개정판

Lecture Notes in

Set Theory

집합론 해설

Charles C. Pinter

박찬녕 엮음

KM 경문사

Preface

This book has been written for junior and senior students for mathematics; it is intended as a basic text for one-semester courses in set theory or the foundations of mathematics.

My chief concern throughout the writing of this work has been to make it accessible to relatively unsophisticated students. The arguments are perfectly rigorous, but I have attempted to make them as simple as possible, balancing the advantages of brevity with the need for sufficient detail. Every definition is accompanied by a fair amount of commentary, whose purpose is to motivate and explain new concepts in simple terms and provide some background information. There is a progressive increase in complexity which is carefully planned; the book begins with a repetition of the familiar arguments of elementary set theory. Then, throughout the first four chapters the level of abstract thinking gradually rises and the arguments become more difficult by small stages. This is designed so that even the less mature student may be brought to the point where he can understand the deeper results of set theory, which are presented, especially, in the last three chapters.

There is a vast literature of elementary textbooks in which set theory is treated from an intuitive point of view. This may be excused, and even be desirable, at the high-school level; but there is no justification for teaching a theory which is known to be inconsistent to university mathematics majors. Set theory must be presented within the framework of an axiomatic system; however, as every lecturer who has taught such a course knows, it is difficult to choose an axiomatization of set theory which is both simple and easy to motivate for undergraduate students. The main disadvantage of the Zermelo-Fraenkel theory is that the major conceptual difficulties appear *at the very beginning*, when the student is least able to cope with them. For example, it needs to be proved that the union and Cartesian product of any two sets *exist*, that the domain and range of an arbitrary function exist, and so forth. On the other hand, although axiomatizations of the von Neumann-Gödel variety are more subtle and abstract, they *can* be made to appear fairly natural to the beginner (if carefully handled) and, most important of all, the chief conceptual difficulties can be pushed back to quite late in the course.

The present text begins with a brief account of the growth of set theory, with special emphasis on the problems which led to the development of the various systems of axiomatic set theory. These systems are compared, and their major advantages and disadvantages are studied. In the next chapter, the undefined notions and elementary axioms of set theory are introduced. The axiomatization used here is an adaptation of the system of von Neumann-Bernays-Gödel. All the objects of the theory are named *classes*; the student is instructed to think of a class as a "collection of objects" in the usual sense. However, for reasons which are carefully explained in the introduction, a distinction must be made between classes and sets: every set is a class, but not all classes are sets. In particular, the reader is advised to think of sets as those collections which are not so all-inclusive as to lead the "paradoxes." The particular properties of sets are not emphasized until later in the book; as a result, Chapter 1 contains a development of the algebra of classes which is hardly distinguishable from the development of intuitive set algebra to be found in an elementary text. The next three chapters, on functions, relations, and partially ordered classes, respectively, give a relatively orthodox treatment of these subjects; the fact that we call our objects *classes* rather than *sets* does not require us to alter in any way the standard development of these ideas.

To recapitulate, then, the axiomatic system used in this book makes it possible to present the elementary properties of classes, functions, relations, and ordering in a way which does not differ substantially from the common intuitive treatment of these topics. After this, the difference between classes and sets is re-affirmed, and the reader is reminded of the things which can be done with sets but cannot be done with classes except at the risk of producing "paradoxes." From this point on, the book deals chiefly with sets. The Axiom of Choice is introduced, and is shown to be equivalent to Zorn's lemma and the well-ordering theorem. The natural numbers are presented, followed by a chapter devoted to the properties of infinite and denumerable sets. Then the transfinite cardinal and ordinal numbers are introduced.

It is fairly common practice, nowadays, to present transfinite arithmetic to upperclass students in an "abstract" fashion, roughly as follows: the class of the ordinal numbers is constructed, followed by a proof of the transfinite recursion theorem; operations on the ordinals are defined via transfinite recursion and the properties of these operations are then proved inductively; finally, the cardinal numbers are introduced as initial ordinals. While this approach is elegant and, in fact, eminently correct, it is in many ways unsound pedagogy, especially in a course for undergraduates. Indeed, the concept of cardinal and ordinal numbers is concrete, tangible, and easy to motivate; it seems a great pity to conceal the underlying simplicity of these objects in a formal system where they are treated, essentially, as abstract symbols.

In the present text, the cardinal numbers are introduced before the ordinals. The concept of cardinal numbers is explained in the following way: the natural numbers, in their capacity as cardinals, serve as a set of standards for measuring the "size" of every finite set. A natural extension of this idea, the class CD of all the cardinal numbers serves as a collection of standards for measuring the "size" of *every* set. In Chapter 8, the class CD of the cardinals is defined to be a class with the following property: for every set A, there is a unique element $a \in \mathrm{CD}$ such that A and a are in one-to-one correspondence. The existence of the class of all the cardinals is assured by means of a provisional axiom (to be removed later) called the Axiom of Cardinality. The arithmetic of the cardinals is then presented using the properties of one-to-one correspondence.

The cardinal numbers are treated in a similar fashion. They are presented as a natural extension of the simple concept of finite ordinality. It is explained to the reader that the class OR of all the ordinal numbers serves as a collection of standards for designating the "position" of any element in any well-ordered set. In the beginning of Chapter 9 the class OR is defined to be a class with the following property: if A is any well-ordered set, there exists a unique $\alpha \in$ OR such that A is similar to α; the existence of the class OR is guaranteed by means of a provisional axiom called the Axiom of Ordinality. The arithmetic of the ordinals is then presented using the properties of similarity of well-ordered sets.

In the last section of Chapter 9 - a section which may be omitted by classes which are not sufficiently advanced - the class OR of the ordinal numbers is constructed using, essentially, the well-known method due to von Neumann. It is shown that the class of the initial ordinals may be used as the class CD of the cardinals. Finally, in Chapter 10, which may be entirely omitted by slower classes, the transfinite recursion theorem is proved, and it is shown that each of the operations on the ordinals has an equivalent definition using transfinite recursion. Several advanced topics are then presented, ending with a discussion of inaccessible ordinals and cardinals.

I am indebted to several anonymous reviewers for their valuable suggestions, which I incorporated into my manuscript. I am grateful to Addision-Wesley for their expert editing and their patience and help in getting this manuscript ready for production.

Lewisburg, Pennsylvania
November 1970

C.C.P.

엮은이의 글

오늘날 수학이 과학의 세계를 떠 받치고 있음은 널리 알려진 사실이다. 현대수학은 집합을 그 대상으로 하고 있다. 집합론은 수학의 세계를 이론적으로 떠받치고 있으며 수학에서 사용하는 언어와 논리를 담고 있다.

Charles C. Pinter의 Set Theory([6], [7])는 집합론을 공리적 방법론으로 전개하고 있다. 공리적 방법론에서 논리는 "이것"이라는 가정으로부터 다른 형태의 "저것"이라는 결론을 이끌어 내는 철저히 기계적인 논리이다. 여기에서 이것이라는 가정은 이것이 참이라는 것이 아니라 이것이 참일 수도 있고 거짓일 수도 있으나 참과 거짓 둘 중 어느 하나만 성립하는 것을 뜻하며 참이면서 거짓인 경우는 배제하고 있다. 다시 말하면 수학에서는 「이렇거나 혹은 이렇지 않는 경우가 공통적으로 일어날 수 없다」는 배중율을 전제로 언어적인 형태의 문장 중에서 참과 거짓의 두 가지로 양분할 수 있는 문장들을 대상으로 "이것이면 저것이다"의 논리를 전개한다.

이것으로 부터 저것을 이끌어내는 수학의 논리는 수학의 약점인 동시에 수학의 강점이라고 말할 수 있다. 즉, 수학의 논리에 의하여 "절대적" 진리를 찾아낼 수는 없을지라도 수학의 논리를 이용하여 증명된 문장은 "그 논리체계 안에서 상대적 진리"로서 객관적인 설득력을 갖게 되고 "과학적인" 것으로 인정받게 된다. 이런 의미에서 수학은 모든 학문분야에서 제기되는 학설들을 과학적으로 설명해 주는 언어로서의 역할도 하게 되는 것이다.

「수를 연구하는 학문인 수학의 이론체계를 어떻게 정립해야 할 것인가?」라는 질문에 답하는 것이 집합론의 중요한 역할 중의 하나이다. 이러한 집합론의 역할을 알아보기 위해 수학의 체계를 간략하게 짚어 보고자 한다.

수학도 말과 글로써 그 뜻을 표현해야 하므로 수학이론을 전개해 나가는데 있어서 사용할 말과 글, 즉 수학용어(數學用語, mathematical notion)를 필요로 한다. 수학용어를 사용할 때에는 주는 자와 받는 자가 있어서 주는 자의 뜻과 받는 자의 뜻이 정확히 일치하여야 한다. 따라서 수학용어에 대하여 완벽한 뜻을 주어야 하는데 이를 낱말 뜻, 즉 정의(定義, definition)라 한다.

그러나 수학용어를 정의하는 내용도 수학용어로써 기술해야하기 때문에,

수학용어를 설명하기 위해서는 맨 처음 정의하지 않고 사용하는 수학용어를 필요로 하게 되는데, 이를 무정의용어(無定義用語, undefined terminology)라 한다. 무정의용어들은 그 수를 최소화하는 것이 바람직하며 무정의용어들을 사용하여 필요한 수학용어들을 정의해 나간다.

수학용어를 사용하여 기술한 문장이 수학적으로 뜻을 가질 때 이를 수학적 문장(數學的 文章, mathematical sentence) 혹은 간단히 문장(文章, sentence)이라고 한다. 여기에서 수학적으로 뜻을 가져야 한다는 말은 주어진 문장이 참이어야 한다는 의미가 아니라 그 문장이 "참"일 수도 있고, "참이 아님" 즉 "거짓"일 수도 있으나 참과 거짓 중 어느 하나로 그 뜻이 정해져야 한다는 것을 뜻한다.

따라서 문장에 대해서는 그것이 참인지 아니면 거짓인지를 판단할 수 있는 근거가 있어야 하는데 이 근거를 다음에 설명할 공리(公理, axiom)로서 설정한다. 이 공리에 따라 문장의 참(眞, true) 혹은 거짓(僞, false)을 판단하는데 이를 주어진 문장의 진리값이라고 한다. 논리에서 가장 중요한 것은 주어진 문장이 참인지 혹은 거짓인지를 판단하는 것, 즉 진리값을 정하는 것이다.

주어진 문장이 참임을 수학적으로 밝히는 것을 증명(證明, proof)한다고 하고 거짓임을 수학적으로 밝히는 것을 반증(反證, disproof)한다고 한다. 어떤 문장이 참임을 밝히는 증명과정에서는 참이라고 알고 있는 다른 문장들을 사용해야 한다. 그러나 처음부터 참이라고 증명되어 있는 문장은 없기 때문에 수학이론을 전개해 나가는데 있어서도 증명 없이 참이라고 인정하는 문장을 필요로 한다. 이와 같이 증명 없이 참인 것으로 받아들이는 문장을 공리(公理, axiom)라 한다. 따라서 수학에서의 문장(文章, sentence)은「설정한 공리들로 부터 참인지 거짓인지를 판단할 수 있는 표현」이다.

수학에서는 참인 문장들을 주로 정리(定理, theorem)라고 한다. 공리와 비교해 보면 정리는 증명에 의하여 참인 것으로 밝혀진 문장이고 공리는 증명 없이 참인 것으로 받아들이는 문장이다. 정리로부터 쉽게 참을 이끌어낼 수 있는 문장을 따름정리(系, corollary)라고 하고 정리를 증명하는데 필요한 참인 문장을 보조정리(補助定理, lemma)라고 한다.

결론적으로 수학이론은 무정의용어들과 공리들을 바탕으로 필요한 용어들과 문장들을 만들고 증명과 반증으로 문장들의 참과 거짓을 판단하여 체계적으로 정리한 것이다.

Charles C. Pinter의 Set Theory([6], [7])가 엮은이가 생각하기에 위에 언급한 수학의 이론체계에 따라 집합이론을 체계적으로 잘 설명한 책이다. Set Theory([6], [7])는 저자의 서문에서 밝혔듯이 대학교 3, 4학년 학생들을 위하여 집합론 혹은 수학기초론에 대한 한 학기 강의용으로 쓰여진 교재이다. 엮은이는 경문사에서 출판한 책([7])으로 경북대학교 수학과에서 주로 2학년 학생들을 대상으로 강의를 하면서 잘못된 부분들을 일부 발견하고 Set Theory([6], [7])의 내용을 충실히 따르면서 일부 내용과 증명을 수정하고 보완하였다. 또한 집합론에 대한 역사적 배경을 설명한 Chapter 0 Historical Introduction에 대한 한글번역을 독자의 이해를 돕기 위하여 번역판 [13]을 참고하여 마지막에 추가하였다.

끝으로, 이 책이 수학을 체계적으로 공부하고자 하는 학생들에게 도움이 될 수 있었으면 한다.

2016년 1월, 경북대학교에서
박찬녕

Contents

Chapter 0

Historical Introduction

0.1 The background of set theory

Although set theory is recognized to be the cornerstone of the "new" mathematics, there is nothing essentially new in the intuitive idea of a set. From the earliest times, mathematicians have been led to consider sets of objects of one kind or another, and the elementary notions of modern set theory are implicit in a great many classical arguments. However, it was not until the latter part of the nineteenth century, in the work of Georg Cantor($1845 \sim 1918$), that sets came into their own as the principal object of a mathematical theory.

Strangely, it was his work in the highly technical field of trigonometric series which first led Cantor to study the properties of sets. At first, he confined himself to certain particular sets of real numbers which occurred in connection with the convergence of series. But Cantor was quick to understand that his discoveries applied to sets quite generally; in a series of remarkable papers, published between 1873 and 1897, he moved progressively further from the concrete problems which had initiated his thinking on sets, and toward the powerful general concepts which underlie set theory today.

The boldest step which Cantor had taken - in the eyes of his contemporaries - was his use of infinite sets, which he considered as no less natural than using finite sets. The question of "infinity" had long been one of the most sensitive problems of mathematics. The reader is undoubtedly acquainted with Zeno's famous "paradox," in which a unit line segment is divided into subintervals by the points $1/2, 1/4, 1/8, 1/16$, etc. Each subinterval - no matter how small - has a definite, nonzero length, and there are infinitely many subintervals; hence, the seemingly paradoxical conclusion that infinitely many nonzero lengths can be added together to produce a finite length. In order to avoid such traps, classical mathematicians made a distinction between the "actual" infinite - in which infinitely many objects are conceived of as existing simultaneously - and the "virtual" infinite, which is simply the potential to exceed any given finite quantity. The "virtual" infinite was regarded as safe, hence admissible, whereas the "actual" infinite was taboo.

It is not surprising then, that Cantor's theory - with its uninhibited use of infinite sets (the notion of infinite was obviously understood here in the "actual" sense) - was not immediately accepted by his contemporaries. It was received at first with skepticism, sometimes even with open hostility. However by the 1980's the more "palatable" parts of Cantor's theory were widely used, for they provided an elegant framework for a wide variety of mathematical theories. And before the turn of the century, even the most revolutionary aspects of set theory had been accepted by a great many mathematicians - chiefly because they turned out to be invaluable tools, particularly in analysis.

Meanwhile, the work of several outstanding mathematicians, in particular Dedekind, was taking a turn which would cast set theory in its most promising role - as the fundamental, "unifying" branch of mathematics. From the earliest times, mathematicians have given thought to the possibility of unifying the entire discipline under a small number of basic principles.

Many of the ancient schools, from Euclid through the Middle Ages, contended that the various branches of mathematics could be subsumed under geometry (numbers might be conceived as geometric proportions); a far more successful attempt at unification came in the nineteenth century, when the work of Weierstrass, Dedekind, and others suggested that all of classical mathematics could be derived from the arithmetic of the natural numbers (positive integers). It was shown that every real number can be regarded as a sequence (called a "Cauchy sequence") of rational numbers; hence the study of the real numbers is reduced to that of the rational numbers. But the rational numbers can easily be regarded as pairs of integers, so finally the mathematics of the real numbers - which includes the calculus and (via analytic geometry) all of geometry - can be based on the natural numbers.

It was at this critical point in the evolution of ideas on the foundations of mathematics that Dedekind, in his little book *Was sind und was sollen die Zahlen* (1888), revealed that the concept of natural numbers can be derived from the basic principles of set theory. A modern way to show this is the following: we let "0" be the empty set (that is, the set with no elements, denoted by the symbol $\emptyset$); "1" is defined to be the set $\{\emptyset\}$, that is, the set (of sets) containing the one element $\emptyset$. Then, "2" is defined to be the set $\{0, 1\}$, "3" is defined to be $\{0, 1, 2\}$, and so on. All the properties of the natural numbers can be proven using these definitions and elementary set theory.

By the turn of the century, then, set theory had not only been accepted as an indispensable tool by a large segment of the mathematical community, but, moreover, it was a serious contender for the position of primacy among the mathematical sciences.

Ironically, at the very time when Cantor's ideas seemed finally to have gained acceptance, the first of certain "paradoxes" were announced, which eventually cast serious doubts as to the basic soundness of set theory in its "Cantorian" form. These paradoxes had such wide repercussions that it is worth looking at them in some detail.

0.2 The paradoxes

Between 1985 and 1910 a number of contradictions were discovered in various parts of set theory. At first, mathematicians paid little attention to them; they were termed "paradoxes" and regarded as little more than mathematical curios. The earliest of the paradoxes was published in 1897 by Burali-Forti, but it had already been discovered, two years earlier, by Cantor himself. Since the Bruali-Forti paradox appeared in a rather technical region of set theory, it was hoped, at first, that a slight alteration of the basic definitions would be sufficient to correct it. However, in 1902 Bertrand Russell gave a version of the paradox which involved the most elementary aspects of set theory, and therefore could not be ignored. In the ensuing years other contradictions were discovered, which seemed to challenge many of the "safest" notions of mathematics.

The "paradoxes" of set theory are of two different kinds, the one called *logical* paradoxes, the other called *semantic* paradoxes. The reason for the names "logical" and "semantic" will become clear to us when we have seen a few examples of these paradoxes; essentially, the "logical" paradoxes arise from faulty logic whereas the "semantic" paradoxes arise from the faulty use of language.

We will devote the remainder of this section to the presentation of two of the most celebrated paradoxes, which involve only elementary concepts of set theory. The first is a "logical" paradox, the second is a "semantic" paradox; both may be considered as typical of their kind.

The simplest of the logical paradoxes is *Russell's paradox*, which can be described as follows:

⌈If A is a set, its elements may themselves be sets; this situation occurs frequently in mathematics - for example, A may be a set of lines, where each line is regarded as a set of points. Now the possibility arises that A may be an element of itself, for example, the set of all sets has this property.

Let S denote the *set of all sets that are not elements of themselves.* Is S an element of itself? Well, if S is an element of S, then, by the very definition of S, S is not an element of S. If S is not an element of S, then (again, because of the way S is defined) S is an element of S. Thus, we have proven that S is an element of S if and only if S is not an element of S - a contradiction of the most fundamental sort.⌋

Usually, in mathematics, when we reach a contradiction of this kind, we are forced to admit that one of our assumptions was in error. In this case, we are led to conclude either that it is meaningless to speak of a set as being an element of itself, or that there is no such thing as a "set of all sets which are not elements of themselves." We will return to this question presently; meanwhile, let us say a few words about the semantic paradoxes.

Typical of the semantic paradoxes is *Berry's paradox*:

⌈For the sake of arguments, let us admit that all the words of the English language are listed in some standard dictionary. Let T be *the set of all the natural numbers that can be described in fewer than twenty words of the English language.* Since there are only a finite number of English words, there are only finitely many combinations of fewer than twenty such words - that is T is a finite set. Quite obviously, then, there are natural numbers which are greater than all the elements of T; hence there is a *least natural number which cannot be described in fewer than twenty words of the English language.* By definition, this number is not in T; yet we have described it in sixteen words, hence it is in T.⌋

Once again, we are faced with a glaring contradiction; since the above argument would be unimpeachable if we admitted the existence of the set T, we are irrevocably led to the conclusion that a set such as T simply cannot exist.

Before the paradoxes, the question of *existence* of sets had never been posed. Cantor "defined" a set to be "a collection of definite, distinguishable objects of our perception which can be conceived as a whole." More specifically, Cantor and his early followers accepted the "common-sense" notion that if we can describe a property of objects, we can also speak of the set of all objects possessing that property. The paradoxes had the singular merit of proving this naive conception of sets to be unacceptable - if only because certain "properties" lead to paradoxical sets.

In the various movements which sprang up, during the early 1900's, with the aim of revising the foundations of set theory, the topic of central concern was the *existence* of sets. What properties legitimately define sets? Under what conditions do properties define sets at all? How can new sets be formed from existing ones?

0.3 The axiomatic method

The appearance of the paradoxes marked the beginning of a crisis in the foundations of mathematics which has been completely resolved to our day. It became abundantly clear that the intuitive conception of a set, as embodied in Cantor's "definition," does not provide a satisfactory basis for set theory - much less for mathematics as a whole. Minor attempts to eliminate the paradoxes by excluding specific types of concepts and definitions were doomed to failure; nothing less than an entirely new approach was needed. Starting about 1905, several ways of dealing with the problem were proposed and developed by their adherents; most of them can be classified into three major groups, called the "axiomatic," the "logistic," and the "intuitionist" schools. The remainder of this chapter is devoted to presenting these three ways of thought. First, however, we shall briefly review the development of the axiomatic method.

The axiomatic method in mathematics emerged in a highly developed form, about 300 B.C., with the appearance of Euclid's *Elements.* Although the method popularized by Euclid has become a characteristic feature of every branch of mathematics today, only in comparatively recent years has it been applied outside of geometry. For this reason, our modern understanding of axiom systems, and of deductive reasoning generally, has to a great extent come out of studies in the field of geometry. It is worth examining a few of the major developments in geometry which influenced the growth of the axiomatic method.

To Euclid and his time, the axioms and postulates represent "truths" whose validity is beyond question. For example, it was this belief in the absolute truth of geometric propositions which led to the millenia-long controversy over Euclid's "parallel postulate." This postulate asserts that if two lines, A and B, intersect a third line C, and if the interior angles which A and B make with C (on a given side of C) add up to less than two right angles, then A and B necessarily intersect. Because this statement appeared to be "obviously true" - yet it lacked the terse simplicity of the other axioms and postulates - geometers from Euclid to the 1700's succeeded one another in vain attempts to prove it from the remaining assumptions. Only in the mid-nineteenth century was the question resolved when Bolyai and Lobaxhevski, each replacing the parallel postulate by an assumption which *contradicted* it, developed "non-Euclidean" geometries. The non-Euclidean geometries were shown to be no less consistent than Euclidean geometry, since they could be given Euclidean interpretations (that is, by suitably reinterpreting "point," "line," "angle," and so forth, the postulates of either Bolyai or Lobachevski can be made to hold in Euclidean geometry). Thus, not only is the parallel postulate independent of the other axioms and postulates of Euclid's system, but alternative, equally consistent geometries can be founded which do not describe the space of our everyday experience. With this came the recognition that axioms are not "universal truths," but are whatever statements we wish to use as premises in an argument.

Perhaps the greatest defect in the *Elements* is the number of tacit assumptions made by Euclid - assumptions not granted by the postulates. For example, in a certain proof it is assumed that two circles, each passing through the center of the other, have a pair of points in common - yet the postulates do not provide for the existence of these points. Elsewhere, Euclid speaks of a point as being *between* two others, yet he does not define "betweenness" or postulate any of its properties. Other arguments in the *Elements* involve the concept of rigid motion - a concept which is not defined or mentioned in the postulates. Thus, throughout Euclid, the orderly chain of logical inferences is frequently broken by tacit appeals to visual evidence. With the discovery of these gaps, mainly in the nineteenth century, grew the understanding that a mathematical argument must be able to proceed without the mediation of spatial or other intuition; that certain objects and relations (such as "point," "line," "betweenness") must be regarded as *undefined notions* and their properties fully specified; that deduction is, in a very essential manner, independent of the *meaning* of concepts. In 1882, M. Pasch published the first formulation of geometry in which the exclusion of any appeal to intuition is clearly stated as a goal and systematically carried out.

By the end of the nineteenth century, then, a modern conception of the axiomatic method began to emerge. In its broad outlines, it did not differ from the ideas held by Euclid: a mathematical theory is "axiomatic" if certain statements are selected to be "axioms", and all the remaining propositions of the theory are derived from the axioms by logical inference. However, there was a new understanding of the *formal* nature of mathematical proof. Inasmuch as possible, the axioms should be sufficiently detailed, and the rules of logical deduction sufficiently explicit, that neither intuition nor intelligence is needed to go through the steps of a proof. Ideally, it should be possible for a computer to verify whether or not a proof is correct.

As long as mathematics is formulated in ordinary languages, such as

English, human understanding is indispensable for interpreting statements and finding the structure of complex sentences. Thus, if intuition is to be completely removed from mathematical proof, an essential prerequisite is the development of a *formal* mathematical language: the "rules" of this language must be strictly codified, so that every statement is unambiguous and its structure clear. The creation of formal, symbolic languages was one of the most important developments of modern mathematics; here is what such a language looks like.

The most basic mathematical statements look like this:

"X is parallel to Y,"

"y lies between x and z,"

"X is an open set." etc.

They are statements about an object, a pair of objects, or more generally, about an ordered n-tuple of objects. These statements are called *elementary predicates*, and the letters X, Y, x, y, z are called their *variables*.

It is convenient to denote a predicate by a single letter followed by the list of its variables. Thus, "X is parallel to Y" may be written $A(X, Y)$, "y lies between x and z" may be written $B(x, y, z)$, and so on. Now the mature student is aware of the fact that the "meaning" of a predicate is immaterial in the process of mathematical reasoning. For example, the "meaning" of the word *parallel* has no bearing on the course of a geometrical argument; all that matters is the relationship between the statement "X is parallel to Y" and other statements such as "X intersects V" and "Y is perpendicular to Z." For this reason, elementary predicates are also called *atomic formulas*; they are integral, "indivisible," not to be analyzed further; they are only to be distinguished from each other.

It is a remarkable fact that every known branch of mathematics requires only a finite number (usually a very small number) of distinct elementary predicates. For example, every statement of plane Euclidean geometry can be expressed in terms of the following basic predicates:

$$\begin{aligned}
P(x):&\quad x \text{ is a point.}\\
L(x):&\quad x \text{ is a line.}\\
B(x,y,z):&\quad y \text{ is between } x \text{ and } z.\\
E(x,y):&\quad x \text{ equals } y.\\
I(x,y):&\quad x \text{ belongs to } y.\\
C(u,v,x,y):&\quad \text{the segment } uv \text{ is congruent to the segment } xy.\\
D(u,v,w,x,y,z):&\quad \text{the angle } uvw \text{ is congruent to the angle } xyz.
\end{aligned} \tag{0.1}$$

Set theory, as we shall see, may be formulated entirely in terms of the one predicate $x \in A$(x is an element of A).

Predicates alone are not sufficient to express all the statements of mathematics, just as nouns alone would be inadequate to write English sentences. For example, we may wish to say that *if* "x is parallel to y" *and* "y is perpendicular to z," *then* "x is perpendicular to z." Such statements consist of predicates joined by means of *logical connectives*. Thus, if P and Q are statements in our language, then so are the following:

$$\begin{aligned}
\neg P:&\quad \text{not } P\\
P \wedge Q:&\quad P \text{ and } Q\\
P \vee Q:&\quad P \text{ or } Q\\
P \to Q:&\quad \text{if } P \text{ then } Q\\
P \leftrightarrow Q:&\quad P \text{ if and only if } Q
\end{aligned}$$

Finally, we may wish to say, for example, that if "x is a point" and "y is a point," then *there exists* a point z such that "z is between x and y." This requires the use of *quantifiers*. Thus, if $P(x)$ is a statement with a variable x, then the following are also statements:

$$\forall x P(x): \qquad \text{for every } x\ P(x).$$

$$\exists\, x \cdot\!\ni\!\cdot^{1}\ P(x): \quad \text{there exists an } x \text{ such that } P(x).$$

This completes our formal mathematical language. All of known mathematics can be expressed in terms of elementary predicates, logical connectives, and quantifiers. To illustrate how this language is used, let us take a simple example.

The sentence "If x and y are distinct points, then there is a point z between x and y" can be symbolized as

$$[P(x) \wedge P(y) \wedge \neg E(x,y)] \rightarrow [\exists\, z \cdot\!\ni\!\cdot\ (P(z) \wedge B(x,z,y))], \qquad (0.2)$$

where the meaning of the predicates is given in (0.1) above.

One of the many benefits to be derived from the use of a formal language is that it is possible to describe precisely and explicitly the process of deduction in this language. A few clear, unambiguous rules decide when a statement T may be inferred from a statement S. A few such rules are the following:

$$\begin{aligned} &\text{Rule } A. \ \text{ from } P \text{ and } (P \rightarrow Q) \text{ we may infer } Q.\\ &\text{Rule } B. \ \text{ from } P \text{ and } Q \text{ we may infer } P \wedge Q.\\ &\text{Rule } C. \ \text{ from } \neg(\neg P) \text{ we may infer } P.\\ &\text{Rule } D. \ \text{ from } P(c) \text{ we may infer } \exists\, x \cdot\!\ni\!\cdot\ P(x). \end{aligned} \qquad (0.3)$$

These and a few other laws[2], are called "rules of inference" in our language. A *formal* argument, from given premises, is a sequence of expressions of the formal language, where each expression is either a premise, or is derived from a preceding expression (or expressions) by applying one of the

[1]The notation $\cdot\!\ni\!\cdot$ is an abbreviation of English words "such that."

[2]The four rules given in (0.3) above, together with seven additional rules, make up the system of natural deduction described in Slupecki and Borkowski [10]; these eleven rules are sufficient for every valid logical argument. Other systems are described in Quine[8] and Suppes[11].

rules of inference.

Example 0.1. Consider a formal language with a predicate $L(x, y)$; let us agree to write $x < y$ for $L(x, y)$. The following is a very simple formal argument in this language.

Premises

Premise 1. $a < b$.

Premise 2. $b < c$.

Premise 3. $[(a < b) \wedge (b < c)] \rightarrow (a < c)$.

Theorem $\exists\, x \cdot\!\ni\!\cdot [(a < x) \wedge (b < x)]$.

proof.

Step 1. $a < b$ (Premise 1)

Step 2. $b < c$ (Premise 2)

Step 3. $[(a < b) \wedge (b < c)]$ (Steps 1, 2 and Rule B)

Step 4. $[(a < b) \wedge (b < c)] \rightarrow (a < c)$ (Premise 3)

Step 5. $a < c$ (Steps 3, 4 and Rule A)

Step 6. $[(a < c) \wedge (b < c)]$ (Steps 5, 2 and Rule B)

Step 7. $\exists\, x \cdot\!\ni\!\cdot [(a < x) \wedge (b < x)]$ (Step 6 and Rule D)

The reader should note that the rules of inference are applied to expression in a perfectly mechanical way. To all intents and purposes, the expressions can be regarded as meaningless arrays of symbols; the fact that they have a meaning *to us* is irrelevant to the task of carrying out the proof. Thus intuition is totally absent from a formal mathematical proof.

An axiomatic theory is said to be *formalized* if its axioms are transcribed in the formal language (for example, the formula (0.2) is an axiom of Hilbert's plane geometry), and all of its proofs are formal proofs. While it is commonly accepted as the ideal, today, that every axiomatic theory be developed formally, it would be far too tedious, in practice, to do so. Symbolic statements are difficult to decipher, and formal proofs tend to be exceedingly long. Thus *mathematicians are usually content to satisfy themselves that an axiomatic theory can be formalized, and then proceed to develop it in an informal manner.* This will be our procedure in this book.

0.4 Axiomatic set theory

To a great many mathematicians in the early 1900's, the answer to the problem posed by the paradoxes was to provide set theory with an axiomatic basis. The term "set" and the relation "is an element of" would be the undefined notions of such a theory, just as "point" and "line" are undefined notions in geometry; their "meaning" would be irrelevant, and their properties would be given formally by the axioms. In particular, *the axioms would be chosen in such a manner that all the useful results of Cantor's theory could be proven, whereas the paradoxes could not.*

The first axiomatization of set theory was given in 1908 by Zermelo. Zermelo's system, with certain modifications due to Skolem and Fraenkel, is widely used up to the present day. Zermelo wrote his work before the time when formal method became widely understood and accepted; thus his set theory is not written in a formal language, but is closer in style to the older axiomatic treatments of geometry.

In Zermelo's system, there is one primitive relation, denoted by the symbol $\in$; the expression $x \in Y$ is to be read "x is an element of Y." The variables x, y, z, X, Y, etc., which we place to the right or to the left of the symbol $\in$, stand for objects which we agree to call "sets."

The reader may feel there ought to be *two* kinds of objects, namely sets and elements. Actually, this distinction is unnecessary: For, on the one hand, the relationship between element and set is a relative one rather than an absolute one (in fact, the element-set relationship is precisely the relation $\in$). On the other hand, almost every set in mathematics is a set *of sets*. For example, in plane analytic geometry, a line is a set of points; a point is a pair of real number (its coordinates); a real number is regarded as a sequence (that is, a set) of rational numbers; etc. Thus a useful simplification which is made in axiomatic set theory is to regard the elements of every set to be sets themselves; in other words, every set is considered to be a set *of sets.* This simplification has no harmful effects, and has the merit of reducing the number of primitive notions and axioms of set theory.

This suggests a comment on notations. Although it is customarily to use small and capital letters as in $x \in Y$, it is in no way necessary. In fact, we will sometimes write things like $x \in y$ and $X \in Y$. All the variables in these expressions denote sets.

Almost every set that arises in our thinking is a set consisting of all the objects of a specified kind - that is, consisting of all the objects which satisfy a given condition. This is the most natural way in which sets occur: we are able to describe a condition on x - let us symbolize this condition by $S(x)$ - and we are led to speak of the set of all objects x which satisfy $S(x)$.

Example 0.2. The set of all objects x which satisfy the condition "x is an irrational number and $0 \leq x \leq 1$" (loosely speaking, the set of all irrational numbers between 0 and 1).

The set of all objects x which can be described by the sentence "x is a man" (loosely speaking, the set of all men).

Since this is the most natural way that sets arise, it is clearly desirable to have a principle in set theory which makes it possible - given any condition $S(x)$ - to form the set of all objects x which satisfy $S(x)$. However, as we noted in Section 2, if such a principle is adopted *without any restrictions*, we are led to the paradoxes (for example, we can form the set of all sets which are not elements of themselves). Thus, we must devise such restrictions on this principle as will eliminate the paradoxes. Zermelo conceived the following restriction: Let $S(x)$ be a condition on x; we *cannot* form the set of all x which satisfy $S(x)$; but if A is a set, we can form the set *of all x in A*

which satisfy $S(x)$. Thus, roughly speaking, a property of objects cannot be used to form a "new" set, but only to "select" from a set A whose existence has already been secured, all the elements which satisfy the given property.

Zermelo introduced this principle as an axiom in his system. Because its role is to *select* elements in sets, he called it the *axiom of selection* and stated it as follows:

⌈Let A be a set, and let $S(x)$ be a statement about x which is meaningful for every object x in A. There exists a set which consists of exactly those elements x in A which satisfy $S(x)$.⌋

The set whose existence is given by the axiom of selection is customarily denoted by

$\{x \in A \mid S(x)\}$ [to be read: "the set of all x in A such that $S(x)$"].

Thus the reader should note that Zermelo's system does not allow us to form $\{x \mid S(x)\}$, [the "set of all x which satisfy $S(x)$"]; but, for any set A, we can form $\{x \in A \mid S(x)\}$.

How does the axiom of selection avoid the paradoxes? First, let us see what happens to Russell's paradox: the crucial set in Russell's argument is the "set of all sets which are not elements of themselves," which can be symbolized as $\{x \mid x \notin x\}$. As we have noted, this set cannot be formed in Zermelo's system; the best we can do is to produce $\{x \in A \mid x \notin x\}$, where A is any set which can be shown to exist. If we substitute $\{x \in A \mid x \notin x\}$ for $\{x \mid x \notin x\}$ in Russell's argument, the outcome changes completely. Indeed, let us go through the steps of the argument, with S denoting $\{x \in A \mid x \notin x\}$:

⌈$S \in S$ is impossible, for $S \in S$ implies $S \notin S$, a contradiction! Thus $S \notin S$. It follows that $S \notin A$, for if S were in A, then (because $S \notin S$) we would have $S \in S$, which would be a contradiction.⌋

Hence Russell's argument merely proves that if A is any set, then the set $\{x \in A \mid x \notin x\}$ cannot be an element of A.

The other logical paradoxes disappear in similar fashion. The crucial sets in all of the logical paradoxes have a common trait: they are overly comprehensive - that is, they are "too large," they include too much. In Russell's paradox it is the "set of all sets which are not elements of themselves"; in Cantor's paradox (which is closely related to that of Russell) it is the "set of all sets." The axiom of selection cannot contribute to the formation of these "excessively large" sets, since it can only be used to form *subsets of existing sets.*

The problem of avoiding the semantic paradoxes is a more difficult one. The crucial sets in paradoxes such as Berry's are not "too large." The trouble seems, rather, to be inherent in the condition $S(x)$ which determines the set; even the restriction imposed by the axiom of selection is not an effective barrier. Thus if $S(x)$ designates the sentence "x can be described in fewer than twenty words of the English language," then the offending set in Berry's paradox is $\{x \in \mathbb{N} \mid S(x)\}$, where $\mathbb{N}$ denotes the set of the natural numbers. This set *can* be formed in Zermelo's system, if we admit $S(x)$ as an acceptable condition on x.

Thus, to prevent the semantic paradoxes, we must place restrictions on the type of "conditions" $S(x)$ which are admissible for determining sets. Zeromelo attempted to do this by stipulating, in the axiom of selection, that $\{x \in A \mid S(x)\}$ can be formed only if $S(x)$ *is meaningful for every element* x *in* A. However, in so doing, he only raised new question: How are we to understand "meaningful"? How do we determine whether $S(x)$ is meaningful?

We are forced, at last, to face a question which the alert reader may already have asked himself: What do we mean by a "condition" $S(x)$, by a "statement about an object x"? We cannot be content to regard the concept of a "statement about x" as intuitively known, since our purpose now is to axiomatize set theory, that is, to free it of all dependence on intuition. Zermelo failed to give a satisfactory answer to this question, because he did not frame his system in a formal language. However, in 1922, Skolem and Fraenkel, both working on formal axiomatization of set theory, saw the natural way out of the dilemma: a "statement about x" is simply a statement *in the formal language* with one "free" variable x. (We say that x is *free* in $S(x)$ if x is not governed by a quantifier $\exists x$ or $\forall x$; thus, in $\exists y \cdot \ni \cdot (x < y)$, x is free whereas y is not).

In Zeromelo's system there is only one elementary predicate, denoted by the symbol $\in$. Thus a statement in the formal language is an expression which can be written using only predicates $x \in Y, u \in V$, etc., logical connectives, and quantifiers.

If we restrict the "statements" $S(x)$ which can be used in the axiom of selection to those which are expressible in the formal language, we immediately eliminate all the semantic paradoxes. For example, there is no way of writing the sentence "x can be described in fewer than twenty words of the English language" in terms of the formal language. This solution - this way of avoiding the semantic paradoxes - is acceptable from the mathematical point of view, though it is hardly an ideal solution philosophically. Mathematically, we can still form all the sets essential for mathematics; from a broader point of view, though, we cannot form anything like the "set of all men," the "set of all Latin verbs," etc. No better solution has been devised to this day.

We have seen how Zermelo's system, with modifications due to Skolem and Fraenkel, manages to avoid paradoxes. The remaining axioms of Zermelo's system are similar to those which will be developed in the following chapters. Essentially, they provide for the existence of the set of all the natural numbers (from which we can construct the other number systems of mathematics), and guarantee the existence of unions, intersections, and products of sets. Before going on, we will briefly review another way of axiomatizing set theory, which is of increasing interest in our day; the essential ideas are due to von Neumann.

Von Neumann noted that two facts combine to produce the logical paradoxes: in the first place, as we have seen, the crucial sets (for example, the set S in Russell's paradox) are "too large"; in the second place, these "large" sets *are allowed to be elements of sets* (for example, it is admitted that Russell's set S may be an element of itself). Of these two facts, Zermelo used the first; he avoided the paradoxes by making it impossible to form the "large" sets. Von Neumann proposed to use the second of these facts: he would permit the excessively large sets to exist, but would not allow them to be elements of sets.

Briefly, von Neumann's system can be described as follows. As in Zermelo's theory, there is only one elementary predicate, namely the predicate $x \in Y$. The variables x, y, X, Y, etc. stand for objects which we agree to call *classes*; however, we distinguish between two kinds of classes, namely *elements* - which are defined to be those classes which are elements of classes - and *classes*, which are not elements of any class. Zermelo's axiom of selection is now replaced by a principle called the *class axiom*, which states the following:

⌈If $S(x)$ is any statement about an object x, there exists a class which consists of all those *elements* x which satisfy $S(x)$.⌋

In other words, if $S(x)$ is any statement about x, we can form the class

$$\{x \mid x \text{ is an element and } S(x)\}.$$

To verify that Russell's paradox does not "work" in this system, let us go through the steps of Russell's argument, with S denoting $\{x \mid x$ is an element and $x \notin x\}$.

⌈ $S \in S$ is impossible, for $S \in S$ implies $S \notin S$, which is a contradiction! Thus $S \notin S$. It follows that S is not an element, for, if S were an element, then we would have $S \in S$, which would be a contradiction.⌋

Thus Russell's argument merely proves that the class S, defined above, is not an element.

The semantic paradoxes are avoided, as in the revised Zermelo system, by admitting in the class axiom only those "statements" which can be written in the formal language.

Variants of von Neumann's system have been developed by Gödel and Bernays. They have an advantage over Zermelo's system in that the class axiom is closer to the spirit of intuitive set theory than the axiom of selection. Indeed, if $S(x)$ is any statement about x, the class axiom guarantees the existence of a *class* containing all the elements x which satisfy $S(x)$. In mathematics, systems of the von Neumann type provide us with the convenience of being able to speak of the "class of all elements" and of being able to operate on classes which are not elements. (Such classes tend to occur at various points in higher mathematics; they can be avoided, but only at a price.) The chief disadvantage of these systems is that the distinction between classes which are elements and classes which are not elements - a highly artificial one - must always be borne in mind; however, this disadvantage is undoubtedly outweighed by the greater flexibility and naturalness of von Neumann type systems. In this text we shall use a slightly modified form of von Neumann's system of axioms.

0.5 Objections to the axiomatic approach and other proposals

What are the chief goals of axiomatic set theory, and to what extent have these goals been successfully attained? In order to answer that question we must remember the circumstances which led mathematicians, in the early years of the twentieth century, to search for an axiomatic basis to set theory. The ideas of Cantor had already thoroughly permeated the fabric of modern mathematics, and had become indispensable tools of the working mathematician. Algebra and analysis were formulated within a framework of set theory, and some of the most elegant, powerful new results in these fields were established by using the methods introduced by Cantor and his followers. Thus, when the paradoxes were discovered and there arose doubts as to the basic validity of Cantor's system, most mathematicians were understandably reluctant to give it up; they trusted that some way would be found to circumvent the contradictions and preserve, if not all, at least most of Cantor's results. Hilbert once wrote in this connection: "We will not be expelled from the paradise into which Cantor has led us."

With the discovery of newer paradoxes, and the failure of all the initial attempts to avoid them, it became increasingly clear that it would not be possible to preserve intuitive set theory in its entirely. Something - possibly quite a lot - would have to be relinquished. The best that one could hope for was to retain as much of intuitive set theory as was needed to save the new results of modern mathematics and provide an adequate framework for classical mathematics.

Briefly, then, axiomatic set theory was created to achieve a limited aim: it had to provide a firm foundation for a system of set theory which - while it did not need to be as comprehensive as intuitive set theory - must include all of Cantor's basic results as well as the constructions (such as the number systems, functions, and relations) needed for classical mathematics.

The systems of both Zermelo and von Neumann were successful in achieving this limited aim. But the amount of intuitive set theory which they had to sacrifice was considerable. For example, in Zermelo's system, as we have already seen, the intuitive way of making sets -by naming a property of objects and forming the set of all objects which have that property- does not take place at all. It is replaced by the axiom of selection, in which properties are allowed only to determine *subsets of given sets.* Furthermore, the only admissible "properties" are those which can be expressed entirely in terms of the seven symbols $\in$, $\wedge$, $\vee$, $\neg$, $\rightarrow$, $\forall$, $\exists$ and variables $x, y, z, \cdots$. As a result, many of the things that we normally think of as sets - for example, the "set of all apples," the "set of all atoms in the universe" - are not admissible as "sets" in axiomatic set theory. In fact, the only "sets" which the axioms provide for are, first the empty set $\emptyset$, and then constructions such as $\{\emptyset\}$, $\{\emptyset, \{\emptyset\}\}, \cdots$, etc., which can be built up from the empty set. It is remarkable fact that all of mathematics can be based upon such a meager concept of set.

While the various axiomatic systems of set theory saved mathematics from its immediate peril, they failed to satisfy a great many people. In particular, many who were sensitive to the elegance and universality of mathematics were quick to point out that the creations of Zermelo and von Neumann must be regarded as provisional solutions - as expedients to solve a temporary problem; they will have to be replaced, sooner or later, by a mathematical theory of broader scope, which treats the concept of "set" in its full, intuitive generality.

This argument against axiomatic set theory - that it deals with an amputated version of our intuitive conception of a set - has important philosophical ramifications; it is part of a far wider debate, on the nature of mathematical "truth." The debate centers around the following question: Are mathematical concepts creations (that is, *inventions*) of the human mind, or do they exist independently of us in a "platonic" realm of concepts, merely to be *discovered* by the mathematician? The latter opinion is often referred to as "platonic realism" and is the dominant viewpoint of classical mathematics. We illustrate these two opposing points of view by showing how they apply tn a particular concept - the notion of natural numbers. From the viewpoint of platonic realism, the concepts "one," "two," "three," and so on, exist in nature and existed before the first man began to count. If intelligent beings exist elsewhere in the universe, then, no matter how different they are from us, they have no doubt discovered the natural numbers and found them to have the same properties they have for us. On the other hand, according to the opposing point of view, while three cows, three stones, or three tress exist on nature, the natural number *three* is a creation of our minds; we have invented a procedure for constructing the natural numbers (by starting from 0 and adding 1 each time, thus producing successively $1, 2, 3$, etc.) and have in this manner fashioned a conceptual instrument of our own making.

How does platonic realism affect the status of axiomatic set theory? From the point of view of platonic realism, mathematical objects are given to us ready-made, with all their features and all their properties. It follows that to say a mathematical theorem is *true* means it expresses a correct statement about the relevant mathematical objects. (For example, the proposition $2 + 2 = 4$ is not merely a formal statement provable in arithmetic; it states an *actual fact* about numbers.) Now - if we admit that mathematical objects are given to us with all their properties, it follows, in particular, that the notion of *set* is a fixed, well-defined concept which we are not free to alter for our own convenience. Thus the "sets" created by Zermelo and von Neumann do not exist, and theorems which purport to describe these nonexistent objects are false! In conclusion, if we were to accept a strict interpretation of platonic realism, we would be forced to reject the systems of Zermelo and von Neumann as mathematically invalid.

Fortunately, the trend, for some time now, has been away from platonism and toward a more flexible, more "agnostic" attitude toward mathematical "truth." For one thing, developments in mathematics have been conforming less and less to the pattern dictated by platonic philosophy. For another, the cardinal requirement of platonism - that every mathematical object correspond to a definite, distinct object of our intuition (just as "point" and "line" refer to well-defined objects of our spatial intuition) - came to be an almost unbearable burden on the work of creative mathematicians by the nineteenth century. They were dealing with a host of new concepts (such as complex numbers, abstract laws of composition, and the general notion of function) which did not lend themselves to a simple interpretation in concrete terms. The case of the complex numbers is a good illustration of what was happening. Classical mathematics never felt at ease with the complex numbers, for it lacked a suitable "interpretation" of them, and as a result there were nagging doubts as to whether such things really "existed." Real numbers may be interpreted as lengths or quantities, but the square root of a negative real number - this did not seem to correspond to anything in the real world or in our intuition of number. Yet the system of the complex numbers arises in a most natural way - as the smallest number system which contains the real numbers and includes the roots of every algebraic equation with real coefficients; whether or not the complex numbers have a physical or psychological counterpart seems irrelevant.

The case of the complex numbers strikes a parallel with the problem of axiomatic set theory. For the "sets" created by Zermelo and von Neumann arise quite naturally in a mathematical context. They give us the simplest notion of set which is adequate for mathematics and yields a consistent axiomatic theory. Whether or not we can interpret them intuitively may be relatively unimportant.

Be that as it may, mathematicians in the early 1900's were reluctant to make so sharp a break with tradition as axiomatic set theory seemed to demand. Furthermore, they felt, on esthetic grounds, that a mathematical theory of sets should describe all the things - and only those thing - which our intuition recognizes to be sets. Among them was Bertrand Russell; in his efforts to reinstate intuitive set theory, Russell was led to the idea that we may consider sets to be ordered in a hierarchy of "levels," where, if A and $\mathscr{B}$ are sets and A is an element of $\mathscr{B}$, then $\mathscr{B}$ is "one level higher" than A. For example, in plane geometry, a *circle* (regarded as a set of points) is one level below a *family of circles*, which, in turn, is one level below a *set of families of circles*. This basic idea was built by Russell into a theory called the *theory of types*, which can be described, in essence, as follows.

⌈Every set has a natural number assigned to it, called its *level*, The simplest sets, those of level 0, are called *individuals* - they do not have elements. A collection of individuals is a sets of level 1; a collection of sets of level 1 is a set of level 2; and so on. In the theory of types the expression $a \in B$ is only meaningful if, for some number n, a is a set of level n and B is a set of level $n+1$. It follows that statement $x \in x$ has no meaning in the theory of types, and as a result, Russell's paradox vanishes for the simple reason that it cannot even be formulated.⌋

Russell's theory of types is built upon a beautifully simple idea. Unfortunately, in order to make it "work," Russell was forced to add a host of new assumptions, until finally the resulting theory became too cumbersome to work with and too complicated to be truly pleasing. For one thing, corresponding to the hierarchy of "levels" of sets, it was necessary to have a hierarchy of "levels" of logical predicates. Then - as a way of avoiding the semantic paradoxes - sets at the same level were further divided into "orders." Finally, Russell had to admit a so-called *Axiom of Reducibility* which was just as arbitrary, just as ungrounded in intuition, as any of the *ad hoc* assumptions made by Zermelo. As a result of these shortcomings, the theory of types has not gained wide acceptance among mathematicians, although it is still an interesting (and maybe promising) area of research.

A far more radical approach was taken by a group of mathematicians calling themselves *intuitionists.* To the intuitionists, much of modern mathematics, including almost all of Cantor's theory of sets, is based on the uncritical use of rules of logic which they consider to be invalid. The intuitionist attitude toward set theory can therefore be summed up very easily: it is one of almost total rejection.

In order to properly understand the philosophy of intuitionism, we must first gain an understanding of its attitude toward logic. As the intuitionist sees it, the rules of logic used by mathematicians have an empirical character. Certain methods of proof came to be commonly used by mathematicians, and, over the years, were codified into a body of rules. These rules were observably correct in their original context, but - after they were codified - they came to be used uncritically in totally different contexts in which they no longer applied. Let us be more specific: in Euclidean geometry, which is the source of most mathematics before the fifteenth century, every theorem involves only a finite number of objects, and each of these objects (geometric figures) is given by an explicit construction. The rules of logic asked by Euclid are perfectly valid in this context, say the intuitionists; it is only when they are transposed to problems involving an infinite domain of objects, or in which the objects are not given by an explicit construction, that the rules are incorrect.

As an example, let us take the *law of the excluded middle.* This is the rule which says that if S is any statement, then either S is true or the denial of S is true. In particular, let A be a set and let $P(x)$ be a statement which is meaningful for every element x in A. By the law of the excluded middle, either *there exists an x in A such that $P(x)$ is true*, or else *for every x in A, $P(x)$ is false.* Now the intuitionists will accept this rule if A is a finite set and if each of its elements can be tested to determine whether $P(x)$ is true or false. In fact, say the intuitionists, this is where the rule originated - our experience tells us that if we examine every element x in A to determine whether or not $P(x)$ is true (to do so, A has to be a finite set), there can be only two possible outcomes: either we have found an x for which $P(x)$ is true, or else, for every x in A we have found $P(x)$ to be false. Our experience, then, confirms the law of the excluded middle in the case of finite sets. But, say the intuitionists, to assume it is true in the case of infinite sets - in an area where we have no experience and experience is impossible - is wholly without foundation. On grounds such as these, the intuitionists deny the law of the excluded middle. Other rules of traditional logic are similarly rejected, because they go beyond the realm of our experience.

The intuitionist points out that mathematics originated as a study of certain mental constructions - chiefly geometric figures and simple constructions involving whole numbers. The theorems of early mathematics were essentially statements to the effect that if certain constructions are carried out, certain results will be achieved. For example, consider the following theorem of geometry: Given two triangles, if two sides and the included angle of one triangle are equal, respectively, to two sides and the included angle of the other triangle, then the two triangles are congruent. What is expressed here is the fact that if we construct two triangles, with two sides and an included angle equal as stated above, we will be able to verify (for example, by using a compass) that the remaining side of one triangle is equal to the remaining side of the other triangle. The proofs of Euclidean geometry, and of early mathematics generally, have a constructive character. For example, the Pythagorean theorem is proven by constructing a figure in which corresponding parts are congruent, hence have the same area. Once the construction is completed, it remains only to point out the congruent parts and thereby reach the desired conclusion. Thus, say the intuitionists, rules of logic were originally intended to describe situations which arose in the context of such constructions and determinations. For example, the rule of the excluded middle was intended simply to note the fact that if we are given a (finite) set of objects and a method (for example, using a ruler and compass) to test each object for some property P, then, if we perform the test on every object, either one of the objects will pass the requirements of the test, or else every object will fail the requirements.

To sum up, then, the intuitionist maintains that a mathematical theorem is nothing more than a factual statement to the effect that a given mental construction will lead to a given result. Every proof must be constructive. If we claim that a mathematical object exists, we must prove it by giving a method for actually constructing the object. If we assert that a relation holds among given pairs of objects, our proof must include a method for testing every pair of objects in question. The "rules of logic" are nothing more than simple observations on the process of performing mathematical constructions; we have no grounds to believe these rules apply outside the context of constructive mathematics - in fact, it is meaningless to apply them outside this context. *Logic is incidental, not essential, to mathematics.*

It is obvious that the intuitionist's notion of set must be quite different from ours. Consider, for example, Cantor's principle that if we can name a property of objects, then there exists a set of all objects, which have that property. Now this principle - as well as the limited versions of it accepted by Zermelo and von Neumann - is anathema to the intuitionists. An object exists only if it can be constructed; hence, a set exists only if we are able to describe a procedure for building it.

A full discussion of intuitionist set theory is outside the scope of this book; however, it is worth mentioning a particular kind of set which is important in intuitionist mathematics; this is called a *spread.* A spread is identified with a rule for producing all of its elements. Thus, a spread is not regarded as an "already formed" totality, but rather as a "process of formation"; each of its elements can be formed if we apply the rule long enough.

The paradoxes do not occur in intuitionist set theory because the crucial sets in the paradoxes cannot be produced in intuitionist mathematics, and the essential arguments cannot be rendered using intuitionist logic.

0.6 Concluding remarks

During the early part of the twentieth century, as we have seen, various ways of building a noncontradictory theory of sets were proposed and developed by different "schools" of mathematicians. We have reviewed the basic principles of axiomatic set theory, Russell's theory of types and the intuitionist (or "constructivist") approach to sets. In addition to these, a great many other ideas were proposed, too various to describe in this brief introduction.

Of all the ways of dealing with sets, the axiomatic method seemed to best suit the needs of modern mathematics. The notion of "set" embodied in the systems of Zermelo and von Neumann is broad enough for the purposes of mathematics, and therefore in a mathematical setting it is virtually indistinguishable from the Cantorian notion of set. The methods of proof, the symbolism, the rigor - all of these correspond to current mathematical usage. Most important of all, axiomatic set theory seems "natural" to most working mathematicians.

Those who reject axiomatic set theory do so on the basis of some philosophical bias. Those philosophical positions, however, which refuse to accept axiomatic set theory also deny the validity of a large part of modern mathematics. For example, the intuitionist school rejects the greater part of contemporary analysis, because it is founded on nonconstructivist principles. While the arguments of these critics present a challenge, and certainly give us food for thought, they are not powerful enough to destroy the achievements of three brilliant generations of mathematicians.

In the course of the past seventy years or so, set theory has come to be widely recognized as the fundamental, "unifying" branch of mathematics. We have already seen how the natural numbers can be constructed, and their properties derived, within the framework of set theory; from there, it is easy to develop the rational numbers, the real and complex numbers, as well as remarkable systems such as Cantor's "transfinite cardinals." The notions of function, relation, operation, and so forth are easily defined in terms of sets, and, as a result, every known branch of mathematics can be formulated within set theory. It is therefore legitimate - and, in fact, vital - to ask the question: "How secure a foundation does set theory provide for the whole edifice of mathematics?" In particular, are we absolutely certain that axiomatic set theory is consistent, that is, free of contradictions? If it is, then everything we develop within it - in other words, all of mathematics

- is consistent; and if it is not, then whatever we build upon it is worthless.

The fact is that there is no known proof of the consistency of axiomatic set theory. This is not too surprising though, in view of some of the results of modern logic. For example, in 1931, K. Gödel proved that it is impossible to give a finitary proof of the consistency of ordinary arithmetic (of the natural numbers). The situation in almost every other branch of mathematics is much the same. Thus the best assurance that we have, at this time, of the consistency of axiomatic set theory is the fact that the familiar contradictions cannot be obtained in the usual way. We cannot do any better at this time.

A result relating to *relative consistency* is of some interest. It has been proven recently that *if* Zermelo's axiomatization of set theory is consistent, *then* von Neumann's axiomatization is also consistent.

It is probable that, in the final analysis, any assurance of the consistency of mathematics will have to rest on some combination of basic intuition and empirical evidence.

Chapter 1

Classes and Sets

1.1 Building sentences

Before introducing the basic notions of set theory, it will be useful to make certain observations on the use of language.

Definition 1.1. By a *sentence*, we will mean a statement which, in a given context, is unambiguously either true or false, but not both simultaneously.

Example 1.2. Here are some examples of sentences.

1. Seoul is the capital of the Republic of Korea.
2. Money grows on trees.
3. Snow is black.

We will use letters P, Q, R, S, etc., to denote sentences.

Sentences may be combined in various ways to form more complicated sentences. The following is a list of essential operations on the sentences.

Definition 1.3. Let P and Q be any sentences.

(1) The *negation* of P is denoted by $\neg P$, to be read "*not* P", which is defined as asserting that "$\neg P$ *is not true if* P *is true*" and "$\neg P$ *is true*

if P is not true". It is convenient to display the relationship between $\neg P$ and P in the following *truth table.*

P	$\neg P$
true	false
false	true

(2) The *conjunction* of P and Q is denoted by $P \wedge Q$, to be read "P *and* Q", which is defined by the following truth table.

P	Q	$P \wedge Q$
true	true	true
true	false	false
false	true	false
false	false	false

(3) The *disjunction* of P and Q is denoted by $P \vee Q$, to be read "P *or* Q", which is defined by the following truth table.

P	Q	$P \vee Q$
true	true	true
true	false	true
false	true	true
false	false	false

(4) The *right arrow* from P to Q is denoted by $P \to Q$, to be read "*if* P, *then* Q", which is defined by the following truth table.

P	Q	$P \to Q$
true	true	true
true	false	false
false	true	true
false	false	true

A word of caution: In ordinary usage, "if P, then Q" is understood to mean that there is a *causual* relationship between P and Q (as in "if John passes the course, then John can graduate"). In mathematics, however, $\rightarrow$ is always understood in the *formal* sense: $P \rightarrow Q$ is true *except if* P is true and Q is false.

(5) The *implication* from P to Q is denoted by $P \Rightarrow Q$, to be read "P *implies* Q", which is defined as asserting that "$P \rightarrow Q$ is true".

(6) We agree that $P \leftrightarrow Q$ is to be an abbreviation for "$(P \rightarrow Q) \wedge (Q \rightarrow P)$". $\leftrightarrow$ is also an abbreviation for *if and only if* (shortened *iff*). Thus we have the following truth table.

P	Q	$P \rightarrow Q$	$Q \rightarrow P$	$P \leftrightarrow Q$
true	true	true	true	true
true	false	false	true	false
false	true	true	false	false
false	false	true	true	true

(7) The *logical equivalence* between P and Q, denoted by $P \Leftrightarrow Q$, will mean that "$P \leftrightarrow Q$ is true". Thus $P \Leftrightarrow Q$ is an abbreviation for $(P \Rightarrow Q) \wedge (Q \Rightarrow P)$.

Remark 1.4. (1) Note that the sentence "$P \rightarrow Q$" is true except if P is true and Q is false and that "$P \leftrightarrow Q$" is true only when P and Q are both true or both false. Thus the sentence $P \rightarrow Q$ is logically equivalent to $(\neg P) \vee Q$ as the following truth table illustrates.

P	Q	$P \rightarrow Q$	$\neg P$	$(\neg P) \vee Q$
true	true	true	false	true
true	false	false	false	false
false	true	true	true	true
false	false	true	true	true

(2) The properties of formal implication "$P \Rightarrow Q$" differ somewhat from the properties we would expect *causal* implication to have, as the following simple examples illustrate.

1. "$1 + 1 = 2 \rightarrow \pi$ is a transcendental number" is true, even though there is no causal relationship between the two component sentences.

2. "$2 + 2 = 5 \rightarrow 4$ is a prime number" is true, even though the two component sentences are false.

(3) The sentence "$P \leftrightarrow Q$" is true only when either both P and Q are true or both P and Q are false.

Certain compound sentences are true regardless of the truth or falsity of their component parts; a typical example is the sentence $P \rightarrow P$. Regardless of whether P is true or false, $P \rightarrow P$ is always true; in other words, no matter what sentence P is, $P \rightarrow P$ is true. For future reference we record a few sentences which have this property.

Theorem 1.5. *For all sentences P and Q, the following statements hold.*

(1) (i) $P \rightarrow (P \vee Q)$. (ii) $Q \rightarrow (P \vee Q)$.

(2) (i) $(P \wedge Q) \rightarrow P$. (ii) $(P \wedge Q) \rightarrow Q$.

(3) $\neg(\neg P) \leftrightarrow P$. (4) $(P \rightarrow Q) \leftrightarrow (\neg Q \rightarrow \neg P)$.

(5) $[(P \rightarrow Q) \wedge P] \rightarrow Q$. (6) $[(P \rightarrow Q) \wedge \neg Q] \rightarrow \neg P$.

(7) (i) $(P \rightarrow Q) \leftrightarrow (\neg P \vee Q)$. (ii) $(P \rightarrow Q) \leftrightarrow [(P \vee Q) \rightarrow Q]$.

(8) $(P \rightarrow Q) \leftrightarrow [(P \wedge \neg Q) \rightarrow (Q \wedge \neg Q)]$.

Proof. (1) (i) We wish to prove the sentence that

"if P and Q are any sentences, then $P \rightarrow (P \vee Q)$ is true."

In other words, we wish to prove that no matter what truth values are assumed by P and Q, $P \to (P \vee Q)$ is always true.

To do this we derive a truth table for $P \to (P \vee Q)$ as follows.

P	Q	$P \vee Q$	$P \to P \vee Q$
true	true	true	true
true	false	true	true
false	true	true	true
false	false	false	true

The basic idea of the derived truth table is this:

For example, in line 1, P and Q both take the value "true".

Thus $P \vee Q$ takes the value "true" by Definition 1.3(3).

Now since P has the value "true" and $P \vee Q$ has the value "true", $P \to (P \vee Q)$ takes the value "true" by Definition 1.3(4).

We do the same for each line.

Thus we find that in every line (that is, for every possible assignment of truth values to P and Q) $P \to (P \vee Q)$ has the value "true".

This is what we had set out to prove that $P \to (P \vee Q)$ is true.

(ii) The derived truth table for $Q \to (P \vee Q)$ is analogous to the one for $P \to (P \vee Q)$ and the conclusion is the same.

(2) (i) In order to prove that $(P \wedge Q) \to P$ is true for all sentences P and Q, we derive a truth table for $(P \wedge Q) \to P$.

P	Q	$P \wedge Q$	$P \wedge Q \to P$
true	true	true	true
true	false	false	true
false	true	false	true
false	false	false	true

In every line, (that is, for every possible assignment of truth values to P and Q), $(P \wedge Q) \to Q$ takes the value "true".

Thus $(P \wedge Q) \to P$ is true irrespective of the truth or falsity of its component sentences P and Q.

(ii) The truth table for $(P \wedge Q) \to Q$ is analogous to the one for $(P \wedge Q) \to P$, and the conclusion is the same.

(3) - (8) We should derive the corresponding truth table for each of the sentences (3) through (8) and verify that the corresponding sentences for (3) through (8) take the truth value "true" in every line of the tables. These are left for the reader. □

Note. For any sentence P, $P \Rightarrow (P \vee P)$ and $P \Rightarrow (P \wedge P)$ by Theorem 1.5.

Remark 1.6. For any sentences P and Q, we have the following truth tables.

P	Q	$\neg\neg P$	$P \to Q$	$P \leftrightarrow Q$	$\neg Q \to \neg P$	$\neg P \wedge \neg Q$
true	true	true	true	true	true	false
true	false	true	false	false	false	false
false	true	false	true	false	true	false
false	false	false	true	true	true	true

P	Q	$P \vee Q$	$Q \to P \vee Q$	$P \vee Q \to Q$	$P \vee Q \leftrightarrow Q$
true	true	true	true	true	true
true	false	true	true	false	false
false	true	true	true	true	true
false	false	false	true	true	true

We can read many interesting facts from the above truth tables.

(1) $P \Leftrightarrow \neg(\neg P)$ for any sentence P.

(2) $(P \to Q) \Leftrightarrow [(P \vee Q) \leftrightarrow Q]$ and $(P \to Q) \Leftrightarrow (\neg Q \to \neg P)$ for any sentences P and Q.

(3) $\neg(P \vee Q) \Leftrightarrow \neg P \wedge \neg Q$ for any sentences P and Q.

Theorem 1.7. *For all sentences P, Q and R, the following statements hold.*

(1) $[(P \to Q) \wedge (Q \to R)] \to (P \to R)$.

(2) $[(P \Rightarrow Q) \wedge (Q \Rightarrow R)] \Rightarrow (P \Rightarrow R)$.

Proof. The reader should derive the truth table for $[(P \to Q) \wedge (Q \to R)] \to (P \to R)$ and verify that the sentence take the truth value "true" in every line of the table. □

Theorem 1.8. *For all sentences P, Q and R, if $Q \to R$ is true, then the following statements hold.*

(1) $(P \vee Q) \to (P \vee R)$. (2) $(P \wedge Q) \to (P \wedge R)$.

Proof. (1) We assume that $Q \to R$ is true, and derive the truth table for $(P \vee Q) \to (P \vee R)$.

P	Q	R	$P \vee Q$	$P \vee R$	$(P \vee Q) \to (P \vee R)$
true	true	true	true	true	true
true	true	false	true	true	true
true	false	true	true	true	true
true	false	false	true	true	true
false	true	true	true	true	true
false	true	false	true	false	false
false	false	true	false	true	true
false	false	false	false	false	true

Since the sentence $Q \to R$ is true, we cannot have simultaneously that Q is true and R is false.

Thus we may disregard the sixth line of the table.

In all of the remaining lines,

$(P \vee Q) \to (P \vee R)$ takes the value "true".

(2) The proof for $(P \wedge Q) \to (P \wedge R)$ is analogous to the above. □

Theorem 1.9. *For all sentences P, Q and R, the following statements hold.*

(1) (i) $(P \vee Q) \leftrightarrow (Q \vee P)$. (ii) $(P \wedge Q) \leftrightarrow (Q \wedge P)$.

(2) (i) $[(P \vee Q) \wedge P] \leftrightarrow P$. (ii) $[(P \wedge Q) \vee P] \leftrightarrow P$.

(3) (i) $(P \vee P) \leftrightarrow P$. (ii) $(P \wedge P) \leftrightarrow P$.

(4) (i) $[P \vee (Q \vee R)] \leftrightarrow [(P \vee Q) \vee R]$. (ii) $[P \wedge (Q \wedge R)] \leftrightarrow [(P \wedge Q) \wedge R]$.

(5) (a) $[P \wedge (Q \vee R)] \leftrightarrow [(P \wedge Q) \vee (P \wedge R)]$.

(b) $[P \vee (Q \wedge R)] \leftrightarrow [(P \vee Q) \wedge (P \vee R)]$.

(c) $[P \vee (Q \vee R)] \leftrightarrow [(P \vee Q) \vee (P \vee R)]$.

(d) $[P \wedge (Q \wedge R)] \leftrightarrow [(P \wedge Q) \wedge (P \wedge R)]$.

(e) $[(P \vee Q) \wedge R] \leftrightarrow [(P \wedge R) \vee (Q \wedge R)]$.

(f) $[(P \wedge Q) \vee R] \leftrightarrow [(P \vee R) \wedge (Q \vee R)]$.

(g) $[(P \vee Q) \vee R] \leftrightarrow [(P \vee R) \vee (Q \vee R)]$.

(h) $[(P \wedge Q) \wedge R] \leftrightarrow [(P \wedge R) \wedge (Q \wedge R)]$.

Proof. It is left as an exercise for the reader. □

Theorem 1.10. *For all sentences P and Q, the following statements hold.*

(1) $\neg(P \vee Q) \leftrightarrow (\neg P \wedge \neg Q)$. (2) $\neg(P \wedge Q) \leftrightarrow (\neg P \vee \neg Q)$.

Proof. It is left as an exercise for the reader. □

Notation 1.11. Here, we list the symbols introduced for our records and give some comments on the usage of the symbols $\Rightarrow$ and $\Leftrightarrow$.

(1) For sentences P, Q, R and S, the following abbreviations were introduced.

$\neg P$: not P. $P \wedge Q$: P and Q. $P \vee Q$: P or Q.

$P \rightarrow Q$: if P then Q. $P \leftrightarrow Q$: if P then Q and if Q then P.

$P \Rightarrow Q$: $P \rightarrow Q$ is true. $P \Leftrightarrow Q$: $P \leftrightarrow Q$ is true.

(2) As is customary, the following abbreviations will also be used if necessary.

$\exists$: there exists. $\forall$: for all. $\cdot\ni\cdot$: such that.

iff: if and only if($\leftrightarrow$). ($\because$): because.

(3) By Theorem 1.7(2), $[(P \Rightarrow Q) \wedge (Q \Rightarrow R)] \Rightarrow [P \Rightarrow R]$. We will write $P \Rightarrow Q \Rightarrow R$ instead of $(P \Rightarrow Q) \wedge (Q \Rightarrow R)$. Also an expression of the form $P_1 \Rightarrow P_2 \Rightarrow P_3 \Rightarrow \cdots \Rightarrow P_n$ should be understood to mean that "$P_1 \Rightarrow P_2$ and $P_2 \Rightarrow P_3 \cdots$ and $P_{n-1} \Rightarrow P_n$." Analogously, $P_1 \Leftrightarrow P_2 \Leftrightarrow P_3 \Leftrightarrow \cdots \Leftrightarrow P_n$ should be understood to mean that "$P_1 \Leftrightarrow P_2$ and $P_2 \Leftrightarrow P_3 \cdots$ and $P_{n-1} \Leftrightarrow P_n$." In other words, $P_1 \Rightarrow P_2 \Rightarrow P_3 \Rightarrow \cdots \Rightarrow P_n$ means $(P_1 \Rightarrow P_2) \wedge (P_2 \Rightarrow P_3) \wedge \cdots \wedge (P_{n-1} \Rightarrow P_n)$, and $P_1 \Leftrightarrow P_2 \Leftrightarrow P_3 \Leftrightarrow \cdots \Leftrightarrow P_n$ means $(P_1 \Leftrightarrow P_2) \wedge (P_2 \Leftrightarrow P_3) \wedge \cdots \wedge (P_{n-1} \Leftrightarrow P_n)$.

(5) An expression "$P_1\ P_2\ \cdots\ P_n$" for sentences $P_1, P_2, \cdots, P_n$, will usually mean the compound sentence $P_1 \wedge P_2 \wedge \cdots P_n$.

Exercise 1.1

1. Prove Theorem 1.5.
2. Prove Theorem 1.7.
3. Prove Theorem 1.9.
4. Prove Theorem 1.10.
5. Prove the following statements for every sentence P.

   (1) $\neg(\neg P) \Rightarrow P$. (2) $P \Rightarrow \neg(\neg P)$.
6. Prove the following statements for all sentences P and Q.

   (1) $(P \to Q) \Leftrightarrow (\neg Q \to \neg P)$. (2) $(P \to Q) \Leftrightarrow (\neg P \vee Q)$.

   (3) $(P \to Q) \Leftrightarrow \neg(P \wedge \neg Q)$. (4) $[P \wedge (P \to Q)] \Rightarrow Q$.

   (5) $[(P \vee Q) \wedge \neg P] \Rightarrow Q$.
7. Prove the following statements for all sentences P, Q and R.

   (1) $[(P \wedge Q) \to R] \Leftrightarrow [P \to (Q \to R)]$.

   (2) $[(P \to R) \wedge (Q \to R)] \Leftrightarrow [(P \vee Q) \to R]$.

(3) $[(P \to R) \vee (Q \to R)] \Leftrightarrow [(P \wedge Q) \to R]$.

(4) $[(P \to Q) \wedge (P \to R)] \Leftrightarrow [P \to (Q \wedge R)]$.

(5) $[(P \to Q) \vee (P \to R)] \Leftrightarrow [P \to (Q \vee R)]$.

8. Assume that $Q \Leftrightarrow R$ for sentences P, Q and R. Prove the following.

(1) $(P \vee Q) \Leftrightarrow (P \vee R)$. (2) $(P \wedge Q) \Leftrightarrow (P \wedge R)$. (3) $(P \to Q) \Leftrightarrow (P \to R)$.

9. Assume that $P \Rightarrow Q$ and $R \Rightarrow S$ for sentences P, Q, R and S. Prove the following statements.

(1) $(P \vee R) \Rightarrow (Q \vee S)$. (2) $(P \wedge R) \Rightarrow (Q \wedge S)$.

1.2 Building classes

We will now begin our development of axiomatic set theory. Every axiomatic system, as we have seen, must start with a certain number of *undefined notions*. For example, in geometry, the words "points" and "lines" and the relation "incidence" are generally taken to be undefined. While we are free in our own minds to attach a "meaning", in the form of a mental picture, to each of these notions, mathematically we must proceed "as if" we did not know what they meant. Now an "undefined" notion has no properties except those which are explicitly assigned to it; therefore, we must state as *axioms* all the elementary properties which we expect our undefined notions to have.

Undefined Notions. In our system of axiomatic set theory, we choose two undefined notions:

the word *class* and the *membership relation* $\in$.

Definition 1.12. A class x is called an *element* if there exists a class A such that $x \in A$.

Remark 1.13. Let us comment briefly on the "meaning" we intend to attach to these notions: elements, classes and the membership relation $\in$.

(1) In the intended interpretation of our axiomatic system, the word *class* means any collection of objects. However as we noted in Section 0.2 of Chapter 0, certain "excessively large" collections can be formed in intuitive set theory (for example, the collection of all x such that $x \notin x$), and if we do not exercise special caution they lead to contradictions such as Russell's paradox.

(2) All the objects of our theory are called classes. Certain classes, to be called *sets*, will be defined in Section 1.7. Every set is a class, but not conversely; a class which is not a set is called a *proper class*. While the word *class* means any collection of objects, the term *proper class* is understood to refer to "*excessively large*" collections and all other collections are *sets* so that a class is either a *proper class* or a *set*.

(3) For classes x and A, the expression $x \in A$ is read "*x is an element of A*" or "*x belongs to A*".

(4) For classes x and A, the notation $x \notin A$ is read "*x is not an element of A*" or "*x does not belong to A*".

(5) The lower-case letters $a, b, c, x, y, \cdots$ will be used only to designate elements and the capital letters $A, B, C, \cdots$ will be used to designate elements or classes.

Definition 1.14. Let A and B be classes.

(1) We say that A is *equal* to B, denoted by $A = B$, if every element of A is an element of B and vice versa. We write $A \neq B$ if A is not equal to B. In symbols,

$$A = B \Leftrightarrow (x \in A \rightarrow x \in B) \wedge (x \in B \rightarrow x \in A).$$

$$A \neq B \Leftrightarrow \neg(A = B).$$

(2) A is called a *subclass* of B, denoted by $A \subseteq B$, if every element of A is an element of B. We write $A \nsubseteq B$ if A is not a subclass of B. In symbols,

$$A \subseteq B \Leftrightarrow (x \in A) \rightarrow (x \in B).$$

$$A \nsubseteq B \Leftrightarrow \neg(A \subseteq B) \Leftrightarrow \exists\, x \cdot\!\ni\!\cdot (x \in A) \wedge (x \notin B).$$

Note that in many other texts, "$\subset$" is used instead of our $\subseteq$.

(3) A is called a *strict subclass* of B, denoted by $A \subset B$, if A is a subclass of B and A is not equal to B. In symbols,

$$A \subset B \Leftrightarrow (A \subseteq B) \wedge (A \neq B).$$

Note that in many other texts, "$\subsetneq$" is used instead of our $\subset$.

We have defined two classes to be equal if and only if they have the same elements. Equal classes have another property: if x and y are equal and x is an element of A, we certainly expect y to be an element of A. This property is stated as our first axiom.

Axiom 1 (Axiom of Extent)**.** Let x, y and A be classes. If $x = y$ and $x \in A$, then $y \in A$.

A few simple properties of equality and inclusion are given in the next theorem.

Theorem 1.15. *Let A, B and C be classes. Then the following are true.*

(1) $A = A$.

(2) $A = B \rightarrow B = A$.

(3) *($A = B$ and $B = C$) $\rightarrow A = C$.*

(4) *($A \subseteq B$ and $B \subseteq A$) $\rightarrow A = B$.*

(5) *($A \subseteq B$ and $B \subseteq C$) $\rightarrow A \subseteq C$.*

Proof. (1) Since $(x \in A \to x \in A)$ is true by Axiom 1 of Extent,

$(x \in A \to x \in A) \wedge (x \in A \to x \in A)$ is true by Theorem 1.9(3)

and hence $A = A$ by Definition 1.14(1).

(2) $A = B$.

$\Rightarrow (x \in A \to x \in B) \wedge (x \in B \to x \in A)$ by Definition 1.14(1).

$\Rightarrow (x \in B \to x \in A) \wedge (x \in A \to x \in B)$ by Theorem 1.9(1).

$\Rightarrow B = A$ by Definition 1.14(1).

(3) $A = B$ and $B = C$.

$\Rightarrow [(x \in A \to x \in B) \wedge (x \in B \to x \in A)] \wedge$

$[(x \in B \to x \in C) \wedge (x \in C \to x \in B)]$ by Definition 1.14.

$\Rightarrow [[(x \in A \to x \in B) \wedge (x \in B \to x \in A)] \wedge (x \in B \to x \in C)]$

$\wedge\ (x \in C \to x \in B)$ by Theorem 1.9(4).

$\Rightarrow (x \in C \to x \in B) \wedge [[(x \in B \to x \in A) \wedge (x \in A \to x \in B)]$

$\wedge\ (x \in B \to x \in C)]$ by Theorem 1.9(1).

$\Rightarrow [(x \in C \to x \in B) \wedge (x \in B \to x \in A)] \wedge$

$[(x \in A \to x \in B) \wedge (x \in B \to x \in C)]$ by Theorem 1.9(4).

$\Rightarrow (x \in C \to x \in A) \wedge (x \in A \to x \in C)$ by Theorem 1.7(2).

$\Rightarrow A = C$ by Definition 1.14(1).

(4) $A \subseteq B$ and $B \subseteq A$.

$\Rightarrow (x \in A \to x \in B)$ and $(x \in B \to x \in A)$ by Definition 1.14(2).

$\Rightarrow A = B$ by Definition 1.14(1).

(5) $A \subseteq B$ and $B \subseteq C$.

$\Rightarrow (x \in A \to x \in B)$ and $(x \in B \to x \in C)$ by Definition 1.14(2).

$\Rightarrow (x \in A \to x \in C)$ by Theorem 1.7(2).

$\Rightarrow A \subseteq C$ by Definition 1.14(2). □

We have seen that the intuitive way of making classes is to name a property of objects and form the class of all the objects which have that property. Our second axiom allows us to make classes in this manner.

Axiom 2 (Axiom of Class Construction)**.** Let $P(x)$ designate a statement about x which can be expressed entirely in terms of symbols $\in$, $\vee$, $\wedge$, $\neg$, $\rightarrow$, $\exists$, $\forall$, brackets, and variables $x, y, z, \cdots, A, B, \cdots$. Then there exists a class C which consists of all the elements x satisfying $P(x)$.

The class C whose existence is asserted by the Axiom 2 will be designated by the symbol $C = \{x \mid P(x)\}$.

Remark 1.16. (1) In Axiom 2 of Class Construction, the expression $\{x \mid P(x)\}$ means the class of all the **elements** x which satisfy $P(x)$. Axiom 2 permits us to form the class of all the **elements** x which satisfy $P(x)$, not the class of all the **classes** x which satisfy $P(x)$. This distinction is sufficient to eliminate the logical paradox. Note that for the class $\mathscr{C}$ defined by the class of all the classes X satisfying $X \notin X$, we meet the logical paradox between $\mathscr{C} \in \mathscr{C}$ and $\mathscr{C} \notin \mathscr{C}$. ($\mathscr{C}$ is a Russell's class. See Section 0.4.)

(2) The semantic paradoxes have been avoided by admitting in Axiom 2 of Class Construction only those statements $P(x)$ which can be written entirely in terms of symbols $\in$, $\vee$, $\wedge$, $\neg$, $\rightarrow$, $\exists$, $\forall$, brackets, and variables $x, y, z, \cdots, A, B, \cdots$.

(3) The use of a small x in the expression $\{x \mid P(x)\}$ is not accident, but quite essential. Note that $C = \{x \mid P(x)\}$ asserts that C is the class of all the **elements** x which satisfy $P(x)$.

(4) Thus $y \in \{x \mid P(x)\}$ asserts that y is an element and that y satisfies the property $P(y)$.

(5) The class equality "=" can be expressed in symbols $\rightarrow$, $\in$ and $\wedge$ by Definition 1.14(1) and Definition 1.3.

Our *set theory* consists of certain statements which built up from the following three types;

the **undefined notions**; the classes $x, y, z, \cdots, A, B, C, \cdots,$

the membership relation $\in$,

the **propositional connectives**; $\wedge$, $\vee$, $\neg$, $\rightarrow$,

the **usual quantifiers**; $\forall$, $\exists$, and the usual brackets.

We will now use Axiom 2 of Class Construction to build some new classes from given classes.

Definition 1.17. Let A and B be classes.

(1) The *union* of A and B, denoted by $A \cup B$, is defined to be the class of all the elements which belong to either to A, or to B, or to both A and B.

The existence of the union $A \cup B$ is a consequence of Axiom 2 because we can express

$$A \cup B = \{x \mid (x \in A) \vee (x \in B)\}.$$

We usually express $A \cup B$ as follows: $A \cup B = \{x \mid x \in A \text{ or } x \in B\}$.

It is convenient to use a *Venn diagram* to represent relations among classes. $A \cup B$ can be expressed as in Figure 1.1. See Remark 1.18(3).

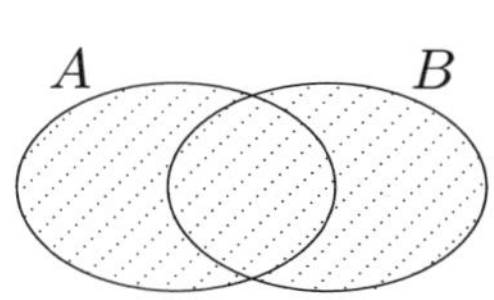

Figure 1.1: $A \cup B$

(2) The *intersection* of A and B, denoted by $A \cap B$, is defined to be the class of all the elements which belong to both A and B.

The existence of the intersection $A \cap B$ is a consequence of Axiom 2 because we can express

$$A \cap B = \{x \mid (x \in A) \wedge (x \in B)\}.$$

We usually express $A \cap B$ as follows: $A \cap B = \{x \mid x \in A \text{ and } x \in B\}$.

In a Venn diagram, $A \cap B$ can be expressed as in Figure 1.2.

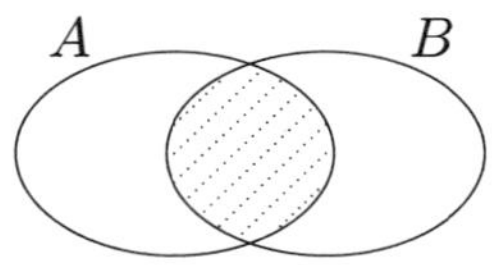

Figure 1.2: $A \cap B$

(3) The *universal class*, denoted by $\mathscr{U}$, is defined to be the class of all elements.

The existence of the universal class $\mathscr{U}$ is a consequence of Axiom 2 because we can express

$$\mathscr{U} = \{x \mid x = x\} = \{x \mid (a \in x) \to (a \in x)\}.$$

(4) The *empty class*, denoted by $\emptyset$, is defined to be the class which has no elements at all.

The existence of the empty class $\emptyset$ is a consequence of Axiom 2 because we can express

$$\emptyset = \{x \mid x \neq x\} = \{x \mid \neg(x = x)\}.$$

(5) A and B are said to be *disjoint* if A and B have no elements in common. In other words,

$$A \text{ and } B \text{ are disjoint } \Leftrightarrow A \cap B = \emptyset.$$

(6) The *complement* of the class A, denoted by A', is defined to be the class of all the elements which do not belong to A. In symbols,

$$A' = \{x \mid x \notin A\} = \{x \mid \neg(x \in A)\}.$$

In a Venn diagram, A' can be expressed as in Figure 1.3.

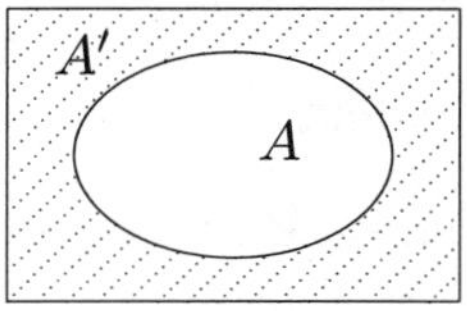

Figure 1.3: A'

(7) The *difference* of A and B, denoted by $A \setminus B$, is defined to be the class of all elements which belong to A, but do not belong to B. In symbols,

$$A \setminus B = \{x \mid (x \in A) \wedge (x \notin B)\} = \{x \mid (x \in A) \wedge \neg(x \in B)\},$$

i.e., $A \setminus B = A \cap B'$.

We usually express $A \setminus B$ as follows: $A \setminus B = \{x \mid x \in A \text{ and } x \notin B\}$.

Remark 1.18. (1) Note that $\{x \mid x = x\} = \{x \mid (a \in x) \to (a \in x)\}$ is a class by Axiom 2 of Class Construction. We claim that the universal class $\mathscr{U}$ can be expressed in symbols as $\mathscr{U} = \{x \mid x = x\}$.

(i) $\mathscr{U} \subseteq \{x \mid x = x\}$.

($\because$) $y \in \mathscr{U}$.

$\Rightarrow$ The statement $y = y$ is true by Theorem 1.15 and y is an element.

$\Rightarrow y \in \{x \mid x = x\}$.

(ii) $\{x \mid x = x\} \subseteq \mathscr{U}$.

($\because$) $y \in \{x \mid x = x\}$.

$\Rightarrow$ y is an element.

$\Rightarrow$ By the definition of the universal class $\mathscr{U}$, $y \in \mathscr{U}$.

Thus $\mathscr{U} = \{x \mid x = x\}$ by Definition 1.14(1).

(2) Note that $\{x \mid \neg(x = x)\}$ is a class by Axiom 2. We claim that the empty class $\emptyset$ can be expressed in symbols as $\emptyset = \{x \mid \neg(x = x)\}$.

($\because$) (i) Since every element x satisfies the property $x = x$ by Theorem 1.15(1), the statement $\lceil y \in \{x \mid \neg(x = x)\}\rfloor$ is not true.

(ii) Since $\emptyset$ consists of no elements at all, the statement $\lceil y \in \emptyset\rfloor$ is not true.

$\Rightarrow \lceil y \in \emptyset \rightarrow y \in \{x \mid \neg(x = x)\}\rfloor$ and $\lceil y \in \{x \mid \neg(x = x)\} \rightarrow y \in \emptyset\rfloor$ are true.

$\Rightarrow \emptyset = \{x \mid \neg(x = x)\}$ by Definition 1.14(1).

(3) Relations among classes can be represented graphically by means of a useful device known as the *Venn diagram.* A class is represented graphically by a simple plane area (circular or oval in shape); if it is desired to show the complement of a class, then the circle or oval is drawn within a rectangle which represents the universal class. Thus, $A \cup B$ is rendered by the shaded area of Figure 1.1, $A \cap B$ by the shaded area of Figure 1.2 and A' by the shaded area of Figure 1.3. The reader will find that Venn diagrams are helpful in guiding his reasoning about classes, and that they give more meaning to set-theoretic formulas by making them more concrete. For example, in Section 1.3 of this chapter we will prove the formula $A \cap (B \cup C) = (A \cap B) \cup (A \cap C)$. This formula is illustrated in Figure 1.4, where the shaded area represents $A \cap (B \cup C)$; one immediately notices that this same shaded area represents $(A \cap B) \cup (A \cap C)$.

Theorem 1.19. *For every class A, the following hold.*

(1) $\emptyset \subseteq A$. (2) $A \subseteq \mathscr{U}$.

Proof. (1) To show that $\emptyset \subseteq A$, we claim that $x \in \emptyset$ implies $x \in A$. Since $\lceil x \in \emptyset\rfloor$ is not true for any element x, $\lceil x \in \emptyset \Rightarrow x \in A\rfloor$

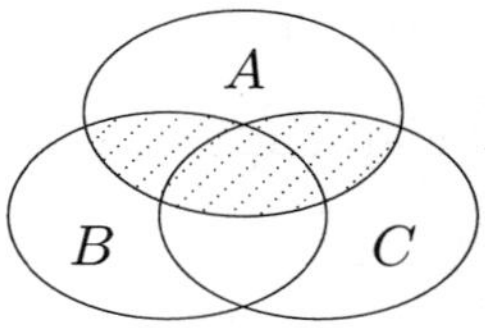

Figure 1.4: $A \cap (B \cup C)$

by Definition 1.3(4). Thus $\emptyset \subseteq A$ by Definition 1.14(2).

(2) To show that $A \subseteq \mathscr{U}$, let $x \in A$.

$\Rightarrow$ x is an element and hence $x \in \mathscr{U}$.

Thus $A \subseteq \mathscr{U}$ by Definition 1.14(2). □

Exercise 1.2

1. Assume that $A \subseteq B$ and $C \subseteq D$. Prove the following.

   (1) $A \cup C \subseteq B \cup D$. (2) $A \cap C \subseteq B \cap D$.

   [Hint Use the result of Exercise 1.1.9.]

2. Assume that $A = B$ and $C = D$. Prove the following.

   (1) $A \cup C = B \cup D$. (2) $A \cap C = B \cap D$.

   [Hint Use the previous Exercise 1.2.1.]

3. Prove that if $A \subseteq B$, then $B' \subseteq A'$.

   [Hint Use the result of Exercise 1.1.6(1).]

4. Prove that if $A = B$, then $B' = A'$.

5. Prove that if $A = B$ and $B \subseteq C$, then $A \subseteq C$.

6. Prove that if $A \subseteq B$ and $B \subseteq C$, then $A \subseteq C$.

7. Prove Theorem 1.15(4).

8. Let $S = \{x \mid x \notin x\}$. Use the Russell's argument to prove that S is not an element.

9. Does Axiom 2 allow us to form the *class of all classes*? Explain your answer.

10. Explain why Russell's paradox and Berry's paradox cannot be produced by using Axiom 2.

1.3 The algebra of classes

One of the most interesting and useful facts about classes is that under the operations of union, intersection, and complementation they satisfy certain algebraic laws from which we can develop an algebra of classes.

Our purpose in this section is to develop the basic laws of the algebra of classes. We remind the reader that the word *class* should be understood to mean *any collection of objects*; thus, the laws we are about to present should be thought of as applying to every collection of objects; in particular, they apply to all *elements* which will be defined later as "sets" in Definition 1.45 of Section 1.7.

Theorem 1.20. *Let A, B and C be any classes. Then the following hold.*

(1) (i) $A \subseteq A \cup B$ *and* $B \subseteq A \cup B$. (ii) $A \cap B \subseteq A$ *and* $A \cap B \subseteq B$.

(2) (i) $(A \subseteq B) \leftrightarrow (A \cup B = B)$. (ii) $(A \subseteq B) \leftrightarrow (A \cap B = A)$.

(3) $(A')' = A$.

(4) (i) $A \cup (A \cap B) = A$. (ii) $A \cap (A \cup B) = A$.

(5) (i) $(A \cup B)' = A' \cap B'$. (ii) $(A \cap B)' = A' \cup B'$.

(6) (i) $A \cup B = B \cup A$. (ii) $A \cap B = B \cap A$.

(7) (i) $A \cup A = A$. (ii) $A \cap A = A$.

(8) (i) $A \cup (B \cup C) = (A \cup B) \cup C$. (ii) $A \cap (B \cap C) = (A \cap B) \cap C$.

(9) (i) $A \cap (B \cup C) = (A \cap B) \cup (A \cap C)$. (ii) $A \cup (B \cap C) = (A \cup B) \cap (A \cup C)$.

(10) (i) $A \cup \emptyset = A$. (ii) $A \cap \emptyset = \emptyset$. (iii) $A \cup \mathscr{U} = \mathscr{U}$. (iv) $A \cap \mathscr{U} = A$.
(v) $\mathscr{U}' = \emptyset$. (vi) $\emptyset' = \mathscr{U}$. (vii) $A \cup A' = \mathscr{U}$. (viii) $A \cap A' = \emptyset$.

Proof. Let A, B and C be any classes.

(1) (i) $x \in A$.

$\Rightarrow (x \in A) \vee (x \in B)$ by Theorem 1.5(1).

$\Rightarrow x \in A$ or $x \in B$ by Definition 1.3(3).

$\Rightarrow x \in A \cup B$ by Definition 1.17(1).

Thus $A \subseteq A \cup B$ by Definition 1.14(2).

Analogously, we can show that $B \subseteq A \cup B$.

(ii) $x \in A \cap B$.

$\Rightarrow x \in A$ and $x \in B$ by Definition 1.17(2).

$\Rightarrow (x \in A) \wedge (x \in B)$ by Definition 1.3(2).

$\Rightarrow x \in A$ by Theorem 1.5(2). Thus $A \cap B \subseteq A$ by Definition 1.14(2).

Analogously, we can show that $A \cap B \subseteq B$.

(2) (i) ($\Rightarrow$) $A \subseteq B$.

$\Rightarrow A \cup B \subseteq B$.

($\because$) $x \in A \cup B$.

$\Rightarrow x \in A$ or $x \in B$ by Definition 1.17(1).

$\Rightarrow$ Since $A \subseteq B$ by assumption, $x \in B$ or $x \in B$ by Definition 1.14(2).

$\Rightarrow (x \in B) \vee (x \in B)$ by Definition 1.3(3).

$\Rightarrow x \in B$ by Theorem 1.9(3). Thus $A \cup B \subseteq B$ by Definition 1.14(2).

$\Rightarrow$ Since $B \subseteq A \cup B$ by (1)(i) above, $B = A \cup B$ by Theorem 1.15(4).

($\Leftarrow$) $B = A \cup B$.

$\Rightarrow A \subseteq B$.

($\because$) $x \in A$.

$\Rightarrow$ Since $A \subseteq A \cup B$ by (1)(i) above, $x \in A \cup B$.

$\Rightarrow$ Since $B = A \cup B$ by assumption, $x \in B$.

Thus $A \subseteq B$ by Definition 1.14(2).

(ii) Analogously, we can show that $A \subseteq B$ iff $A \cap B = A$.

(3) Since $[x \in (A')'] \leftrightarrow [x \notin A'] \leftrightarrow (x \in A)$ is true by Definition 1.17(6),

$(A')' \subseteq A$ and $A \subseteq (A')'$ by Definition 1.14(2).

Thus $(A')' = A$ by Theorem 1.15(4).

(4) (i) Since $A \cap B \subseteq A$ by (1)(ii) above,

$A \cup (A \cap B) = A$ by (2)(i) above.

(ii) Analogously, we can show that $A \cap (A \cup B) = A$.

(5) (i) $x \in (A \cup B)'$.

$\Leftrightarrow x \notin A \cup B$ by Definition 1.17(6).

$\Leftrightarrow \neg(x \in A \cup B)$ by Definition 1.3(1).

$\Leftrightarrow \neg(x \in A$ or $x \in B)$ by Definition 1.17(1).

$\Leftrightarrow \neg[(x \in A) \vee (x \in B)]$ by Definition 1.3(3).

$\Leftrightarrow [\neg(x \in A)] \wedge [\neg(x \in B)]$ by Theorem 1.10(1).

$\Leftrightarrow [\neg(x \in A)]$ and $[\neg(x \in B)]$ by Definition 1.3(2).

$\Leftrightarrow x \notin A$ and $x \notin B$ by Definition 1.3(1).

$\Leftrightarrow x \in A'$ and $x \in B'$ by Definition 1.17(6).

$\Leftrightarrow x \in A' \cap B'$ by Definition 1.17(2).

Thus $(A \cup B)' = A' \cap B'$ by Definition 1.14(1).

(ii) Analogously, we can show that $(A \cap B)' = A' \cup B'$.

(6) (i) $x \in A \cup B$.

$\Leftrightarrow x \in A$ or $x \in B$ by Definition 1.17(1).

$\Leftrightarrow (x \in A) \vee (x \in B)$ by Definition 1.3(3).

$\Leftrightarrow (x \in B) \vee (x \in A)$ by Theorem 1.9(1).

$\Leftrightarrow x \in B$ or $x \in A$ by Definition 1.3(3).

$\Leftrightarrow x \in B \cup A$ by Definition 1.17(1).

Thus $A \cup B = B \cup A$ by Definition 1.14(1).

(ii) Analogously, we can show that $A \cap B = B \cap A$.

(7) (i) $x \in A \cup A$.

$\Leftrightarrow x \in A$ or $x \in A$ by Definition 1.17(1).

$\Leftrightarrow (x \in A) \vee (x \in A)$ by Definition 1.3(3).

$\Leftrightarrow x \in A$ by Theorem 1.9(3).

Thus $A \cup A = A$ by Definition 1.14(1).

(ii) Analogously, we can show that $A \cap A = A$.

(8) (i) $x \in A \cup (B \cup C)$.

$\Leftrightarrow (x \in A)$ or $(x \in B$ or $x \in C)$ by Definition 1.17(1).

$\Leftrightarrow (x \in A) \vee [(x \in B) \vee (x \in C)]$ by Definition 1.3(3).

$\Leftrightarrow [(x \in A) \vee (x \in B)] \vee (x \in C)$ by Theorem 1.9(4).

$\Leftrightarrow (x \in A$ or $x \in B)$ or $(x \in C)$ by Definition 1.3(3).

$\Leftrightarrow x \in (A \cup B) \cup C$ by Definition 1.17(1).

Thus $A \cup (B \cup C) = (A \cup B) \cup C$ by Definition 1.14(1).

(ii) Analogously, we can show that $A \cap (B \cap C) = (A \cap B) \cap C$.

(9) (i) $x \in A \cap (B \cup C)$.

$\Leftrightarrow (x \in A)$ and $(x \in B$ or $x \in C)$ by Definition 1.17(1) and (2).

$\Leftrightarrow (x \in A) \wedge [(x \in B) \vee (x \in C)]$ by Definition 1.3(2) and (3).

$\Leftrightarrow [(x \in A) \wedge (x \in B)] \vee [(x \in A) \wedge (x \in C)]$ by Theorem 1.9(5).

$\Leftrightarrow [(x \in A)$ and $(x \in B)]$ or $[(x \in A)$ and $(x \in C)]$ by Definition 1.3(2) and (3).

$\Leftrightarrow x \in (A \cap B) \cup (A \cap C)$ by Definition 1.17(1) and (2).

Thus $A \cap (B \cup C) = (A \cap B) \cup (A \cap C)$ by Definition 1.14(1).

(ii) Analogously, we can show that $A \cup (B \cap C) = (A \cup B) \cap (A \cup C)$.

(10) (i) Since $\emptyset \subseteq A$ by Theorem 1.19(1), $A \cup \emptyset = A$ by (2)(i) above.

(iv) Since $A \subseteq \mathscr{U}$ by Theorem 1.19(2), $A \cap \mathscr{U} = A$ by (2)(ii) above.

(v) Note that $\mathscr{U} = \{x \mid x = x\}$ and $\emptyset = \{x \mid x \neq x\}$ by Remark 1.18.

For an element x, $x \in \mathscr{U}' \Leftrightarrow \neg(x = x) \Leftrightarrow x \neq x \Leftrightarrow x \in \emptyset$,

$\mathscr{U}' = \emptyset$ by Definition 1.14(1).

(vii) $x \in \mathscr{U}$.

$\Rightarrow$ Since $x \in A$ or $x \in A'$, $x \in A \cup A'$.

Thus $\mathscr{U} \subseteq A \cup A'$ by Definition 1.14(2).

$\Rightarrow$ Since $A \cup A' \subseteq \mathscr{U}$ by Theorem 1.19(2), $\mathscr{U} = A \cup A'$ by Theorem 1.15(4).

The proofs for the remaining parts are left for students. □

Remark 1.21. By using the laws of class algebra in Theorem 1.20, we can prove all the elementary properties of classes without referring to the definitions of the symbols $\cup$, $\cap$, $'$ and $\subseteq$.

Example 1.22. Let A and B be any classes. Prove the following.

(1) $A \cap (A' \cup B) = A \cap B$. (2) $A \setminus B = B' \setminus A'$.

Proof. Let A and B be any classes.

(1) $A \cap (A' \cup B) = (A \cap A') \cup (A \cap B) = \emptyset \cup (A \cap B) = A \cap B$.

(2) $A \setminus B = A \cap B' = B' \cap A = B' \cap (A')' = B' \setminus A'$. □

Exercise 1.3

1. Prove Theorem 1.20(2)(ii), (4)(ii).
2. Prove Theorem 1.20(5)(ii).
3. Prove Theorem 1.20(6)(ii), (7)(ii), (8)(ii), (9)(ii).
4. Prove Theorem 1.20(10)(ii)(iii)(vi)(viii).
5. Use class algebra to prove the following.

   (1) $(A \cap B) \cup C = (A \cup C) \cap (B \cup C)$. (2) $(A \cup B) \cap C = (A \cap C) \cup (B \cap C)$.
6. Use class algebra to prove the following.

   (1) If $A \cap C = \emptyset$, then $A \cap (B \cup C) = A \cap B$.

   (2) If $A \cap B = \emptyset$, then $A \setminus B = A$.

   (3) If $A \cap B = \emptyset$ and $A \cup B = C$, then $A = C \setminus B$.
7. Using class algebra, prove each of the following.

   (1) $A \cap (B \setminus C) = (A \cap B) \setminus C$. (2) $(A \cup B) \setminus C = (A \setminus C) \cup (B \setminus C)$.

   (3) $A \setminus (B \cup C) = (A \setminus B) \cap (A \setminus C)$. (4) $A \setminus (B \cap C) = (A \setminus B) \cup (A \setminus C)$.
8. We define the operation $+$ on classes as follows: For classes A and B, $A + B = (A \setminus B) \cup (B \setminus A)$. Prove each of the following.

   (1) $A + B = B + A$. (2) $A + (B + C) = (A + B) + C$.

   (3) $A \cap (B + C) = (A \cap B) + (A \cap C)$. (4) $A + A = \emptyset$. (5) $A + \emptyset = A$.

9. Prove each of the following.

   (1) If $A \cup B = \emptyset$, then $A = \emptyset$ and $B = \emptyset$.

   (2) $A \cap B' = \emptyset$ if and only if $A \subseteq B$.

   (3) $A + B = \emptyset$ if and only if $A = B$.

10. Prove each of the following.

    (1) $A \cup C = B \cup C$ if and only if $A + B \subseteq C$.

    (2) $(A \cup C) + (B \cup C) = (A + B) \setminus C$.

11. Use class algebra to prove that if $A \subseteq B$ and $C = B \setminus A$, then $A = B \setminus C$.

1.4 Ordered pairs and Cartesian products

Definition 1.23. (1) For an element a, by Axiom 2 of Class Construction, $\{x \mid x = a\}$ forms a class, which contains a single element. A class containing a single element is called a *singleton*. We write $\{a\} = \{x \mid x = a\}$.

(2) For elements a and b, by Axiom 2 of Class Construction, $\{x \mid (x = a) \vee (x = b)\}$ forms a class which contains either only one element or exactly two elements. We write $\{a, b\} = \{x \mid x = a \text{ or } x = b\}$. In like fashion, we can form the classes $\{a, b, c\}, \{a, b, c, d\}$ and so on.

A class containing exactly two elements is called an *unordered pair* or simply a *doubleton*.

(3) Let a and b be elements. Then the *ordered pair* (a, b) is defined to be the class

$$(a, b) = \{\{a\}, \{a, b\}\}.$$

(4) Let A and B be classes. Then the *Cartesian product* $A \times B$ of A and B is defined to be the class of all ordered pairs (x, y) where $x \in A$ and $y \in B$. In symbols,

$$A \times B = \{(x, y) \mid (x \in A) \wedge (y \in B)\}.$$

Remark 1.24. (1) In the expression $\{a, b\}$ of Definition 1.23(2), there is no restriction on a and b. Thus both cases $a = b$ and $a \neq b$ are possible.

(2) By Axiom 2 of Class Construction, for classes A and B, the Cartesian product $A \times B$ forms a class.

(3) Let $A = \{x \mid P(x)\}$ and $B = \{x \mid Q(x)\}$. Then by Definition 1.17(1) and (2),

$$A \cup B = \{x \mid x \in A \text{ or } x \in B\} = \{x \mid P(x) \vee Q(x)\} \text{ and}$$

$$A \cap B = \{x \mid x \in A \text{ and } x \in B\} = \{x \mid P(x) \wedge Q(x)\}, \text{ respectively.}$$

Note that for elements $x_1, x_2, \cdots, x_n$, the notation $\{x_1, x_2, \cdots x_n\}$ is defined by the following class equality

$$\{x_1, x_2, \cdots, x_n\} = \{x \mid x = x_1, \text{ or } x = x_2, \text{ or } \cdots, \text{ or } x = x_n\}.$$

Thus $\{x_1, x_2, \cdots, x_n\} = \{x \mid x = x_1, \text{ or } x = x_2, \text{ or } \cdots, \text{ or } x = x_n\}$

$$= \{x_1\} \cup \{x_2\} \cup \cdots \cup \{x_n\}.$$

Frequently, in mathematics, we need to form classes whose elements are doubletons. In order to be able to do this legitimately, we need a new axiom which will guarantee that if a and b are elements, then the doubleton $\{a, b\}$ is an element. This motivates our next axiom, which is often called the *Axiom of Pair*:

Axiom 3 (Axiom of Pair). If a and b are elements, then $\{a, b\}$ is an element.

Theorem 1.25. (1) *If a is an element, then $\{a\}$ is an element.*

(2) *If $\{a, b\} = \{c, d\}$, then ($a = c$ and $b = d$) or ($a = d$ and $b = c$).*

(3) *If $(a, b) = (c, d)$, then $a = c$ and $b = d$.*

Proof. (1) a is an element.

$\Rightarrow$ $\{a, a\}$ is an element by Axiom 3 of Pair and $\{a\} = \{a, a\}$ by Definition 1.14(1).

$\Rightarrow \{a\}$ is an element.

(2) Assume that $\{a,b\} = \{c,d\}$. Consider the two cases $a = b$ and $a \neq b$.

(Case 1) $a = b$.

$\Rightarrow c = a = b$ and $d = a = b$ because $c, d \in \{c,d\} = \{a,b\}$).

$\Rightarrow a = b = c = d$, which satisfies both statements of the conclusion.

(Case 2) $a \neq b$.

$\Rightarrow$ Since $\{a,b\} = \{c,d\}$, $c \neq d$.

($\because$) $c = d$ implies $a = b = c = d$ which contradicts to $a \neq b$.

Note that $c = a$ or $c = b$ because $c \in \{c,d\} = \{a,b\}$.

If $c = a$, then $a = c$ and $b = d$, which is the first statement of the conclusion.

($\because$) $c = a$.

$\Rightarrow b \in \{a,b\} = \{c,d\} = \{a,d\}$.

$\Rightarrow b = d$ because $a \neq b$.

If $c = b$, then $a = d$ and $b = c$, which is the second statement of the conclusion.

($\because$) $c = b$.

$\Rightarrow a \in \{a,b\} = \{c,d\} = \{b,d\}$.

$\Rightarrow a = d$ because $a \neq b$.

(3) $(a,b) = (c,d)$.

$\Rightarrow \{\{a\},\{a,b\}\} = \{\{c\},\{c,d\}\}$ by Definition 1.23(3).

$\Rightarrow$ Either ($\{a\} = \{c\}$ and $\{a,b\} = \{c,d\}$)

or ($\{a\} = \{c,d\}$ and $\{a,b\} = \{c\}$) by (2) above.

$\Rightarrow$ If $\{a\} = \{c\}$ and $\{a,b\} = \{c,d\}$, then $a = c$, and $b = d$ by (2) above.

If $\{a\} = \{c,d\}$ and $\{a,b\} = \{c\}$, then $a = b = c = d$.

($\because$) $\{a\} = \{c,d\}$ and $\{a,b\} = \{c\}$.

$\Rightarrow$ Since $c \in \{c,d\} = \{a\}$, $c = a$.

Since $d \in \{c,d\} = \{a\}$, $d = a$.

Since $b \in \{a,b\} = \{c\}$, $b = c$.

$\Rightarrow a = b = c = d$.

$\Rightarrow a = c$ and $b = d$. □

Remark 1.26. (1) An important notion in mathematics is that of an *ordered pair* of elements. Intuitively, an *ordered pair* is a class consisting of two elements in a specified order. In fact, the order is not really essential; what is essential is that ordered pairs have the property (3) of Theorem 1.25. The ordered pair in Definition 1.23(4) is defined in such a way to avoid introducing a new undefined notion of "order".

(2) For elements a and b, the ordered pair (a, b) exists and (a, b) is an element.

($\because$) Let a and b be elements.

$\Rightarrow$ The singleton $\{a\}$ is an element by Theorem 1.25, and

$\{a, b\}$ is an element by Axiom 3 of Pairing.

$\Rightarrow$ The ordered pair $(a, b) = \{x \mid (x = \{a\}) \vee (x = \{a, b\})\}$ is a class

by Axiom 2 of class construction and

$(a, b) = \{\{a\}, \{a, b\}\}$ is an element by Axiom 3 of Pair.

(3) Since $(b, a) = \{\{b\}, \{b, a\}\} = \{\{b\}, \{a, b\}\}$, there is a clear distinction between the two possible "orders" (a, b) and (b, a) as explained in Theorem 1.25(3).

Theorem 1.27. *Let A, B, C and D be classes. Then the following hold.*

(1) $A \times (B \cap C) = (A \times B) \cap (A \times C)$.

(2) $A \times (B \cup C) = (A \times B) \cup (A \times C)$.

(3) $(A \times B) \cap (C \times D) = (A \cap C) \times (B \cap D)$.

Proof. Let A, B, C and D be classes.

(1) $(x, y) \in A \times (B \cap C)$.

$\Leftrightarrow x \in A$ and $y \in B \cap C$ by Definition 1.23(4).

$\Leftrightarrow x \in A$ and ($y \in B$ and $y \in C$) by Definition 1.17(2).

$\Leftrightarrow (x \in A) \wedge [(y \in B) \wedge (y \in C)]$ by Definition 1.3(2).

$\Leftrightarrow [(x \in A) \wedge (y \in B)] \wedge [(x \in A) \wedge (y \in C)]$ by Theorem 1.9(5).

$\Leftrightarrow (x \in A$ and $y \in B)$ and $(x \in A$ and $y \in C)$ by Definition 1.3(2).

$\Leftrightarrow (x, y) \in A \times B$ and $(x, y) \in A \times C$ by Definition 1.23(4).

$\Leftrightarrow (x, y) \in (A \times B) \cap (A \times C)$ by Definition 1.17(2).

Thus $A \times (B \cap C) = (A \times B) \cap (A \times C)$ by Definition 1.14(1).

(2) $(x, y) \in A \times (B \cup C)$.

$\Leftrightarrow x \in A$ and $y \in B \cup C$ by Definition 1.23(4).

$\Leftrightarrow x \in A$ and $(y \in B$ or $y \in C)$ by Definition 1.17(1).

$\Leftrightarrow (x \in A) \wedge [(y \in B) \vee (y \in C)]$ by Definition 1.3(2),(3).

$\Leftrightarrow [(x \in A) \wedge (y \in B)] \vee [(x \in A) \wedge (y \in C)]$ by Theorem 1.9(5).

$\Leftrightarrow (x \in A$ and $y \in B)$ or $(x \in A$ and $y \in C)$ by Definition 1.3(2),(3).

$\Leftrightarrow (x, y) \in A \times B$ or $(x, y) \in A \times C$ by Definition 1.23(4).

$\Leftrightarrow (x, y) \in (A \times B) \cup (A \times C)$ by Definition 1.17(1).

Thus $A \times (B \cup C) = (A \times B) \cup (A \times C)$ by Definition 1.14(1).

(3) $(x, y) \in (A \times B) \cap (C \times D)$.

$\Leftrightarrow (x, y) \in A \times B$ and $(x, y) \in C \times D$ by Definition 1.17(2).

$\Leftrightarrow (x \in A$ and $y \in B)$ and $(x \in C$ and $y \in D)$ by Definition 1.23(4).

$\Leftrightarrow [(x \in A) \wedge (y \in B)] \wedge [(x \in C) \wedge (y \in D)]$ by Definition 1.3(2).

$\Leftrightarrow [(x \in A) \wedge (x \in C)] \wedge [(y \in B) \wedge (y \in D)]$ by using Theorem 1.9(1) and (4), repeatedly.

$\Leftrightarrow (x \in A$ and $x \in C)$ and $(y \in B$ and $y \in D)$ by Definition 1.3(2).

$\Leftrightarrow x \in A \cap C$ and $y \in B \cap D$ by Definition 1.17(2).

$\Leftrightarrow (x, y) \in (A \cap C) \times (B \cap D)$ by Definition 1.23(4).

Thus $(A \times B) \cap (C \times D) = (A \cap C) \times (B \cap D)$ by Definition 1.14(1). □

Remark 1.28. Just as we found it instructive to represent relations between classes by means of Venn diagram, it is often convenient to illustrate relations between products of classes by using a graphic device known as a *coordinate diagram.*

A coordinate diagram is analogous to the familiar Cartesian coordinate plane. There are two axes - a vertical one and a horizontal one - but we consider only one "quadrant". If we wish to represent a class $A \times B$, then a segment of the horizontal axis is marked off to represent A and a segment of the vertical axis is marked off to represent B; $A \times B$ is the rectangle determined by these two segments (See Figure 1.5). As an example of the use of coordinate diagrams, Theorem 1.27(3) is illustrated in Figure 1.6.

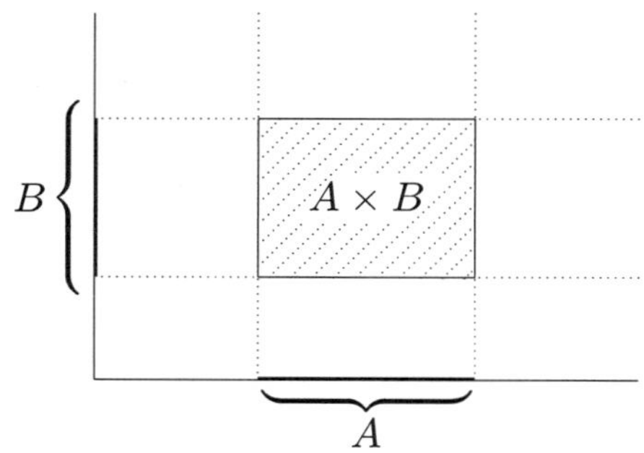

Figure 1.5: the Cartesian product of A and B

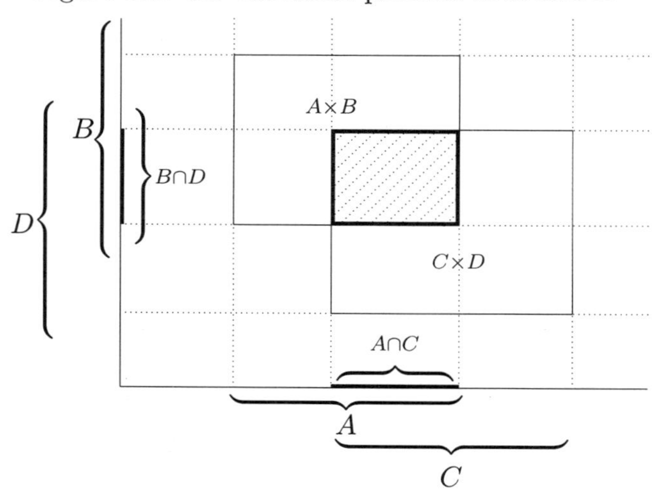

Figure 1.6: $(A \times B) \cap (C \times D) = (A \cap C) \times (B \cap D)$

Exercise 1.4

1. Let $A = \{a, b, c, d\}, B = \{1, 2, 3\}$, and $C = \{x, y, z\}$. Find $A \times B, B \times A$, $C \times (B \times A)$, $(A \cup B) \times C$, $(A \times C) \cup (B \times C)$, and $(A \cup B) \times (B \cup C)$.

2. Prove that $A \times (B \setminus D) = (A \times B) \setminus (A \times D)$.

3. Prove that $(A \times B) \cap (C \times D) = (A \times D) \cap (C \times B)$.

4. If A, B and C are classes, prove the following.

   (1) $(A \times A) \cap (B \times C) = (A \cap B) \times (A \cap C)$.

   (2) $(A \times B) \setminus (C \times C) = ((A \setminus C) \times B) \cup (A \times (B \setminus C))$.

   (3) $(A \times A) \setminus (B \times C) = ((A \setminus B) \times A) \cup (A \times (A \setminus C))$.

5. Prove that A and B are disjoint if and only if, for any nonempty class C, $A \times C$ and $B \times C$ are disjoint.

6. If A and C are nonempty classes, prove that $A \subseteq B$ and $C \subseteq D$ if and only if $A \times C \subseteq B \times D$.

7. Let A, B, C and D be nonempty classes. Prove that $A \times B = C \times D$ if and only if $A = C$ and $B = D$.

8. If A, B and C are classes, prove the following.

   (1) $A \times B$ and $A' \times C$ are disjoint. (2) $B \times A$ and $C \times A'$ are disjoint.

9. Prove that $A \times B = \emptyset$ if and only if $A = \emptyset$ or $B = \emptyset$.

10. Prove each of the following.

    (1) If $a = \{b\}$, then $b \in a$. (2) $x = y$ if and only if $\{x\} = \{y\}$.

    (3) $x \in a$ if and only if $\{x\} \subseteq a$. (4) $\{a, b\} = \{a\}$ if and only if $a = b$.

11. We give the following alternative definition of ordered pairs:

    $(x, y) = \{\{x, \emptyset\}, \{y, \{\emptyset\}\}\}$. Using this definition, prove that

    $$(a, b) = (c, d) \Rightarrow a = c \text{ and } b = d.$$

1.5 Graphs

Definition 1.29. (1) A *graph* is defined to be a class of ordered pairs. Thus, G is a graph if and only if G is a subclass of $\mathscr{U} \times \mathscr{U}$. i.e., $G \subseteq \mathscr{U} \times \mathscr{U}$.

(2) For a graph G, the *inverse graph* of G, denoted by G^{-1}, is the graph defined by

$$G^{-1} = \{(y, x) \mid (x, y) \in G\}.$$

(3) For two graphs G and H, the *composite graph* of G followed by H, denoted by $G \circ H$, is the graph defined by

$$G \circ H = \{(x, z) \mid \text{there exists } y \text{ such that } (x, y) \in H \text{ and } (y, z) \in G\}.$$

(4) For a graph G, the *domain of* G, denoted by $\mathrm{dom}(G)$, and the *range of* G, denoted by $\mathrm{ran}(G)$, are defined to be the classes

$$\mathrm{dom}(G) = \{x \mid \text{there exists } y \text{ such that } (x, y) \in G\},$$

$$\mathrm{ran}(G) = \{y \mid \text{there exists } x \text{ such that } (x, y) \in G\}.$$

Remark 1.30. The domain of a graph G is the class of all the *first components* of elements of G and the range of a graph G is the class of all the *second components* of elements of G.

Theorem 1.31. *Let G, H and J be graphs. Then the following hold.*

(1) $(G \circ H) \circ J = G \circ (H \circ J)$. (2) $(G^{-1})^{-1} = G$.

(3) $(G \circ H)^{-1} = H^{-1} \circ G^{-1}$.

Proof. Let G, H and J be graphs.

(1) $(x, y) \in (G \circ H) \circ J$.

$\Leftrightarrow$ There exists z such that $(x, z) \in J$ and $(z, y) \in G \circ H$ by Definition 1.29(3).

$\Leftrightarrow$ There exist z and w such that $(x, z) \in J$ and $[(z, w) \in H$ and $(w, y) \in G]$ by Definition 1.29(3).

$\Leftrightarrow$ There exist z and w such that $[(x,z) \in J$ and $(z,w) \in H]$ and $(w,y) \in G$ by applying Theorem 1.9(4).

$\Leftrightarrow$ There exists w such that $(x,w) \in H \circ J$ and $(w,y) \in G$ by Definition 1.29(3).

$\Leftrightarrow (x,y) \in G \circ (H \circ J)$ by Definition 1.29(3).

Thus $(G \circ H) \circ J = G \circ (H \circ J)$ by Definition 1.14(1).

(2) Since $(x,y) \in (G^{-1})^{-1} \Leftrightarrow (y,x) \in G^{-1} \Leftrightarrow (x,y) \in G$ by Definition 1.29(2), $(G^{-1})^{-1} = G$ by Definition 1.14(1).

(3) $(x,y) \in (G \circ H)^{-1}$.

$\Leftrightarrow (y,x) \in G \circ H$ by Definition 1.29(2).

$\Leftrightarrow$ There exists z such that $(y,z) \in H$ and $(z,x) \in G$ by Definition 1.29(3).

$\Leftrightarrow$ There exists z such that $(x,z) \in G^{-1}$ and $(z,y) \in H^{-1}$ by Definition 1.29(2).

$\Leftrightarrow (x,y) \in H^{-1} \circ G^{-1}$ by Definition 1.29(3).

Thus $(G \circ H)^{-1} = H^{-1} \circ G^{-1}$ by Definition 1.14(1). □

Theorem 1.32. *For graphs G and H, the following hold.*

(1) $\mathrm{dom}(G) = \mathrm{ran}(G^{-1})$. (2) $\mathrm{ran}(G) = \mathrm{dom}(G^{-1})$.

(3) $\mathrm{dom}(G \circ H) \subseteq \mathrm{dom}(H)$. (4) $\mathrm{ran}(G \circ H) \subseteq \mathrm{ran}(G)$.

Proof. Let G and H be graphs.

(1) $x \in \mathrm{dom}(G)$.

$\Leftrightarrow$ There exists y such that $(x,y) \in G$ by Definition 1.29(4).

$\Leftrightarrow$ There exists y such that $(y,x) \in G^{-1}$ by Definition 1.29(2).

$\Leftrightarrow x \in \mathrm{ran}(G^{-1})$ by Definition 1.29(4).

Thus $\mathrm{dom}(G) = \mathrm{ran}(G^{-1})$ by Definition 1.14(1).

(2) $y \in \mathrm{ran}(G)$.

$\Leftrightarrow$ There exists x such that $(x,y) \in G$ by Definition 1.29(4).

$\Leftrightarrow$ There exists x such that $(y,x) \in G^{-1}$ by Definition 1.29(2).

$\Leftrightarrow y \in \mathrm{dom}(G^{-1})$ by Definition 1.29(4).

Thus $\operatorname{ran}(G) = \operatorname{dom}(G^{-1})$ by Definition 1.14(1).

(3) $x \in \operatorname{dom}(G \circ H)$.

$\Rightarrow$ There exists z such that $(x, z) \in G \circ H$ by Definition 1.29(4).

$\Rightarrow$ There exists y such that $(x, y) \in H$ and $(y, z) \in G$ by Definition 1.29(3).

$\Rightarrow$ $x \in \operatorname{dom}(H)$ by Definition 1.29(4) and hence $\operatorname{dom}(G \circ H) \subseteq \operatorname{dom}(H)$ by Definition 1.14(2).

(4) $z \in \operatorname{ran}(G \circ H)$.

$\Rightarrow$ There exists x such that $(x, z) \in G \circ H$ by Definition 1.29(4).

$\Rightarrow$ There exists y such that $(x, y) \in H$ and $(y, z) \in G$ by Definition 1.29(3).

$\Rightarrow$ $z \in \operatorname{ran}(G)$ by Definition 1.29(4) and hence $\operatorname{ran}(G \circ H) \subseteq \operatorname{ran}(G)$ by Definition 1.14(2). □

Corollary 1.33. *Let G and H be graphs.*
If $\operatorname{ran}(H) \subseteq \operatorname{dom}(G)$, *then* $\operatorname{dom}(G \circ H) = \operatorname{dom}(H)$.

Proof. Let G and H be graphs such that $\operatorname{ran}(H) \subseteq \operatorname{dom}(G)$.

$\Rightarrow$ $\operatorname{dom}(H) \subseteq \operatorname{dom}(G \circ H)$.

($\because$) $x \in \operatorname{dom}(H)$.

$\Rightarrow$ There exists y such that $(x, y) \in H$ by Definition 1.29(4).

$\Rightarrow$ $y \in \operatorname{ran}(H)$ by Definition 1.29(4).

$\Rightarrow$ Since $\operatorname{ran}(H) \subseteq \operatorname{dom}(G)$, $y \in \operatorname{dom}(G)$ by Definition 1.14(2).

$\Rightarrow$ There exists z such that $(y, z) \in G$ by Definition 1.29(4).

$\Rightarrow$ Since $(x, y) \in H$, $(x, z) \in G \circ H$ by Definition 1.29(3).

$\Rightarrow$ $x \in \operatorname{dom}(G \circ H)$ by Definition 1.29(4) and hence $\operatorname{dom}(H) \subseteq \operatorname{dom}(G \circ H)$ by Definition 1.14(2).

$\Rightarrow$ Since $\operatorname{dom}(G \circ H) \subseteq \operatorname{dom}(H)$ by Theorem 1.32(3), $\operatorname{dom}(H) = \operatorname{dom}(G \circ H)$ by Definition 1.14(1). □

Exercise 1.5

1. Let $G = \{(b, b), (b, c), (c, c), (c, d)\}$ and $H = \{(b, a), (c, b), (d, c)\}$.
   Find G^{-1}, H^{-1}, $G \circ H$, $H \circ G$, $(G \circ H)^{-1}$, $(G \cup H)^{-1}$, $H^{-1} \circ G$.

2. If G, H and J are graphs, prove each of the following.
   (1) $(H \cup J) \circ G = (H \circ G) \cup (J \circ G)$. (2) $(G \setminus H)^{-1} = G^{-1} \setminus H^{-1}$.
   (3) $G \circ (H \cap J) \subseteq (G \circ H) \cap (G \circ J)$. (4) $(G \circ H) \setminus (G \circ J) \subseteq G \circ (H \setminus J)$.

3. If G and H are graphs, prove each of the following.
   (1) $(G \cap H)^{-1} = G^{-1} \cap H^{-1}$. (2) $(G \cup H)^{-1} = G^{-1} \cup H^{-1}$.

4. If G, H, J and K are graphs, prove each of the following.
   (1) If $G \subseteq H$ and $J \subseteq K$, then $G \circ J \subseteq H \circ K$.
   (2) $G \subseteq H$ if and only if $G^{-1} \subseteq H^{-1}$.

5. If A, B and C are classes, prove each of the following.
   (1) $(A \times B)^{-1} = B \times A$.
   (2) If $A \cap B \neq \emptyset$, then $(A \times B) \circ (A \times B) = A \times B$.
   (3) If A and B are disjoint, then $(A \times B) \circ (A \times B) = \emptyset$.
   (4) If $B \neq \emptyset$, then $(B \times C) \circ (A \times B) = A \times C$.

6. Let G and H be graphs. Prove each of the following.
   (1) If $G \subseteq A \times B$, then $G^{-1} \subseteq B \times A$.
   (2) If $G \subseteq A \times B$ and $H \subseteq B \times C$, then $H \circ G \subseteq A \times C$.

7. If G and H are graphs, prove each of the following.
   (1) $\mathrm{dom}(G \cup H) = \mathrm{dom}(G) \cup \mathrm{dom}(H)$. (2) $\mathrm{ran}(G \cup H) = \mathrm{ran}(G) \cup \mathrm{ran}(H)$.
   (3) $\mathrm{dom}(G) \setminus \mathrm{dom}(H) \subseteq \mathrm{dom}(G \setminus H)$. (4) $\mathrm{ran}(G) \setminus \mathrm{ran}(H) \subseteq \mathrm{ran}(G \setminus H)$.

8. Let G be a graph and let B and C be subclasses of the domain of G. By the *restriction* of G to B, we mean the graph

   $$G|_B = \{(x, y) \mid (x, y) \in G \text{ and } x \in B\}.$$

   Prove each of the following.
   (1) $G|_B = G \cap (B \times \mathrm{ran}(G))$. (2) $G|_{B \cup C} = G|_B \cup G|_C$.
   (3) $G|_{B \cap C} = G|_B \cap G|_C$. (4) $(G \circ H)|_B = G \circ (H|_B)$.

9. Let G be a graph and let B be a subclass of the domain of G. We use the symbol $G(B)$ to designate the class

$$G(B) = \{y \mid \text{there exists } x \in B \text{ such that } (x, y) \in G\}.$$

Prove each of the following.

(1) $G(B) = \text{ran}(G|_B)$.

(2) $G(B \cup C) = G(B) \cup G(C)$.

(3) $G(B \cap C) = G(B) \cap G(C)$.

(4) If $B \subseteq C$, then $G(B) \subseteq G(C)$.

1.6 Generalized union and intersection

Consider the class $\{A_1, A_2, \cdots, A_n\}$ whose elements are indexed by the numbers $1, 2, \cdots, n$. The class $\{A_1, A_2, \cdots, A_n\}$ is often called an *indexed family* of classes and the numbers $1, 2, \cdots, n$ are called *indices* and the class $\{1, 2, \cdots, n\}$ is called the *index class.*

More generally, we are frequently led to think of a class I whose elements $i, j, k, \cdots$ serve as indices to designate the elements of a class $\{A_i, A_j, A_k, \cdots\}$. The class $\{A_i, A_j, A_k, \cdots\}$ is called an *indexed family of classes*, I is called its *index class*, and the elements of I are called *indices.* A compact notation which is often used to designate the class $\{A_i, A_j, A_k, \cdots\}$ is $\{A_i\}_{i \in I}$.

Thus speaking informally, $\{A_i\}_{i \in I}$ is the class of all the classes A_i, as i ranges over I.

Remark 1.34. The definition of an indexed family of classes which we have just given is, admittedly, an intuitive one; it relies on the intuitive notion of *indexing.* This intuitive definition is adequate at the present time; however, for future reference and to avoid such an intuitive notion, we now give a *formal* definition of the same concept:

Definition 1.35. By an *indexed family of classes*, $\{A_i\}_{i \in I}$, indexed by a class I, we mean a graph G whose domain is I defined by
$A_i = \{x \mid (i, x) \in G\}$ for each $i \in I$.

Remark 1.36. The correspondence between the intuitive definition $\{A_i\}_{i\in I}$ in the beginning of this section and the graph G in the formal definition(Definition 1.35) is as follows;

$$A_i = \{x \mid (i,x) \in G\} \;\longleftrightarrow\; \{(i,x) \mid x \in A_i\}(\subseteq G) \text{ for each } i \in I.$$

Example 1.37. Let $I = \{1,2\}, A_1 = \{a,b\}, A_2 = \{c,d\}$. Then $\{A_i\}_{i\in I}$ is intuitively $\{\{a,b\},\{c,d\}\}$ and formally is the graph $G = \{(1,a)$, $(1,b)$, $(2,c)$, $(2,d)\}$.

If $\{A_i\}_{i\in I}$ is an indexed family of classes such that for each $i \in I$, A_i is an element, then we let $\{A_i \mid i \in I\}$ designate the class whose elements are all the A_i, that is, $\{A_i \mid i \in I\} = \{x \mid x = A_i \text{ for some } i \in I\}$. However, we shall follow current mathematical usage and use the two expressions, $\{A_i\}_{i\in I}$ and $\{A_i \mid i \in I\}$, interchangeably.

Definition 1.38. Let $\{A_i\}_{i\in I}$ be an indexed family of classes and let $\mathscr{A}$ be a class whose elements are classes.

(1) The *union of the classes* A_i consists of all the elements which belong to at least one class A_i of the family. In symbols,

$$\bigcup_{i\in I} A_i = \{x \mid (\exists j \in I) \wedge (x \in A_j)\}.$$

The *union* of $\mathscr{A}$, or *union of the elements of* $\mathscr{A}$ is defined to be the union of all the classes which are elements of $\mathscr{A}$. In symbols,

$$\bigcup_{A\in\mathscr{A}} A = \{x \mid (\exists A \in \mathscr{A}) \wedge (x \in A)\}.$$

(2) The *intersection of the classes* A_i consists of all the elements which belong to every class A_i of the family. In symbols,

$$\bigcap_{i\in I} A_i = \{x \mid x \in A_i \ \forall i \in I\}.$$

The *intersection* of $\mathscr{A}$, or *intersection of the elements of* $\mathscr{A}$ is defined to be the union of all the classes which are elements of $\mathscr{A}$. In symbols,

$$\bigcap_{A\in\mathscr{A}} A = \{x \mid x \in A \ \forall A \in \mathscr{A}\}.$$

Notation 1.39. Let $\mathscr{A}$ be a class whose elements are classes and let A_i be a class for each index $i \in I$. We use the following notations.

(1) $\{A_i\}_{i \in I} = \{A_i \mid i \in I\}$.

(2) $\bigcup \mathscr{A} = \bigcup\limits_{A \in \mathscr{A}} A = \{x \mid x \in A \text{ for some } A \in \mathscr{A}\}$.

(3) $\bigcap \mathscr{A} = \bigcap\limits_{A \in \mathscr{A}} A = \{x \mid x \in A \text{ for all } A \in \mathscr{A}\}$.

Example 1.40. Let $\mathscr{A} = \{K, L, M\}$ where $K = \{a, b, d\}, L = \{a, c, d\}$ and $M = \{d, e\}$. Then $\bigcup\limits_{A \in \mathscr{A}} A = \{a, b, c, d, e\}$ and $\bigcap\limits_{A \in \mathscr{A}} A = \{d\}$.

Theorem 1.41. *Let $\{A_i\}_{i \in I}$ be an indexed family of classes. Then the following hold.*

(1) *If $A_i \subseteq B$ for all $i \in I$, then $\bigcup\limits_{i \in I} A_i \subseteq B$.*
(2) *If $B \subseteq A_i$ for all $i \in I$, then $B \subseteq \bigcap\limits_{i \in I} A_i$.*

Proof. Let $\{A_i\}_{i \in I}$ be an indexed family of classes.

(1) Assume that $A_i \subseteq B$ for all $i \in I$ and let $x \in \bigcup\limits_{i \in I} A_i$.

$\Rightarrow x \in A_i$ for some $i \in I$ by Definition 1.38(1).

$\Rightarrow x \in B$ by Definition 1.14(2) because $A_i \subseteq B$.

Thus $\bigcup\limits_{i \in I} A_i \subseteq B$ by Definition 1.14(2).

(2) Assume that $B \subseteq A_i$ for all $i \in I$ and let $x \in B$.

$\Rightarrow x \in A_i$ for all $i \in I$ by Definition 1.14(2).

$\Rightarrow x \in \bigcap\limits_{i \in I} A_i$ by Definition 1.38(2).

Thus $B \subseteq \bigcap\limits_{i \in I} A_i$ by Definition 1.14(2). □

Theorem 1.42 (Generalized DeMorgan's Laws).
Let $\{A_i\}_{i \in I}$ be an indexed family of classes. Then the following hold.

(1) $(\bigcup\limits_{i \in I} A_i)' = \bigcap\limits_{i \in I} A_i{}'$. (2) $(\bigcap\limits_{i \in I} A_i)' = \bigcup\limits_{i \in I} A_i{}'$.

Proof. Let $\{A_i\}_{i\in I}$ be an indexed family of classes.

(1) $x \in (\bigcup_{i\in I} A_i)'$.

$\Leftrightarrow x \notin \bigcup_{i\in I} A_i$ by Definition 1.17(6).

$\Leftrightarrow x \notin A_i$ for all $i \in I$ by applying Theorem 1.5(4) and Definition 1.38(1).

$\Leftrightarrow x \in A_i'$ for all $i \in I$ by Definition 1.17(6).

$\Leftrightarrow x \in \bigcap_{i\in I} A_i'$ by Definition 1.38(2).

Thus $(\bigcup_{i\in I} A_i)' = \bigcap_{i\in I} A_i'$ by Definition 1.14(1).

(2) $x \in (\bigcap_{i\in I} A_i)'$.

$\Leftrightarrow x \notin \bigcap_{i\in I} A_i$ by Definition 1.17(6).

$\Leftrightarrow$ There exists $i \in I$ such that $x \notin A_i$ by applying Theorem 1.5(4) and Definition 1.17(2).

$\Leftrightarrow$ There exists $i \in I$ such that $x \in A_i'$ by Definition 1.17(6).

$\Leftrightarrow x \in \bigcup_{i\in I} A_i'$ by Definition 1.38(1).

Thus $(\bigcap_{i\in I} A_i)' = \bigcup_{i\in I} A_i'$ by Definition 1.14(1). □

Theorem 1.43 (Generalized Distributive Laws). *Let $\{A_i\}_{i\in I}$ and $\{B_j\}_{j\in J}$ be indexed families of classes. Then the following hold.*

(1) $(\bigcup_{i\in I} A_i) \bigcap (\bigcup_{j\in J} B_j) = \bigcup_{(i,j)\in I\times J} (A_i \bigcap B_j)$.

(2) $(\bigcap_{i\in I} A_i) \bigcup (\bigcap_{j\in J} B_j) = \bigcap_{(i,j)\in I\times J} (A_i \bigcup B_j)$.

Proof. Let $\{A_i\}_{i\in I}$ and $\{B_j\}_{j\in J}$ be indexed families of classes.

(1) $x \in (\bigcup_{i\in I} A_i) \bigcap (\bigcup_{j\in J} B_j)$.

$\Leftrightarrow x \in \bigcup_{i\in I} A_i$ and $x \in \bigcup_{j\in J} B_j$ by Definition 1.17(2).

$\Leftrightarrow x \in A_h$ for some $h \in I$ and $x \in B_k$ for some $k \in J$ by Definition 1.38(1).

$\Leftrightarrow x \in A_h \cap B_k$ for some $(h, k) \in I \times J$ by Definition 1.17(2) and Definition 1.23(4).

$\Leftrightarrow x \in \bigcup_{(i,j)\in I\times J} (A_i \bigcap B_j)$ by Definition 1.38(1).

Thus $(\bigcup_{i\in I} A_i) \bigcap (\bigcup_{j\in J} B_j) = \bigcup_{(i,j)\in I\times J} (A_i \bigcap B_j)$ by Definition 1.14(1).

(2) $x \in (\bigcap_{i\in I} A_i) \bigcup (\bigcap_{j\in J} B_j)$.

$\Leftrightarrow x \in \bigcap_{i\in I} A_i$ or $x \in \bigcap_{j\in J} B_j$ by Definition 1.17(1).

$\Leftrightarrow (x \in A_h$ for all $h \in I)$ or $(x \in B_k$ for all $k \in J)$ by Definition 1.38(2).

$\Leftrightarrow x \in A_h \cup B_k$ for all $(h,k) \in I \times J$ by Definition 1.38(2) and Definition 1.23(4).

$\Leftrightarrow x \in \bigcap_{(i,j)\in I\times J} (A_i \bigcup B_j)$ by Definition 1.38(2).

Thus $(\bigcap_{i\in I} A_i) \bigcup (\bigcap_{j\in J} B_j) = \bigcap_{(i,j)\in I\times J} (A_i \bigcup B_j)$ by Definition 1.14(1). □

Theorem 1.44. *Let $\{G_i\}_{i\in I}$ be an indexed family of graphs. Then the following hold.*

(1) $\mathrm{dom}(\bigcup_{i\in I} G_i) = \bigcup_{i\in I} \mathrm{dom}(G_i)$. (2) $\mathrm{ran}(\bigcup_{i\in I} G_i) = \bigcup_{i\in I} \mathrm{ran}(G_i)$.

Proof. Let $\{G_i\}_{i\in I}$ be an indexed family of graphs.

(1) $x \in \mathrm{dom}(\bigcup_{i\in I} G_i)$.

$\Leftrightarrow$ There exists y such that $(x,y) \in \bigcup_{i\in I} G_i$ by Definition 1.29(4).

$\Leftrightarrow$ There exists y such that $(x,y) \in G_j$ for some $j \in I$ by Definition 1.38(1).

$\Leftrightarrow x \in \mathrm{dom}(G_j)$ for some $j \in I$ by Definition 1.29(4).

$\Leftrightarrow x \in \bigcup_{i\in I} \mathrm{dom}(G_i)$ by Definition 1.38(1).

Thus $\mathrm{dom}(\bigcup_{i\in I} G_i) = \bigcup_{i\in I} \mathrm{dom}(G_i)$ by Definition 1.14(1).

(2) $y \in \mathrm{ran}(\bigcup_{i\in I} G_i)$.

$\Leftrightarrow$ There exists x such that $(x,y) \in \bigcup_{i\in I} G_i$ by Definition 1.29(4).

$\Leftrightarrow$ There exists x such that $(x,y) \in G_j$ for some $j \in I$ by Definition 1.38(1).

$\Leftrightarrow y \in \mathrm{ran}(G_j)$ for some $j \in I$ by Definition 1.29(4).

$\Leftrightarrow y \in \bigcup_{i\in I} \mathrm{ran}(G_i)$ by Definition 1.38(1).

Thus $\mathrm{ran}(\bigcup_{i\in I} G_i) = \bigcup_{i\in I} \mathrm{ran}(G_i)$ by Definition 1.14(1). □

Exercise 1.6

1. Let $\{A_i\}_{i\in I}$ and $\{B_i\}_{i\in I}$ be two families of classes with the same index class I. Assume that $A_i \subseteq B_i$ for all $i \in I$. Prove the following.

   (1) $\bigcup_{i\in I} A_i \subseteq \bigcup_{i\in I} B_i$. (2) $\bigcap_{i\in I} A_i \subseteq \bigcap_{i\in I} B_i$.

2. Let $\{A_i\}_{i\in I}$ and $\{B_j\}_{j\in J}$ be indexed families of classes. Prove the following.

   (1) $(\bigcap_{i\in I} A_i) \times (\bigcap_{j\in J} B_j) = \bigcap_{(i,j)\in I\times J} (A_i \times B_j)$.

   (2) $(\bigcup_{i\in I} A_i) \times (\bigcup_{j\in J} B_j) = \bigcup_{(i,j)\in I\times J} (A_i \times B_j)$.

3. Let $\{A_i\}_{i\in I}$ and $\{B_j\}_{j\in J}$ be indexed families of classes. Assume that for each $i \in I$, there exists $j \in J$ such that $B_j \subseteq A_i$. Prove that $\bigcap_{j\in J} B_j \subseteq \bigcap_{i\in I} A_i$.

4. Let $\{A_i\}_{i\in I}$ and $\{B_j\}_{j\in J}$ be indexed families of classes. Prove the following.

   (1) $(\bigcup_{i\in I} A_i) \setminus (\bigcup_{j\in J} B_j) = \bigcup_{i\in I} (\bigcap_{j\in J} (A_i \setminus B_j))$.

   (2) $(\bigcap_{i\in I} A_i) \setminus (\bigcap_{j\in J} B_j) = \bigcap_{i\in I} (\bigcup_{j\in J} (A_i \setminus B_j))$.

5. We say that an indexed family $\{B_i\}_{i\in I}$ is a *covering* of A if $A \subseteq \bigcup_{i\in I} A_i$. Assume that $\{B_i\}_{i\in I}$ and $\{C_j\}_{j\in J}$ are two coverings of A. Prove that the family $\{B_i \cap C_j\}_{(i,j)\in I\times J}$ is a covering of A.

6. Let $a = \{u, v, w\}$, $b = \{w, x\}$, $c = \{w, y\}$, $r = \{a, b\}$, $s = \{b, c\}$, and $p = \{r, s\}$. Find the classes $\cup(\cup p)$, $\cap(\cap p)$, $\cup(\cap p)$, $\cap(\cup p)$.

7. Let $\mathscr{A}$ and $\mathscr{B}$ be classes. Prove that $\cap(\mathscr{A} \cup \mathscr{B}) = (\cap\mathscr{A}) \cap (\cap\mathscr{B})$.

8. Let $\mathscr{A}$ and $\mathscr{B}$ be classes. Prove each of the following.

   (1) If $A \in \mathscr{B}$, then $A \subseteq \bigcup \mathscr{B}$ and $\bigcap \mathscr{B} \subseteq A$.

   (2) $\mathscr{A} \subseteq \mathscr{B}$ if and only if $\bigcup \mathscr{A} \subseteq \bigcup \mathscr{B}$.

   (3) If $\emptyset \in \mathscr{A}$, then $\bigcap \mathscr{A} = \emptyset$.

1.7 Sets

Undoubtedly, everything we have said in the preceding pages is fairly familiar to the reader. Even though we said *class* where the reader is more accustomed to hearing *set*, it is obvious that the "union" and "intersection" defined in this chapter are precisely the familiar union and intersection of sets, the "Cartesian product" is exactly the usual Cartesian product of sets, and similarly for the other concepts introduced in this chapter. At this point, it appears as though everything we are accustomed to doing with sets can be done with classes. Thus, the reader may very well ask. "Why bother distinguishing between classes and sets? Why not develop all of mathematics in terms of classes?" Since, as we have said, a class means "any collection of elements," why not simply call a class a set, and be done with it? The answer to this question is of great importance; the chief purpose of this section is to explain why we *do* want to distinguish between classes and sets.

First, we note that the axiom of class construction (Axiom 2) permits us to form the *class of all elements* which satisfy a given property; it does not allow us to form the *class of all classes* which satisfy a given property. The reason for this limitation is obvious: if we were enabled to form the class of all classes which satisfy any given property, then we could form "Russell's class" of all classes which are not elements of themselves, and this would give us Russell's paradox.

Next, we note that in mathematics we often need to form particular *sets of sets.* A few examples which come to mind are the following:

The set of all closed intervals $[a, b]$ of real numbers.

The set of all convergent sequences of real numbers.

The set of all the lines in the plane (where each line is regarded as a set of points).

Let us look more closely at the first example; in a discussion in elementary calculus, we would feel perfectly free to say "let $\mathscr{A}$ consist of all the sets which are closed intervals $[a, b]$ of real numbers". Now if "sets" were no different from "classes" then, by the proceeding paragraph, we would not be allowed to form $\mathscr{A}$. This would be an intolerable restriction upon our freedom of operating with sets in mathematics.

Let us recapitulate: The notion of *class* is appealing because of its intuitive simplicity and generality; however, there is a serious drawback to dealing with classes, namely that it is not permissible [for an arbitrary property $P(X)$] to form the "class of all classes X which satisfy $P(X)$". This would be an intolerable restriction on our mathematical freedom of action if we were to base mathematics upon classes. Instead, we base mathematics upon *sets*; the concept of a set is somewhat narrower than that of a class; sets will be defined in such a way, however, that for any property $P(X)$, *it is legitimate to form the class of all sets X which satisfy $P(X)$.* Thus, the freedom we require is restored to us, provided we are willing to deal with sets rather than the broader notion of classes. We are more than willing to do so, because, as we will be able to show, all the classes we deal with in mathematics *are sets.*

Now, how should sets be defined? In order to answer this question, it is essential to remind ourselves, once again, that we are seeking a way of defining sets which makes it legitimate to form the class of all *sets* X which satisfy any property $P(X)$. One obvious answer is to identify sets with elements: let a set be the same thing as an element. Then Axiom 2 of Class Construction will certainly permit us to form the class of all sets X which satisfy any property $P(X)$.

This simple answer is, in fact, the one which has been adopted by most mathematicians. We will use it here; thus,

Definition 1.45. A *set* is defined to be the class which is an element of a class. A subclass of a set is called a *subset* and a class which is not a set is called a *proper class.*

This definition is supported by our intuitive perception of what a set should be. For if $A, B, C, \cdots$ are sets, it is perfectly reasonable that we should be able to form the class $\{A, B, C, \cdots\}$ whose elements are $A, B, C, \cdots$. In other words, we would quite certainly expect every set to be an element. The converse is equally reasonable: for if A is *not* a set, then A is a proper class, and we have already seen that in order to avoid contradictions, proper classes should not be elements of anything. Thus, if A is not a set, then A is not an element.

In the remainder of this section, we will state the basic axioms dealing with sets. The main purpose of these axioms is to guarantee that when the usual set-theoretic operations are performed on sets, the result, each time, is a set.

First the Axiom 3 of Pair may be restated in terms of sets as follows;

Axiom 3(Axiom of Pair) If a and b are sets, then $\{a, b\}$ is a set.

Now, if B is a set and $A \subseteq B$, one would reasonably expect A to be a set. This is the content of our next axiom, called the *Axiom of Subset.*

Axiom 4 (Axiom of Subset)**.** Every subclass of a set is a set.

Remark 1.46. The universal class $\mathscr{U}$ consisting of all sets is not a set.

Proof. Suppose that the universal class $\mathscr{U}$ is a set.

Consider $B = \{x \in \mathscr{U} \mid x \notin x\}$.

$\Rightarrow$ Since $B \subseteq \mathscr{U}$, B is a set by Axiom 4 of Subset.

$\Rightarrow$ One of the statements $B \in B$ or $B \notin B$ must be true.

$\Rightarrow$ If $B \in B$, then $B \notin B$ by the definition of B, which is a contradiction.

If $B \notin B$, then $B \in B$ by the definition of B, which is a contradiction.

$\Rightarrow$ Both of the statements $B \in B$ and $B \notin B$ are not true and we arrive at a contradiction! Thus the universal class $\mathscr{U}$ is not a set. □

The union of "not too many" sets should be a set. This is guaranteed by our next axiom, called the *Axiom of Union.*

Axiom 5 (Axiom of Union)**.** If $\mathscr{A}$ is a set of sets, then $\bigcup_{A\in\mathscr{A}} A$ is a set.

Theorem 1.47. *Let A and B be sets. Then*

(1) *$A\cap B$ is a set.* (2) *$A\cup B$ is a set.*

Proof. Let A and B be sets.

(1) Since $A\cap B$ is a subclass of the set A by Theorem 1.20(1), $A\cap B$ is a set by Axiom 4 of Subset.

(2) Since A and B are sets, $\{A,B\}$ is a set by Axiom 3 of Pair.

$\Rightarrow$ Since $\bigcup_{X\in\{A,B\}} X = A\cup B$ by Definition 1.38(1), $A\cup B$ is a set by Axiom 5 of Union. □

Remark 1.48. Since every doubleton containing two elements is a set by Axiom 3 of Pair, every singleton containing a single element is a set by letting $a=b$ in Axiom 3 of Pair. Since the union of two sets is a set, every class of three elements, which is a union of a doubleton and a singleton, is a set, and every class of four elements is a set, and so on. Thus in an intuitive sense, every finite class is a set.

Definition 1.49. For a set A, the *power set* of A, $\mathscr{P}(A)$, is defined to be the class of all subsets of A. In symbols, $\mathscr{P}(A)=\{B\mid B\subseteq A\}$.

Example 1.50. If $A=\{a,b\}$, then $\mathscr{P}(A)=\{\emptyset,\{a\},\{b\},\{a,b\}\}$.

Now we will establish that if A is a set, then the class of all the subsets of A is a set. This is the content of our next axiom, called the *Axiom of Power Set.*

Axiom 6 (Axiom of Power Set)**.** If A is a set, then $\mathscr{P}(A)$ is a set.

Remark 1.51. Recall that the existence of the empty *class* $\emptyset$ was guaranteed by Axiom 2 in Remark 1.18(2). From all that we have said so far, it does not yet follow that there exist any sets at all. To fill this vacuum, we assume that the empty class $\emptyset$ is a set. This will be followed from Axiom 9 of Infinity later in chapter 6. Henceforth on, we will refer to the empty class $\emptyset$ as the empty *set*.

From the empty set, together with Axiom 3 and Axiom 5, we may infer the existence of great many sets. We have the empty set $\emptyset$; we have singletons such as $\{\emptyset\}, \{\{\emptyset\}\}$, etc.; we have doubletons such as $\{\emptyset, \{\emptyset\}\}$, formed by any two of the above. Similarly, taking unions of the above repeatedly, we may form sets with *any finite number* of elements.

Theorem 1.52. *Let A be a set and let $P(X)$ be a property of X. Then the class of all the subsets of A which satisfy $P(X)$ is a set.*

Proof. Let A be a set and let $P(X)$ be a property of X and let B denote the class of all the subsets X of A which satisfy $P(X)$.
i.e., $B = \{X \mid (X \subseteq A) \wedge P(X)\}$.
$\Rightarrow$ B forms a class by Axiom 2 of Class Construction.
$\Rightarrow$ $B \subseteq \mathscr{P}(A)$. ($\because$ $X \in B \Rightarrow X$ is a subset of $A \Rightarrow X \in \mathscr{P}(A)$.)
$\Rightarrow$ Since $\mathscr{P}(A)$ is a set by Axiom 6 of Power Set,
B is a set by Axiom 4 of Subset. □

Theorem 1.53. *If A and B are sets, then $A \times B$ is a set.*

Proof. Let A and B be sets.
(i) $\mathscr{P}(\mathscr{P}(A \cup B))$ is a set.
($\because$) Since A and B are sets, $A \cup B$ is a set by Theorem 1.47.
$\Rightarrow$ $\mathscr{P}(A \cup B)$ is a set by Axiom 6 of Power Set.
$\Rightarrow$ $\mathscr{P}(\mathscr{P}(A \cup B))$ is a set by Axiom 6 of Power Set again.
(ii) $A \times B \subseteq \mathscr{P}(\mathscr{P}(A \cup B))$.
($\because$) $(x, y) \in A \times B$.
$\Rightarrow$ $(x, y) = \{\{x\}, \{x, y\}\}$, $x \in A$ and $y \in B$ by Definition 1.23.
$\Rightarrow$ Since $x \in A \cup B$, $\{x\} \subseteq A \cup B$ and hence $\{x\} \in \mathscr{P}(A \cup B)$.
Similarly, since $\{x, y\} \subseteq A \cup B$, $\{x, y\} \in \mathscr{P}(A \cup B)$.
$\Rightarrow$ $\{\{x\}, \{x, y\}\} \subseteq \mathscr{P}(A \cup B)$ by Definition 1.49.

$\Rightarrow \{\{x\}, \{x, y\}\} \in \mathscr{P}(\mathscr{P}(A \cup B))$ by Definition 1.49.

$\Rightarrow (x, y) \in \mathscr{P}(\mathscr{P}(A \cup B))$ by Axiom 1 of Extent.

$\Rightarrow A \times B$ is a set by Axiom 4 of Subset by (i) and (ii) above. □

Corollary 1.54. *Let A and B be sets and let G and H be graphs.*

(1) *If $G \subseteq A \times B$, then G,* $\operatorname{dom}(G)$ *and* $\operatorname{ran}(G)$ *are sets.*

(2) *If a graph G is a set, then* $\operatorname{dom}(G)$ *and* $\operatorname{ran}(G)$ *are also sets.*

(3) *If graphs G and H are sets, then $G \circ H$ and G^{-1} are also sets.*

Proof. Let A and B be sets and let G and H be graphs.

(1) Assume that $G \subseteq A \times B$.

(i) G is a set.

($\because$) Since A and B are sets, $A \times B$ is a set by Theorem 1.53.

$\Rightarrow$ Since $G \subseteq A \times B$, G is a set by Axiom 4 of Subset.

(ii) $\operatorname{dom}(G)$ and $\operatorname{ran}(G)$ are sets.

($\because$) Since $\operatorname{dom}(G) \subseteq A$ and $\operatorname{ran}(G) \subseteq B$,

$\operatorname{dom}(G)$ and $\operatorname{ran}(G)$ are sets by Axiom 4 of Subset.

(2) Assume that G is a set.

(i) $\cup(\cup G)$ is a set.

($\because$) Since G is a set, $\cup G$ is a set by Axiom 5 of Union.

$\Rightarrow \cup(\cup G)$ is a set by Axiom 5 of Union again.

(ii) $\operatorname{dom}(G)$ and $\operatorname{ran}(G)$ are subclasses of the set $\cup(\cup G)$.

($\because$) $x \in \operatorname{dom}(G)$.

$\Rightarrow$ There exists y such that $(x, y) \in G$ by Definition 1.29(4).

$\Rightarrow$ Since $(x, y) = \{\{x\}, \{x, y\}\}$, $\{\{x\}, \{x, y\}\} \in G$.

$\Rightarrow \{x\} \in \cup G$ and hence $x \in \cup(\cup G)$.

$\Rightarrow \operatorname{dom}(G) \subseteq \cup(\cup G)$ by Definition 1.14(2).

Similarly, we can show that $\operatorname{ran}(G) \subseteq \cup(\cup G)$.

$\Rightarrow \operatorname{dom}(G)$ and $\operatorname{ran}(G)$ are sets by Axiom 4 of Subset.

(3) G and H are sets.

$\Rightarrow$ dom(G), ran(G), dom(H) and ran(H) are sets by (2) above.

$\Rightarrow$ dom(H) $\times$ ran(G) and ran(G) $\times$ dom(G) are sets by Theorem 1.53.

$\Rightarrow$ (i) $G \circ H$ is a set.

($\because$) Note that $G \circ H \subseteq \text{dom}(H) \times \text{ran}(G)$ because if $(x, y) \in G \circ H$, then there exists z such that $(x, z) \in H$ and $(z, y) \in G$ and hence $(x, y) \in \text{dom}(H) \times \text{ran}(G)$.

$\Rightarrow$ Since dom(H) $\times$ ran(G) is a set, $G \circ H$ is a set by Axiom 4 of Subset.

(ii) G^{-1} is a set.

($\because$) Since $G^{-1} \subseteq \text{ran}(G) \times \text{dom}(G)$, G^{-1} is a set by Axiom 4 of Subset. □

Exercise 1.7

1. If A and B are sets, prove that $A \setminus B$ and $A + B$ are sets. (See Exercise 1.3.8)

2. If A is a proper class and $A \subseteq B$, prove that B is a proper class. Conclude that the union of two proper classes is a proper class.

3. Prove that the Russell class and the universal class are proper classes. [Hint: Use the result of Exercise 1.2.8.]

4. Let $\{A_i\}_{i \in I}$ be an indexed family of sets. Prove that $\bigcap_{i \in I} A_i$ is a set.

5. Let $r = \{a, b\}$, $s = \{b, c\}$, and $p = \{r, s\}$. Find the sets $\mathscr{P}(r)$, $\mathscr{P}(\mathscr{P}(r))$, and $\mathscr{P}(\cup p)$.

6. Let A and B be sets. Prove that the following statements are true.

   (1) $A \subseteq B$ if and only if $\mathscr{P}(A) \subseteq \mathscr{P}(B)$.

   (2) $A = B$ if and only if $\mathscr{P}(A) = \mathscr{P}(B)$.

   (3) $\mathscr{P}(A) \cap \mathscr{P}(B) = \mathscr{P}(A \cap B)$.

   (4) $\mathscr{P}(A) \cup \mathscr{P}(B) \subseteq \mathscr{P}(A \cup B)$.

   (5) $A \cap B = \emptyset$ if and only if $\mathscr{P}(A) \cap \mathscr{P}(B) = \{\emptyset\}$.

7. If A and $\mathscr{B}$ are sets, prove that the following statements are true.

   (1) $\cup(\mathscr{P}(\mathscr{B})) = \mathscr{B}$. (2) $\cap(\mathscr{P}(\mathscr{B})) = \emptyset$.

   (3) If $\mathscr{P}(A) \in \mathscr{P}(\mathscr{B})$, then $A \in \mathscr{B}$.

8. Exhibit the set $\mathscr{P}\big(\mathscr{P}(\emptyset)\big)$ and $\mathscr{P}\left(\mathscr{P}\big(\mathscr{P}(\emptyset)\big)\right)$.

9. For any proper class $\mathscr{C}$, prove that the collection $\mathscr{P}\big(\mathscr{C}\big)$ of all subclasses of $\mathscr{C}$ is not a class.

Chapter 2

Functions

2.1 Introduction

The concept of a function is one of the most basic mathematical ideas and enters into almost every mathematical discussion. We start with an unofficial definition of a function and its graph.

(Unofficial Definition of a Function and Its Graph) Let A and B be classes.

(1) A *function* from A to B, denoted by $f : A \to B$, is a rule which to every element $x \in A$ assign a unique element $y \in B$. To indicate this connection between x and y we usually write $y = f(x)$.

(2) The *graph* of a function $f : A \to B$ is defined to be the class of all ordered pairs (x, y) such that $y = f(x)$.

Example 2.1. (1) Consider the function $y = \sin x$. Then it is easy to see that $y = \sin x$ is a rule which, to every number $x \in \mathbb{R}$, assigns a unique number $y \in [-1, 1]$ and $\sin : \mathbb{R} \to [-1, 1]$ is a function.

(2) Let $A = \{a, b, c\}$ and $B = \{d, e\}$, and let f be the function defined by the following table.

x	$f(x)$
a	d
b	e
c	d

Then the graph of f is $\{(a,d),(b,e),(c,d)\}$.

Remark 2.2. (1) The usuage of the notation $\to$ for a function $f : A \to B$ is different from that of $\to$ for $P \to Q$ in Definition 1.3(4).

(2) In Example 2.1(2), we may use the information contained in the table to construct the graph of f, and conversely, we may also use the information contained in the graph to construct the table of f. Thus a function f completely determines its graph, and conversely, its graph completely determines f. Hence there is no need to distinguish between a function and its graph.

(3) Since a function and its graph are essentially one and the same thing, we may *define* a function to be a graph. There is an advantage to be gained by doing this. Namely, we avoid having to introduce the word *rule* as a new undefined concept of set theory. For this reason it is customary, in rigorous treatments of mathematics, to introduce the notion of *function* via that of *graph*. We shall follow that procedure here. And an *official* definition of a function will be given in this way in the next section.

2.2 Fundamental concepts and definitions

We begin by giving our ⌜*official*⌟ definition of a function.

Definition 2.3. Let A and B be classes.

(1) A *function* from A to B is a triple of objects $\langle f, A, B\rangle$, where f is a subclass of $A \times B$ with the following properties.

F1. For all $x \in A$, there exists $y \in B$ such that $(x, y) \in f$.

F2. If $(x, y_1) \in f$ and $(x, y_2) \in f$, then $y_1 = y_2$.

(2) Note that the function $\langle f, A, B\rangle$ is defined as a special type of a graph. Thus as defined in Definition 1.29, for a function $\langle f, A, B\rangle$,

the class $A = \{x \mid \text{there exists } y \text{ such that } (x, y) \in f\}$ is called the *domain* of $\langle f, A, B\rangle$, which is denoted by $\mathrm{dom}(f)$,

the class $\mathrm{ran}(f) = \{y \in B \mid \text{there exists } x \text{ such that } (x, y) \in f\}$ is called the *range* of $\langle f, A, B\rangle$, which is denoted by $\mathrm{ran}(f)$ and

the class B is called the *codomain* of $\langle f, A, B\rangle$, which is denoted by $\mathrm{cod}(f)$.

(3) It is customary to write $f : A \to B$ instead of $\langle f, A, B\rangle$. Even though there is a distinction between the function $f : A \to B$ and the subclass $f \subseteq A \times B$ we write simply f, as a function $f : A \to B$, if there is no ambiguity on its domain and codomain.

For a function $f : A \to B$, if $(x, y) \in f$, then we say that y is the *image* of x (with respect to f) and that x is the pre-image of y (with respect to f) or that f *maps* x *onto* y.

If y is the image of x (with respect to f), we write $x \overset{f}{\mapsto} y$ or $y = f(x)$. Thus, $x \overset{f}{\mapsto} y$ or $y = f(x)$ has the same meaning as $(x, y) \in f$.

Remark 2.4. (1) In ordinary mathematical applications, every function $\langle f, A, B\rangle$ is a function from a set A to a set B. However, the intuitive concept of a function from A to B is meaningful for any two collections A and B, whether A and B be sets or proper classes. Hence it is natural to give the definition of a function in its most general form, letting A and B be any classes. Once again, every set is a class, hence everything we have to say about functions from a class A to a class B applies, in particular, to functions from a set A to a set B.

(2) For three elements x, y, z, the *ordered triple* can be defined formally by either $(x, y, z) = \{\{x\}, \{x, y\}, \{x, y, z\}\}$ or $(x, y, z) = \{\{x \times \{\emptyset\}\}, \{y \times \{\{\emptyset\}\}\}, \{z \times \{\{\{\emptyset\}\}\}\}\}$. In general for n elements $x_1, x_2, \cdots, x_n$, the *ordered n-tuple* can be defined by either
$(x_1, x_2, \cdots, x_n) = \{\{x_1\}, \{x_1, x_2\}, \cdots, \{x_1, x_2, \cdots, x_n\}\}$ or
$(x_1, x_2, \cdots, x_n) = \{\{x_1 \times \{\emptyset\}\}, \{x_2 \times \{\{\emptyset\}\}\}, \cdots, \{x_n \times \{\cdots \{\{\{\emptyset\}\}\} \cdots \}\}\}$. Note that all these classes are sets.

(3) For three classes X, Y, Z, the *ordered triple* can be defined formally by $\langle X, Y, Z\rangle = \{\{X \times \{\emptyset\}\}, \{Y \times \{\{\emptyset\}\}\}, \{Z \times \{\{\{\emptyset\}\}\}\}\}$. In general for n classes $X_1, X_2, \cdots, X_n$, the *ordered n-tuple* can be defined by $(X_1, X_2, \cdots, X_n) = \{\{X_1 \times \{\emptyset\}\}, \{X_2 \times \{\{\emptyset\}\}\}, \cdots, \{X_n \times \{\cdots \{\{\emptyset\}\} \cdots \}\}\}$. Note that these are classes by Axiom 2 of Class Construction, but may not be sets.

Remark 2.5. The conditions F1 and F2 in Definition 2.3 can be restated in many different ways.

(1) The conditions F1 and F2 can be restated as follows:

F1. Every element $x \in A$ has an image $y \in B$.

F2. If $x \in A$, then the image of x is unique.

($\because$ If $(x, y_1) \in f$ and $(x, y_2) \in f$, then $y_1 = y_2$ by F2.)

(2) When we write $y = f(x)$ instead of $(x, y) \in f$, the conditions F1 and F2 take the following forms.

F1. For all $x \in A$, there exists $y \in B$ such that $y = f(x)$.

F2. If $y_1 = f(x)$ and $y_2 = f(x)$ for $x \in A$, then $y_1 = y_2$.

(3) It is often convenient to write F2 in a slightly different way.

F2. If $x_1 \overset{f}{\mapsto} f(x_1)$, $x_2 \overset{f}{\mapsto} f(x_2)$, and $x_1 = x_2$, then $f(x_1) = f(x_2)$.

(4) Thus, F2 may be written in the following form.

F2°. If $x_1 = x_2$ and $x_1 \in A$, then $f(x_1) = f(x_2)$.

Theorem 2.6. *Let A and B be classes and let $f : A \to B$ be a function. Then every element $x \in A$ has a uniquely determined image $y \in B$.*

Proof. It follows directly from Definition 2.3 or Remark 2.5. □

Theorem 2.7. *Let A and B be classes and let f be a graph. Then $f : A \to B$ is a function if and only if the following conditions are satisfied.*

(1) F2. (2) $\text{dom}(f) = A$. (3) $\text{ran}(f) \subseteq B$.

Proof. Let A and B be classes and let f be a graph.

(only if) Assume that $f : A \to B$ is a function.

(1) F2 holds for f by Definition 2.3.

(2) $\text{dom}(f) = A$.

(i) $\text{dom}(f) \subseteq A$.

($\because$) $x \in \text{dom}(f)$.

$\Rightarrow$ There exists y such that $(x, y) \in f$ by Definition 1.29(4).

$\Rightarrow$ Since $f \subseteq A \times B$ by Definition 2.3, $(x, y) \in A \times B$.

$\Rightarrow$ $x \in A$ by Definition 1.23(4) and hence $\text{dom}(f) \subseteq A$ by Definition 1.14(2).

(ii) $A \subseteq \text{dom}(f)$.

($\because$) $x \in A$.

$\Rightarrow$ There exists y such that $(x, y) \in f$ by F1.

$\Rightarrow$ $x \in \text{dom}(f)$ by Definition 1.29 and hence $A \subseteq \text{dom}(f)$ by Definition 1.14(2).

(3) $\text{ran}(f) \subseteq B$.

($\because$) $y \in \text{ran}(f)$.

$\Rightarrow$ There exists x such that $(x, y) \in f$ by Definition 1.29.

$\Rightarrow$ Since $f \subseteq A \times B$ by Definition 2.3, $(x, y) \in A \times B$.

$\Rightarrow$ $y \in B$ by Definition 1.23(4) and hence $\text{ran}(f) \subseteq B$ by Definition 1.14(2).

(if) Conversely, assume that (1), (2), and (3) hold.

(i) $f \subseteq A \times B$.

($\because$) $(x, y) \in f$.

$\Rightarrow x \in \text{dom}(f)$ and $y \in \text{ran}(f)$ by Definition 1.29.

$\Rightarrow x \in A$ and $y \in B$ by the assumptions (2) and (3).

$\Rightarrow (x, y) \in A \times B$ by Definition 1.23(4) and hence $f \subseteq A \times B$ by Definition 1.14(2).

(ii) f satisfies F1.

($\because$) $x \in A$.

$\Rightarrow x \in \text{dom}(f)$ by the assumption (2).

$\Rightarrow$ There exists an element y such that $(x, y) \in f$.

$\Rightarrow y \in \text{ran}(f)$ and $(x, y) \in f$.

$\Rightarrow y \in B$ by the assumption (3), and $(x, y) \in f$.

$\Rightarrow$ F1 holds for f.

$\Rightarrow$ Since F2 holds by the assumption (1), $f : A \to B$ is a function by Definition 2.3. □

Remark 2.8. From Theorem 2.7 we conclude that if $f : A \to B$ is a function, then A is the domain of f and B contains the range of f. We will state and prove it in the following corollary.

Corollary 2.9. *Let $f : A \to B$ be a function. If C is any class such that $\text{ran}(f) \subseteq C$, then $f : A \to C$ is a function.*

Proof. Let $f : A \to B$ be a function and let C be a class with $\text{ran}(f) \subseteq C$.

$\Rightarrow$ By Theorem 2.7, F2 holds for f and $\text{dom}(f) = A$.

$\Rightarrow$ Since $\text{ran}(f) \subseteq C$, $f : A \to C$ is a function by Theorem 2.7. □

Theorem 2.10. *Let $f : A \to B$ and $g : A \to B$ be functions. Then $f = g$ if and only if $f(x) = g(x)$ for all $x \in A$.*

Proof. **(only if)** $f = g$.

$\Rightarrow$ For all $x \in A$, $y = f(x)$ iff $(x, y) \in f$ iff $(x, y) \in g$ iff $y = g(x)$.

$\Rightarrow$ $f(x) = g(x)$ for all $x \in A$.

(if) $f(x) = g(x)$ for all $x \in A$.

$\Rightarrow$ For all $x \in A$, $(x, y) \in f$ iff $y = f(x)$ iff $y = g(x)$ iff $(x, y) \in g$.

$\Rightarrow$ Since $f : A \to B$ and $g : A \to B$ are functions, $f = g$ by Definition 2.3. □

The following definition is of great importance in the study of functions.

Definition 2.11. Let $f : A \to B$ be a function. Then

(1) $f : A \to B$ is said to be *injective* if it has the following property:

INJ. If $(x_1, y) \in f$ and $(x_2, y) \in f$, then $x_1 = x_2$.

(2) $f : A \to B$ is said to be *surjective* if it has the following property:

SRJ. For all $y \in B$, there exists $x \in A$ such that $y = f(x)$

(3) $f : A \to B$ is said to be *bijective* if it is both injective and surjective. In this case, we say that A *and* B *are in one-to-one correspondence.*

Remark 2.12. Let $f : A \to B$ be a function.

(1) The condition INJ in Definition 2.11(1) can be restated in many different ways.

INJ. If $y \in B$, then y has no more than one pre-image.

($\because$) If x_1, x_2 are pre-images of y, then $(x_1, y), (x_2, y) \in f$ and hence $x_1 = x_2$ means that y has no more than one pre-image.

INJ. Every element of B is the image of *no more than one* element of A.

INJ°. If $x_1 \in A$, $x_2 \in A$ and $f(x_1) = f(x_2)$, then $x_1 = x_2$.

(2) The condition SRJ in Definition 2.11(2) can be restated in many different ways.

SRJ. $\operatorname{ran}(f) = B$.

($\because$) The condition SRJ of Definition 2.11(2) holds.

$\Leftrightarrow$ Every element of B is the image of some element of A.

$\Leftrightarrow B \subseteq \operatorname{ran}(f)$.

$\Leftrightarrow \operatorname{ran}(f) = B$ because $\operatorname{ran}(f) \subseteq B$ by Theorem 2.7(3).

SRJ. Every element of B is the image of *at least one* element of A.

(3) $f : A \to B$ is bijective if and only if every element of B is the image of *exactly one* element of A. Thus if $f : A \to B$ is bijective, then every element of A has exactly one image in B and every element of B is the image of exactly one element of A, and hence all the elements of A and all the elements of B are associated in pairs.

Example 2.13. Let A and B be classes.

(1) **Identity Function.** By the *identity function on* A we mean the function $1_A : A \to A$ given by

$$1_A(x) = x \text{ for all } x \in A, \quad \text{i.e., } 1_A = \{(x, x) \mid x \in A\}.$$

Note the following.

(i) 1_A is injective.

($\because$) $1_A(x) = 1_A(y)$ for some $x, y \in A$.

$\Rightarrow$ Since $1_A(x) = x$ and $1_A(y) = y$, $x = y$.

$\Rightarrow$ INJ$^\circ$ holds.

(ii) 1_A is surjective because $\operatorname{ran}(1_A) = A$.

(iii) 1_A is bijective because it is injective and bijective.

(2) **Constant Function.** Let $b \in B$. By the *constant function* c_b at b from A to B we mean the function $c_b : A \to B$ given by

$$c_b(x) = b \text{ for all } x \in A, \quad \text{i.e., } c_b = \{(x, b) \mid x \in A\}.$$

Note that if B has more than one element, then c_b is not surjective.

(3) **Inclusion Function.** If B is a subclass of A, then the *inclusion function* of B in A is defined to be the function $i_B : B \to A$ given by

$$i_B(x) = x \text{ for all } x \in B, \quad \text{i.e., } i_B = \{(x, x) \mid x \in B\}.$$

Note the following

(i) $i_B = 1_A$ if and only if $A = B$ and

(ii) i_B is not surjective if $B \neq A$.

(4) **Characteristic Function.** Let 2 designate a class of two elements, say the class $\{0, 1\}$. If B is a subclass of A, then the *characteristic function of* B *in* A is the function $\chi_B : A \to 2$ given by

$$\chi_B(x) = \begin{cases} 0 & \text{if } x \in B, \\ 1 & \text{otherwise} \end{cases} \quad \text{for all } x \in A,$$

i.e., $\chi_B = \{(x, 0) \mid x \in B\} \cup \{(x, 1) \mid x \in A \setminus B\}$.

Thus χ_B maps every element of B onto 0 and every element of $A \setminus B$ onto 1.

(5) **Restriction Function.** Let $f : A \to B$ be a function and let C be a subclass of A. By the *restriction of* f *to* C we mean the function $f|_C : C \to B$ given by $(f|_C)(x) = f(x)$ for all $x \in C$.

i.e., $f|_C = \{(x, f(x)) \mid x \in C\} = \{(x, y) \mid (x, y) \in f \text{ and } x \in C\}$.

Note that $f|_C \subseteq f$.

Restrictions of functions have the following properties, which will be useful to us later.

Theorem 2.14. *If $f : B \cup C \to A$ is a function, then $f = f|_B \cup f|_C$.*

Proof. $f : B \cup C \to A$ is a function.

$\Rightarrow$ (1) $f|_B \cup f|_C \subset f$ because $f|_B = \{(x, y) \mid (x, y) \in f \text{ and } x \in B\}$ and $f|_C = \{(x, y) \mid (x, y) \in f \text{ and } x \in C\}$.

(2) $f \subset f|_B \cup f|_C$.

($\because$) $(x, y) \in f$.

$\Rightarrow x \in B \cup C$ and ($y \in A$ and $y = f(x)$) by Definition 2.3.

$\Rightarrow$ ($x \in B$ or $x \in C$) and ($y \in A$ and $y = f(x)$) by Definition 1.17(1).

$\Rightarrow$ [$x \in B$ and ($y \in A$ and $y = f(x)$)] or [$x \in C$ and ($y \in A$ and $y = f(x)$)] by Theorem 1.9(5).

$\Rightarrow$ ($x \in B$ and $y \in A, y = f|_B(x)$) or ($x \in C$ and $y \in A, y = f|_C(x)$).

$\Rightarrow (x, y) \in f|_B$ or $(x, y) \in f|_C$ by Definition 2.3.

$\Rightarrow (x, y) \in f|_B \cup f|_C$ by Definition 1.17(1).

$\Rightarrow f = f|_B \cup f|_C$ by (1) and (2) above. □

Theorem 2.15. *Let A, B and C be classes. Assume that B and C are disjoint. Let $f_1 : B \to A$ and $f_2 : C \to A$ be functions and let $f = f_1 \cup f_2$. Then the following hold.*

(1) *$f : B \cup C \to A$ is a function.*

(2) *$f_1 = f|_B$ and $f_2 = f|_C$.*

(3) *If $x \in B$ then $f(x) = f_1(x)$, and if $x \in C$ then $f(x) = f_2(x)$.*

Proof. (1) We will show that $f : B \cup C \to A$ is a function.

(i) $f = f_1 \cup f_2 \subseteq (B \times A) \cup (C \times A) = (B \cup C) \times A$.

(ii) $f : B \cup C \to A$ satisfies the conditions F1.

($\because$) $x \in B \cup C$.

$\Rightarrow x \in B$ or $x \in C$.

$\Rightarrow$ If $x \in B$, since $f_1 : B \to A$ is a function,

there exists a unique element $y_1 \in A$ such that $(x, y_1) \in f_1$.

If $x \in C$, since $f_2 : C \to A$ is a function,

there exists a unique element $y_2 \in A$ such that $(x, y_2) \in f_2$.

Let $y = \begin{cases} y_1 & \text{if } x \in B, \\ y_2 & \text{if } x \in C. \end{cases}$

$\Rightarrow$ Since $B \cap C = \emptyset$ and $f = f_1 \cup f_2$, y is a unique element of A such that $(x, y) \in f_1 \cup f_2 = f$.

Thus the condition F1 for f is established.

(iii) $f : B \cup C \to A$ satisfies the condition F2.

($\because$) $(x, y_1) \in f$ and $(x, y_2) \in f$.

$\Rightarrow$ Since $f = f_1 \cup f_2$, there are four cases to consider.

Case 1: $(x, y_1) \in f_1$ and $(x, y_2) \in f_1$.

Case 2: $(x, y_1) \in f_1$ and $(x, y_2) \in f_2$.

Case 3: $(x, y_1) \in f_2$ and $(x, y_2) \in f_1$.

Case 4: $(x, y_1) \in f_2$ and $(x, y_2) \in f_2$.

$\Rightarrow$ Since B and C are disjoint, Cases 2 and 3 do not happen.

For Cases 1 and 4, since f_1 and f_2 are functions respectively, we have $y_1 = y_2$ in each case.

Thus the condition F2 for f is established.

$\Rightarrow$ Since the conditions (i), (ii), and (iii) for $f : B \cup C \to A$ are satisfied, $f : B \cup C \to A$ is a function by Definition 2.3.

(2) $f_1 = f|_B$ and $f_2 = f|_C$.

($\because$) $(x, y) \in f_1$.

$\Leftrightarrow x \in B$ and $y = f_1(x)$.

$\Leftrightarrow x \in B$ and $y = f(x)$.

$\Leftrightarrow (x, y) \in f|_B$. Thus $f_1 = f|_B$.

Similarly, we can show that $f_2 = f|_C$.

(3) If $x \in B$ then $f(x) = f_1(x)$, and if $x \in C$ then $f(x) = f_2(x)$.

($\because$) $x \in B$.

$\Rightarrow$ Since $f : B \cup C \to A$ and $f_1 : B \to A$ are functions,

$(x, f(x)) \in f$ and $(x, f_1(x)) \in f_1$.

$\Rightarrow$ Since $f_1 = f|_B$ by (2) above, $f(x) = f_1(x)$ by Theorem 2.10.

Similarly we can show that $f(x) = f_2(x)$ for all $x \in C$. $\square$

Exercise 2.2

1. Prove that the functions introduced in Example 2.13 qualify as functions under Definition 2.3.

2. Prove that if $f : A \to B$ is an injective function and $C \subseteq A$, then $f|_C : C \to B$ is an injective function.

3. Let A be a class and let $f = \{(x, (x, x)) \mid x \in A\}$. Show that f is a bijective function from A to 1_A.

4. Let $f : A \to B$ and $g : A \to B$ be functions. Prove that if $f \subseteq g$, then $f = g$.

5. Let $f : A \to B$ and $g : C \to D$ be functions. The *product* of f and g is the function $f \cdot g : A \times C \to B \times D$ defined to as follows:

$$(f \cdot g)(x, y) = (f(x), g(y)) \text{ for all } (x, y) \in A \times C.$$

   Prove the following.

   (1) $f \cdot g$ is a function from $A \times C$ to $B \times D$.

   (2) If f and g are injective, then $f \cdot g$ is injective.

   (3) If f and g are surjective, then $f \cdot g$ is surjective.

   (4) $\mathrm{ran}(f \cdot g) = \mathrm{ran}(f) \times \mathrm{ran}(g)$.

6. Let $f_1 : A \to B$ and $f_2 : C \to D$ be bijective functions, where $A \cap C = \emptyset$ and $B \cap D = \emptyset$. Let $f = f_1 \cup f_2$. Prove that $f : A \cup C \to B \cup D$ is a bijective function.

7. Let $f : B \to A$ and $g : C \to A$ be functions and assume that $f|_{B\cap C} = g|_{B\cap C}$. Let $h = f \cup g$. Prove the following.

   (1) $h : B \cup C \to A$ is a function. (2) $f = h|_B$. (3) $g = h|_C$.

8. Let $f : A \to B$ be a function. Prove that f is in one-to-one correspondence with A.

9. By a *functional graph*, we mean a graph which satisfies the condition F2. Thus G is a functional graph if and only if $(x, y_1) \in G$ and $(x, y_2) \in G$ imply $y_1 = y_2$.

   If G is a functional graph, show that every subclass of G is a functional graph.

10. Let G be a graph. Prove that G is a functional graph if and only if for arbitrary graphs J and H, $(H \cap J) \circ G = (H \circ G) \cap (J \circ G)$.

11. Let G be a functional graph. Prove that G is injective if and only if for arbitrary graphs J and H, $G \circ (H \cap J) = (G \circ H) \cap (G \circ J)$.

2.3 Properties of composite functions and inverse functions

Theorem 2.16. *Let $f : A \to B$ and $g : B \to C$ be functions. Then the composite graph $g \circ f$ is a function $g \circ f : A \to C$ which is called the composite function of g followed by f.*

Proof. $f : A \to B$ and $g : B \to C$ are functions.

$\Rightarrow$ (i) $\text{dom}(g \circ f) = A$.

$(\because)$ $\text{dom}(g \circ f) = \text{dom}(f)$ by Corollary 1.33.

$\Rightarrow$ Since $\text{dom}(f) = A$, $\text{dom}(g \circ f) = A$.

(ii) $\text{ran}(g \circ f) \subseteq C$.

$(\because)$ $\text{ran}(g \circ f) \subseteq \text{ran}(g)$ by Theorem 1.32(4).

$\Rightarrow$ Since $\text{ran}(g) \subseteq C$, $\text{ran}(g \circ f) \subseteq C$.

(iii) $g \circ f : A \to C$ satisfies F2.

$(\because)$ $(x, z_1) \in g \circ f$ and $(x, z_2) \in g \circ f$.

$\Rightarrow$ There exists y_1 such that $(x, y_1) \in f$ and $(y_1, z_1) \in g$ and there exists y_2 such that $(x, y_2) \in f$ and $(y_2, z_2) \in g$.

$\Rightarrow$ $y_1 = y_2$ by applying F2 to f.

$\Rightarrow$ $(y_1, z_1) \in g$ and $(y_1, z_2) \in g$.

$\Rightarrow$ $z_1 = z_2$ by applying F2 to g.

$\Rightarrow$ $g \circ f : A \to C$ is a function by Theorem 2.7. □

Definition 2.17. A function $f : A \to B$ is said to be *invertible* if the inverse graph f^{-1} is a function $f^{-1} : B \to A$. In this case $f^{-1} : B \to A$ is called the *inverse function* of f.

Remark 2.18. (1) $f : A \to B$ and $g : B \to C$ are functions.

$\Rightarrow (x, z) \in g \circ f$ iff there exists y such that $(x, y) \in f$ and $(y, z) \in g$, by Definition 1.29(3).

$\Rightarrow x \xmapsto{g \circ f} z$ iff $x \xmapsto{f} y$ and $y \xmapsto{g} z$ for some y. (See Figure 2.1.)

$\Rightarrow z = (g \circ f)(x)$ iff $y = f(x)$ and $z = g(y)$ for some y.

$\Rightarrow (g \circ f)(x) = g(f(x))$.

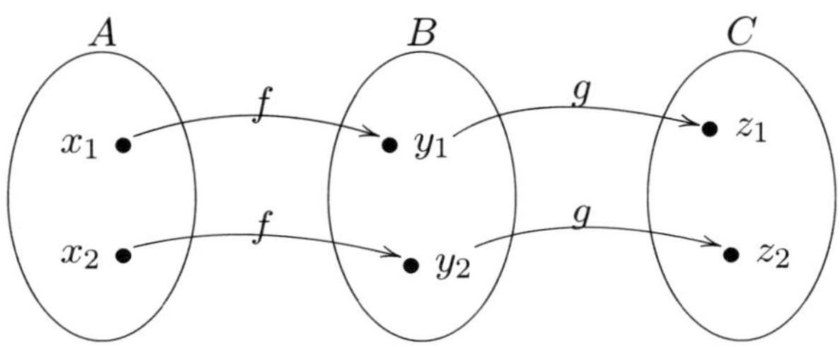

$z_1=(g\circ f)(x_1)=g(f(x_1))$

$z_2=(g\circ f)(x_2)=g(f(x_2))$

Figure 2.1: composite function

(2) $f : A \to B$ is an invertible function.

$\Rightarrow (x, y) \in f$ iff $(y, x) \in f^{-1}$ by Definition 1.29.

$\Rightarrow x \xmapsto{f} y$ iff $y \xmapsto{f^{-1}} x$. (See Figure 2.2.)

$\Rightarrow y = f(x)$ iff $x = f^{-1}(y)$.

The next theorem gives a necessary and sufficient condition for a function to be invertible.

Theorem 2.19. (1) *If $f : A \to B$ is a bijective function, then $f^{-1} : B \to A$ is also a bijective function.*

(2) *If $f : A \to B$ is an invertible function, then $f : A \to B$ is bijective.*

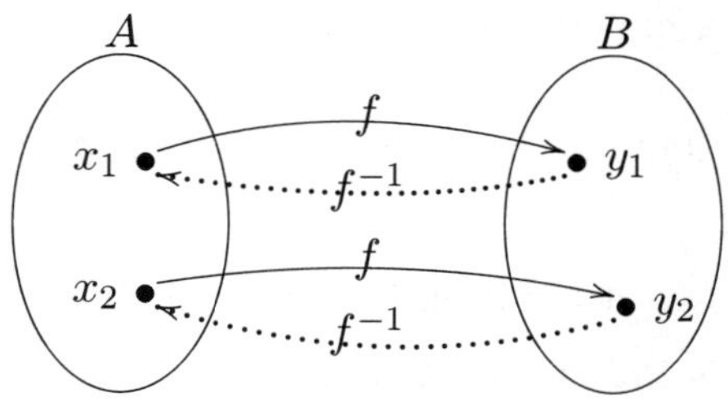

Figure 2.2: invertible function

Proof. (1) $f : A \to B$ is a bijective function.

$\Rightarrow$ (i) Since f is a graph, f^{-1} is a graph and $\text{dom}(f^{-1}) = \text{ran}(f) = B$ and $\text{ran}(f^{-1}) = \text{dom}(f) = A$ by Theorem 1.32.

(ii) f^{-1} satisfies the condition F2.

($\because$) $(y, x_1) \in f^{-1}$ and $(y, x_2) \in f^{-1}$.

$\Rightarrow$ $(x_1, y) \in f$ and $(x_2, y) \in f$ by Definition 1.29(2).

$\Rightarrow$ Since f is injective, $x_1 = x_2$ by Definition 2.11(1).

(iii) f^{-1} satisfies the condition INJ.

($\because$) $(y_1, x) \in f^{-1}$ and $(y_2, x) \in f^{-1}$.

$\Rightarrow$ $(x, y_1) \in f$ and $(x, y_2) \in f$ by Definition 1.29(2).

$\Rightarrow$ $y_1 = y_2$ by Definition 2.3.

(iv) f^{-1} satisfies the condition SRJ because $\text{ran}(f^{-1}) = A$.

$\Rightarrow$ $f^{-1} : B \to A$ is a bijective function.

(2) $f : A \to B$ is an invertible function.

$\Rightarrow$ $f^{-1} : B \to A$ is a function by Definition 2.17.

(i) $f : A \to B$ is surjective because $\text{ran}(f) = \text{dom}(f^{-1}) = B$ by Theorem 1.32(2).

(ii) $f : A \to B$ is injective.

($\because$) $(x_1, y) \in f$ and $(x_2, y) \in f$.

$\Rightarrow$ $(y, x_1) \in f^{-1}$ and $(y, x_2) \in f^{-1}$ by Definition 1.29(2).

$\Rightarrow$ By applying the condition F2 to f^{-1}, we have $x_1 = x_2$.

Thus $f : A \to B$ is injective.

$\Rightarrow f : A \to B$ is a bijective function. □

Remark 2.20. For a function $f : A \to B$, Theorem 2.19 may be summarized as follows:

(1) $f : A \to B$ is invertible if and only if it is bijective.

(2) If $f : A \to B$ is invertible, then $f^{-1} : B \to A$ is bijective.

The next two theorems give another useful characterization of invertible functions.

Theorem 2.21. *Let $f : A \to B$ be an invertible function. Then*

(1) $f^{-1} \circ f = 1_A$, (2) $f \circ f^{-1} = 1_B$.

Proof. Let $f : A \to B$ be an invertible function so that $f^{-1} : B \to A$ is a function and hence $f^{-1} \circ f : A \to A$ and $f \circ f^{-1} : B \to B$ are functions.

(1) $(f^{-1} \circ f)(x) = 1_A(x)$ for all $x \in A$.

($\because$) $x \in A$ and $y = f(x)$.

$\Rightarrow x = f^{-1}(y)$ by Remark 2.18(2).

$\Rightarrow (f^{-1} \circ f)(x) = f^{-1}(f(x)) = f^{-1}(y) = x = 1_A(x)$.

$\Rightarrow$ Since $\text{dom}(f^{-1} \circ f) = \text{dom}(f) = A = \text{dom}(1_A)$ and since $\text{ran}(f^{-1} \circ f) = \text{ran}(f^{-1}) = \text{dom}(f) = A = \text{ran}(1_A)$, $f^{-1} \circ f = 1_A$ (as functions) by Theorem 2.10.

The proof of (2) is analogous to that of (1), and is left for the reader. □

Theorem 2.22. *Let $f : A \to B$ and $g : B \to A$ be functions. If $g \circ f = 1_A$ and $f \circ g = 1_B$, then $f : A \to B$ is bijective (hence invertible), and $g = f^{-1}$.*

Proof. $f : A \to B$ and $g : B \to A$ are functions satisfying the assumption that $g \circ f = 1_A$ and $f \circ g = 1_B$.

$\Rightarrow$ (i) $f : A \to B$ is injective.

($\because$) $f(x_1) = f(x_2)$ for some $x_1, x_2 \in A$.

$\Rightarrow g(f(x_1)) = g(f(x_2))$ by applying the condition F2° to g.

$\Rightarrow (g \circ f)(x_1) = (g \circ f)(x_2)$ by Remark 2.18(1).

$\Rightarrow x_1 = 1_A(x_1) = (g \circ f)(x_1) = (g \circ f)(x_2) = 1_A(x_2) = x_2.$

(ii) $f : A \to B$ is surjective.

($\because$) $y \in B$.

$\Rightarrow y = 1_B(y) = (f \circ g)(y) = f\,(g\,(y)).$

$\Rightarrow y = f(x)$, where $x = g(y) \in A$.

(iii) $g = f^{-1}$.

($\because$) $(y, x) \in g.\ \Leftrightarrow x = g(y).\ \Leftrightarrow f(x) = y.\ \Leftrightarrow (x, y) \in f.$

$\Leftrightarrow (y, x) \in f^{-1}.$

$\Rightarrow f : A \to B$ is a bijective function by (i) and (ii) above and since $\mathrm{dom}(g) = B = \mathrm{dom}(f^{-1})$ and $\mathrm{ran}(g) = A = \mathrm{ran}(f^{-1})$, from (iii) $g = f^{-1}$ (as functions) by Theorem 2.10. □

Theorems 2.21 and 2.22 may be summarized as follows:

Remark 2.23. $f : A \to B$ is an invertible function if and only if there exists a function $g : B \to A$ such that $g \circ f = 1_A$ and $f \circ g = 1_B$. The function g, if it exists, is the inverse function of f.

Our next theorem gives an important characterization of injective functions.

Theorem 2.24. *Let $f : A \to B$ be a function and let $A \neq \emptyset$, say, $a \in A$. Then $f : A \to B$ is injective if and only if there exists a function $g : B \to A$ such that $g \circ f = 1_A$.*

Proof. Let $f : A \to B$ be a function and let $a \in A$.

(if) Assume that there exists a function $g : B \to A$ such that $g \circ f = 1_A$. To prove that $f : A \to B$ is injective, we just repeat the part (i) of the proof of Theorem 2.22.

(only if) Assume that $f : A \to B$ is injective. Let $C = \mathrm{ran}(f)$ and consider a function $f' : A \to C$ defined by the graph $f' = f$, which is guaranteed by Corollary 2.9.

The functions $f : A \to B$ and $f' : A \to C$ differ by their codomains.

$\Rightarrow$ Since $f : A \to B$ is injective and since $C = \mathrm{ran}(f) = \mathrm{ran}(f')$,

$f' : A \to C$ is a bijective function.

$\Rightarrow$ $(f')^{-1} : C \to A$ is a bijective function by Theorem 2.19(1).

Now to define a function $g : B \to A$, consider the constant function $c_a : (B \setminus C) \to A$ which maps every element of $(B \setminus C)$ onto a.

Let $g = (f')^{-1} \cup c_a$. (See Figure 2.3.)

$\Rightarrow$ (i) Since $C \cup (B \setminus C) = B$ and $C \cap (B \setminus C) = \emptyset$, $g : B \to A$ is a function by Theorem 2.15(1).

(ii) $(g \circ f)(x) = 1_A(x)$ for all $x \in A$.

($\because$) $x \in A$.

$\Rightarrow$ Since $f' : A \to C$ is a function, there exists $y \in C$ such that $y = f'(x)$.

$\Rightarrow$ $y \in C = \mathrm{cod}(f') = \mathrm{dom}((f')^{-1})$ and $(f')^{-1}(y) = x$.

$\Rightarrow$ Since $g(y) = (g|_C)(y) = (f')^{-1}(y)$ by Theorem 2.15(3), $(g \circ f)(x) = g(f(x)) = g(y) = (f')^{-1}(y) = x = 1_A(x)$.

$\Rightarrow$ $(g \circ f)(x) = x = 1_A(x)$.

$\Rightarrow$ $g : B \to A$ is a function, and $g \circ f = 1_A$ by Theorem 2.10. $\square$

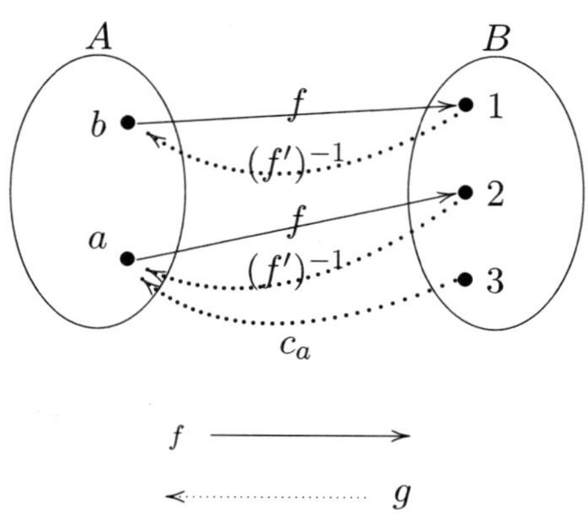

Figure 2.3: the function $g = (f')^{-1} \cup c_a$

Remark 2.25. (1) In Chapter 5 we will prove a companion theorem to Theorem 2.24, which will state the following: $f : A \to B$ is surjective if and only if there exits a function $g : B \to A$ such that $f \circ g = 1_B$. For functions $f : A \to B$ and $g : B \to A$, f is called a *left inverse* of g if $f \circ g = 1_B$ and g is called a *right inverse* of f if $g \circ f = 1_A$.

(2) Theorem 2.24 and its companion are often paraphrased as follows.

Let $f : A \to B$ be a function. Then

(i) $f : A \to B$ is injective if and only if it has a *left inverse.*

(ii) $f : A \to B$ is surjective if and only if it has a *right inverse.*

Theorem 2.26. *Let $f : A \to B$ and $g : B \to C$ be functions.*

(1) *If f and g are injective, then $g \circ f$ is injective.*

(2) *If f and g are surjective, then $g \circ f$ is surjective*

(3) *If f and g are bijective, then $g \circ f$ is bijective.*

(4) *If $g \circ f$ is injective, then f is injective.*

(5) *If $g \circ f$ is surjective, then g is surjective.*

(6) *If $g \circ f$ is bijective, then f is injective and g is surjective.*

Proof. Let $f : A \to B$ and $g : B \to C$ be functions.

(1) $f : A \to B$ and $g : B \to C$ are injective.

$\Rightarrow g \circ f : A \to C$ is injective.

($\because$) $(g \circ f)(x_1) = (g \circ f)(x_2)$ for some $x_1, x_2 \in A$.

$\Rightarrow g(f(x_1)) = g(f(x_2))$.

$\Rightarrow$ Since $g : B \to C$ is injective, $f(x_1) = f(x_2)$.

$\Rightarrow$ Since $f : A \to B$ is injective, $x_1 = x_2$.

Thus $g \circ f$ satisfies INJ° and hence $g \circ f : A \to C$ is injective.

(2) $f : A \to B$ and $g : B \to C$ are surjective.

$\Rightarrow g \circ f : A \to C$ is surjective.

$(\because)$ $z \in C$.

$\Rightarrow$ Since $g : B \to C$ is surjective, there exists $y \in B$ such that $z = g(y)$.

$\Rightarrow$ Since $f : A \to B$ is surjective, there exists $x \in A$ such that $y = f(x)$.

$\Rightarrow z = g(f(x)) = (g \circ f)(x)$.

Thus $g \circ f$ satisfies SRJ and hence $g \circ f : A \to C$ is surjective.

(3) follows immediately from (1) and (2) above.

(4) $g \circ f$ is injective.

$\Rightarrow f : A \to B$ is injective.

$(\because)$ $f(x_1) = f(x_2)$ for some $x_1, x_2 \in A$.

$\Rightarrow g(f(x_1)) = g(f(x_2))$. i.e., $(g \circ f)(x_1) = (g \circ f)(x_2)$.

$\Rightarrow$ Since $g \circ f$ is injective, $x_1 = x_2$ by Definition 2.11(1).

Thus $f : A \to B$ is injective by Definition 2.11(1).

(5) $g \circ f : A \to C$ is surjective.

$\Rightarrow g : B \to C$ is surjective.

$(\because)$ $z \in C$.

$\Rightarrow$ There exists $x \in A$ such that $(g \circ f)(x) = z$. Let $y = f(x)$.

$\Rightarrow y \in B$ and $g(y) = g(f(x)) = (g \circ f)(x) = z$.

Thus $g : B \to C$ is surjective by Definition 2.11(2).

(6) follows immediately from (4) and (5). □

Remark 2.27. If $f : A \to B$ and $g : B \to C$ are invertible functions, then $g \circ f : A \to C$ is invertible by Theorem 2.26(3) and $(g \circ f)^{-1} = f^{-1} \circ g^{-1}$ by Theorem 1.31.

Exercise 2.3

1. Let $f : A \to B$ be a function. Prove that $1_B \circ f = f$ and $f \circ 1_A = f$.
2. Determine whether the converse of each statement of Theorem 2.26 is true or not. Justify your answer.

3. Let $f : A \to B$ and $g : B \to A$ be functions. Assume that $y = f(x)$ if and only if $x = g(y)$. Prove that f is invertible and that $g = f^{-1}$.

4. Let $g : B \to C$ and $h : B \to C$ be functions. Assume that $g \circ f = h \circ f$ for every function $f : A \to B$. Prove that $g = h$.

5. Let $g : A \to B$ and $h : A \to B$ be functions and let C be a set with more than one element. Assume that $f \circ g = f \circ h$ for every function $f : B \to C$. Prove that $g = h$.

6. Let $f : B \to C$ be a function. Prove that f is injective if and only if, for every pair of functions $g : A \to B$ and $h : A \to B$, $f \circ g = f \circ h$ implies $g = h$.

7. Let $f : A \to B$ be a function. Prove that f is surjective if and only if, for every pair of functions $g : B \to C$ and $h : B \to C$, $g \circ f = h \circ f$ implies $g = h$.

8. Let $f : A \to C$ and $g : A \to B$ be functions. Prove the following.

   (1) There exists a function $h : B \to C$ such that $f = h \circ g$

   if and only if for all $x, y \in A$, $g(x) = g(y)$ implies $f(x) = f(y)$.

   (2) Such an $h : B \to C$ is unique.

9. Let $f : C \to A$ and $g : B \to A$ be functions and assume that g is bijective. Prove the following.

   (1) There exists a function $h : C \to B$ such that $f = g \circ h$

   if and only if $\operatorname{ran}(f) \subseteq \operatorname{ran}(g)$.

   (2) Such an $h : C \to B$ is unique.

10. Let $f : A \to B$ be a function and let $C \subseteq A$. Prove that $f|_C = f \circ i_C$ where $i_C : C \to A$ is the inclusion function.

2.4 Direct images and inverse images under functions.

Definition 2.28. Let $f : A \to B$ be a function. Let C be a subclass of A and let D be a subclass of B.

(1) The *direct image* of C under f, which we write $f(C)$, is the following subclass of B.

$$\begin{aligned} f(C) &= \{y \in B \mid \text{ there exists } x \in C \text{ such that } y = f(x)\} \\ &= \{f(x) \in B \mid x \in C\}. \end{aligned}$$

That is, $f(C)$ is the class of all the images of elements in C.

If $C = \{a\}$ is a singleton, we will write $f(a)$ for $f(\{a\})$. Thus $f(a) = f(\{a\}) = \{f(a)\}$ even though there is some ambiguity between a set and its element.

(2) The *inverse image* of D under f, which we write $f^{-1}(D)$, is the following subclass of A.

$$f^{-1}(D) = \{x \in A \mid f(x) \in D\}.$$

That is, $f^{-1}(D)$ is the class of all the pre-images of elements in D.

If $D = \{b\}$ is a singleton, we will write $f^{-1}(b)$ for $f^{-1}(\{b\})$. Thus

$$f^{-1}(b) = f^{-1}(\{b\}) = \{x \in A \mid f(x) = b\}.$$

Remark 2.29. Let $f : A \to B$ be a function. Consider the graphs

$$\overline{f} = \{(C, f(C)) \mid C \in \mathscr{P}(A)\} \text{ and } \overline{\overline{f}} = \{(D, f^{-1}(D)) \mid D \in \mathscr{P}(B)\}.$$

Note that $\overline{f} \subseteq \mathscr{P}(A) \times \mathscr{P}(B)$ and $\overline{\overline{f}} \subseteq \mathscr{P}(B) \times \mathscr{P}(A)$ while $f \subseteq A \times B$ and $f^{-1} \subseteq B \times A$. In fact, $\overline{f} : \mathscr{P}(A) \to \mathscr{P}(B)$ and $\overline{\overline{f}} : \mathscr{P}(B) \to \mathscr{P}(A)$ are functions which will be shown in the following Theorem 2.30.

Theorem 2.30. *Let $f : A \to B$ be a function. Define $\overline{f} : \mathscr{P}(A) \to \mathscr{P}(B)$ and $\overline{\overline{f}} : \mathscr{P}(B) \to \mathscr{P}(A)$ by*

$$\overline{f} = \{(C, f(C)) \mid C \in \mathscr{P}(A)\} \textit{ and } \overline{\overline{f}} = \{(D, f^{-1}(D)) \mid D \in \mathscr{P}(B)\}.$$

Then we have the following.

(1) *If* $A_1, A_2 \subseteq A$ *and* $A_1 = A_2$, *then* $f(A_1) = f(A_2)$.

Thus $\overline{f} : \mathscr{P}(A) \to \mathscr{P}(B)$ *is a function.*

(2) *If* $B_1, B_2 \subseteq B$ *and* $B_1 = B_2$, *then* $f^{-1}(B_1) = f^{-1}(B_2)$.

Thus $\overline{\overline{f}} : \mathscr{P}(B) \to \mathscr{P}(A)$ *is a function.*

Proof. (1) If $A_1, A_2 \subseteq A$ and $A_1 = A_2$, then $f(A_1) = f(A_2)$.

($\because$) $y \in f(A_1)$.

$\Leftrightarrow$ There exists $x \in A_1$ such that $y = f(x)$ by Definition 2.28.

$\Leftrightarrow$ There exists $x \in A_2$ such that $y = f(x)$ because $A_1 = A_2$.

$\Leftrightarrow$ $y \in f(A_2)$ by Definition 2.28(1).

$\Rightarrow$ Since $\text{dom}(\overline{f}) = \mathscr{P}(A)$ and $\text{ran}(\overline{f}) \subseteq \mathscr{P}(B)$, $\overline{f} : \mathscr{P}(A) \to \mathscr{P}(B)$ is a function by Theorem 2.7.

(2) If $B_1, B_2 \subseteq B$ and $B_1 = B_2$, then $f^{-1}(B_1) = f^{-1}(B_2)$.

($\because$) $x \in f^{-1}(B_1)$.

$\Leftrightarrow$ $f(x) \in B_1$ by Definition 2.28(2).

$\Leftrightarrow$ $f(x) \in B_2$ because $B_1 = B_2$.

$\Leftrightarrow$ $x \in f^{-1}(B_2)$ by Definition 2.28(2).

$\Rightarrow$ Since $\text{dom}(\overline{\overline{f}}) = \mathscr{P}(B)$ and $\text{ran}(\overline{\overline{f}}) \subseteq \mathscr{P}(A)$, $\overline{\overline{f}} : \mathscr{P}(B) \to \mathscr{P}(A)$ is a function by Theorem 2.7. □

Example 2.31. (1) $f(A_1) = f(A_2)$ does not always imply that $A_1 = A_2$. For a simple counterexample, see Figure 2.4.

(2) $f^{-1}(B_1) = f^{-1}(B_2)$ does not always imply $B_1 = B_2$. For a simple counterexample, see Figure 2.5.

Theorem 2.32. *Let* $f : A \to B$ *be a function. Let* $\{C_i\}_{i \in I}$ *be a family of subclasses of* A, *and let* $\{D_i\}_{i \in I}$ *be a family of subclasses of* B. *Then the following hold.*

(1) $f(\bigcup_{i \in I} C_i) = \bigcup_{i \in I} f(C_i)$.

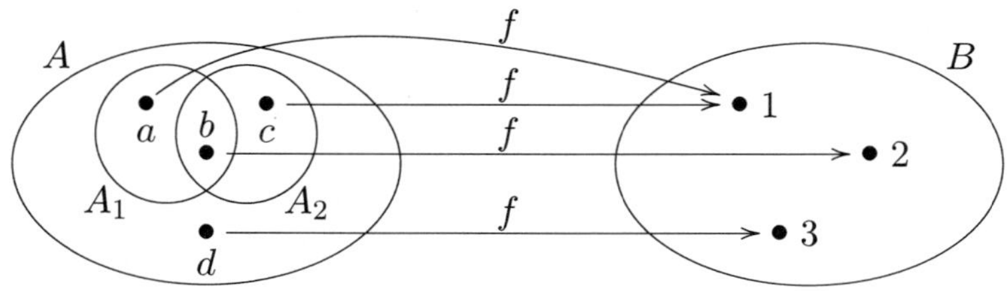

Figure 2.4: $f(A_1) = f(A_2) = \{1, 2\}$ but $A_1 \neq A_2$

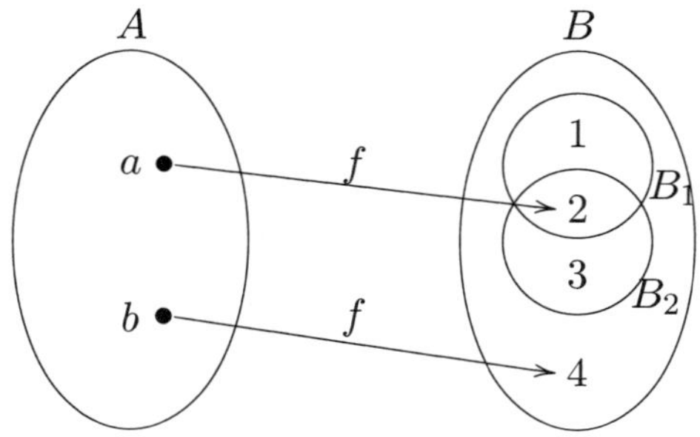

Figure 2.5: $f^{-1}(B_1) = f^{-1}(B_2) = \{a\}$ but $B_1 \neq B_2$

(2) $f^{-1}\big(\bigcup_{i \in I} D_i\big) = \bigcup_{i \in I} f^{-1}(D_i)$.

(3) $f^{-1}\big(\bigcap_{i \in I} D_i\big) = \bigcap_{i \in I} f^{-1}(D_i)$.

Proof. (1) $y \in f\big(\bigcup_{i \in I} C_i\big)$.

$\Leftrightarrow$ There exists $x \in \bigcup_{i \in I} C_i$ such that $y = f(x)$ by Definition 2.28(1).

$\Leftrightarrow$ For some $j \in I$, there exists $x \in C_j$ such that $y = f(x)$ by Definition 1.38(1).

$\Leftrightarrow$ For some $j \in I, y \in f(C_j)$ by Definition 2.28(1).

$\Leftrightarrow y \in \bigcup_{i \in I} f(C_i)$ by Definition 1.38(1).

Thus $f\big(\bigcup_{i \in I} C_i\big) = \bigcup_{i \in I} f(C_i)$ by Definition 1.14(2).

(2) $x \in f^{-1}\big(\bigcup_{i \in I} D_i\big)$.

$\Leftrightarrow f(x) \in \bigcup_{i \in I} D_i$ by Definition 2.28(2).

$\Leftrightarrow$ For some $j \in I, f(x) \in D_j$ by Definition 1.38(1).

$\Leftrightarrow$ For some $j \in I, x \in f^{-1}(D_j)$ by Definition 2.28(2).

$\Leftrightarrow x \in \bigcup_{i \in I} f^{-1}(D_i)$ by Definition 1.38(1).

Thus $f^{-1}\big(\bigcup_{i \in I} D_i\big) = \bigcup_{i \in I} f^{-1}(D_i)$ by Definition 1.14(2).

(3) It is left for the reader. □

Note. It is important to note that there is no counterpart of Theorem 2.32(3) for f. More precisely, we have

$$f\big(\bigcap_{i \in I} C_i\big) \subseteq \bigcap_{i \in I} f(C_i) \text{ but } \bigcap_{i \in I} f(C_i) \nsubseteq f\big(\bigcap_{i \in I} C_i\big) \text{ in general.}$$

For a simple counter example, see Figure 2.4, where $f(A_1 \cap A_2) = f(b) = \{2\} \neq \{1, 2\} = f(A_1) \cap f(A_2)$.

For this reason, a variety of theorems which are true for inverse images of sets fail to hold for direct images of sets.

Exercise 2.4

1. Let $f : A \to B$ be a function and let $C \subseteq A$ and $D \subseteq B$. Prove the following.

   (1) $C \subseteq f^{-1}(f(C))$. (2) $f(f^{-1}(D)) \subseteq D$.

2. Let $f : A \to B$ be a function and let $C \subseteq A$ and $D \subseteq B$. Prove the following.

   (1) If f is injective, then $C = f^{-1}(f(C))$.

   (2) If f is surjective, then $f(f^{-1}(D)) = D$.

3. Let $f : A \to B$ be a function and let $A_1, A_2 \subseteq A$ and $B_1, B_2 \subseteq B$. Prove the following.

   (1) If f is injective, then $f(A_1) = f(A_2)$ implies $A_1 = A_2$.

   (2) If f is surjective, then $f^{-1}(B_1) = f^{-1}(B_2)$ implies $B_1 = B_2$.

   [Hint: Use the result of Exercise 2.4.2 above.]

4. Let $f : A \to B$ be a function. Prove the following.

   (1) If f is injective, then $\overline{\overline{f}} \circ \overline{f} : \mathscr{P}(A) \to \mathscr{P}(A)$ is bijective.

   (2) If f is surjective, then $\overline{f} \circ \overline{\overline{f}} : \mathscr{P}(B) \to \mathscr{P}(B)$ is bijective.

5. Let $f : A \to B$ be a function and let $C \subseteq A$. Prove the following.

   (1) $f(f^{-1}(f(C))) = f(C)$. (2) $\overline{f} \circ \overline{\overline{f}} \circ \overline{f} = \overline{f} : \mathscr{P}(A) \to \mathscr{P}(B)$.

6. Let $f : A \to B$ be a function. Prove the following.

   (1) If f is injective, then $\overline{f} : \mathscr{P}(A) \to \mathscr{P}(B)$ is injective.

   (2) If f is surjective, then $\overline{f} : \mathscr{P}(A) \to \mathscr{P}(B)$ is surjective.

   (3) If f is bijective, then $\overline{f} : \mathscr{P}(A) \to \mathscr{P}(B)$ is bijective.

7. Let $f : A \to B$ be a function. Prove the following.

   (1) If f is injective, then $\overline{\overline{f}} : \mathscr{P}(B) \to \mathscr{P}(A)$ is surjective.

   (2) If f is surjective, then $\overline{\overline{f}} : \mathscr{P}(B) \to \mathscr{P}(A)$ is injective.

   (3) If f is bijective, then $\overline{\overline{f}} : \mathscr{P}(B) \to \mathscr{P}(A)$ is bijective.

8. Let $f : A \to B$ be a function. Prove that $f(C \cap D) = f(C) \cap f(D)$ for every pair of subclasses $C \subseteq A$ and $D \subseteq A$ if and only if f is injective.

9. Let $f : A \to B$ be a function and let $C \subseteq B$ and $D \subseteq B$. Prove that $f^{-1}(C \setminus D) = f^{-1}(C) \setminus f^{-1}(D)$.

10. Let $f : A \to B$ be a function. Prove each of the following.

    (1) If $C \subseteq A$ and $D \subseteq A$, then $f(C) \setminus f(D) \subseteq f(C \setminus D)$.

    (2) $f(C) \setminus f(D) = f(C \setminus D)$ for every pair of subclasses $C \subseteq A$ and $D \subseteq A$ if and only if f is injective.

2.5 Product of a family of classes

In this section we generalize the notion of Cartesian product, introduced in Section 4 of Chapter 1, and define the Cartesian product of an arbitrary family of sets. Note that we confine our attention, here, to sets. We need to do this because, as the reader will recall, we may form arbitrary class of sets, but not arbitrary classes of classes.

The product of two sets A and B has been defined to be the set $A \times B$ of all ordered pairs (x, y), where $x \in A$ and $y \in B$. This definition may be extended, in a natural way, to a finite number of sets $A_1, A_2, \ldots, A_n$.

We may define the product $A_1 \times A_2 \times \cdots \times A_n$ to be the set of all *ordered n-tuples* $(a_1, a_2, \ldots, a_n)$, where $a_i \in A_i$ for each index $i = 1, 2, \ldots, n$. Now, we wish to extend this concept to the case of an indexed family of sets, $\{A_i\}_{i \in I}$, where the index set I is any set whatsoever. Evidently we cannot speak of *I-tuples* of elements, because I may be an infinite set and may not be ordered. Therefore, we must alter our approach to the problem.

Let us take another look at the product $A_1 \times A_2 \times \cdots \times A_n$. Clearly, $\{A_1, A_2, \ldots, A_n\}$ is a family whose index set is $I = \{1, 2, \ldots, n\}$. Now an ordered n-tuple may be regarded as a function whose domain is I, which maps an element $i \in I$ onto an element a_i in A_i. Indeed, if f is such a function, then f is described by the following table.

x	$f(x)$
1	a_1
2	a_2
$\vdots$	$\vdots$
n	a_n

Using the table, we may construct the ordered n-tuple $(a_1, a_2, \ldots, a_n)$. Conversely, if we are given the ordered n-tuple $(a_1, a_2, \ldots, a_n)$, then we may construct the table. (In fact, the ordered n-tuple is simply the table presented as a horizontal array). Thus, the function f and the ordered n-tuple $(a_1, a_2, \ldots, a_n)$ are, essentially, one and the same thing. This simple observation leads to the following definition of the product of a family of sets.

Definition 2.33. Let $\{A_i\}_{i \in I}$ be a family of sets, where I is a set.

(1) The *product* of the sets A_i is defined to be the class

$$\prod_{i \in I} A_i = \left\{ f \mid f : I \to \bigcup_{i \in I} A_i \text{ is a function, and } f(i) \in A_i \text{ for all } i \in I \right\}.$$

The legitimacy of this definition is established in the next section. (See Proposition 2.39.)

(2) For arbitrary sets A and B, the symbol B^A refers to the class of all functions from A to B. In particular, if 2 denotes a set of two elements, say $\{0,1\}$, then 2^A denotes the class of all functions from A to 2.

(3) If a is an element of $\prod_{i\in I} A_i$ and $j \in I$, we write $a(j) = a_j$ and a_j is called the *jth coordinate* of a.

(4) For each $i \in I$, let $x_i \in A_i$. The symbol $(x_i)_{i\in I}$ designates the element in $\prod_{i\in I} A_i$ whose ith-coordinate is x_i for each $i \in I$.

(5) For each index $i \in I$, we define a function $p_i : \prod_{i\in I} A_i \to A_i$ by

$$p_i(a) = a_i, \text{ for all } a \in \prod_{i\in I} A_i.$$

The function $p_i : \prod_{i\in I} A_i \to A_i$ is called the *ith projection* of $\prod_{i\in I} A_i$ to A_i.

Example 2.34. Let $I = \{1,2\}, A_1 = \{a,b\}$, and $A_2 = \{c,d\}$. By Definition 2.33, $\prod_{i\in I} A_i$ consists of all the functions $f : \{1,2\} \to \{a,b,c,d\}$ such that $f(1) \in A_1$ and $f(2) \in A_2$. There are four such functions, given by the following tables.

x	$f(x)$
1	a
2	c

x	$f(x)$
1	a
2	d

x	$f(x)$
1	b
2	c

x	$f(x)$
1	b
2	d

We may identify these four functions with the four ordered pairs (a,c), (a,d), (b,c) and (b,d), respectively. Thus $\prod_{i\in I} A_i$ is exactly $A_1 \times A_2$.

The following is an important result which will be used in a later chapter.

Theorem 2.35. *If A is a set, then 2^A and $\mathscr{P}(A)$ are in one-to-one correspondence, where 2 denotes the set of two elements 0 and 1.*

Proof. We will define a bijective function $\gamma : \mathscr{P}(A) \to 2^A$.

For arbitrary $B \in \mathscr{P}(A)$, define $\gamma(B) = \chi_B$, where $\chi_B : A \to 2$ is the characteristic function of B in A defined in Example 2.13(4).

$\Rightarrow$ (i) $\gamma : \mathscr{P}(A) \to 2^A$ is a well-defined function.

($\because$) For each $B \in \mathscr{P}(A)$, χ_B is a uniquely determined element of 2^A, $\text{dom}(\gamma) = \mathscr{P}(A)$ and $\text{ran}(\gamma) \subseteq 2^A$.

$\Rightarrow$ γ satisfies the condition F2.

$\Rightarrow$ $\gamma : \mathscr{P}(A) \to 2^A$ is a function by Theorem 2.7.

(ii) $\gamma : \mathscr{P}(A) \to 2^A$ is injective.

($\because$) $\gamma(B) = \gamma(D)$ for some $B, D \in \mathscr{P}(A)$.

$\Rightarrow$ $\chi_B = \chi_D$.

$\Rightarrow$ $\chi_B(x) = \chi_D(x)$ for all $x \in A$ by Theorem 2.10.

$\Rightarrow$ $\{x \in A \mid \chi_B(x) = 0\} = \{x \in A \mid \chi_D(x) = 0\}$.

$\Rightarrow$ $B = D$.

$\Rightarrow$ $\gamma : \mathscr{P}(A) \to 2^A$ satisfies the condition INJ°.

$\Rightarrow$ Since $\gamma : \mathscr{P}(A) \to 2^A$ is a function, $\gamma : \mathscr{P}(A) \to 2^A$ is injective.

(iii) $\gamma : \mathscr{P}(A) \to 2^A$ is surjective.

($\because$) $f \in 2^A$ and $B = f^{-1}(0)$.

$\Rightarrow$ If $x \in B$, then $f(x) = 0 = \chi_B(x)$.

If $x \in A \setminus B$, then $f(x) = 1 = \chi_B(x)$.

$\Rightarrow$ $f(x) = \chi_B(x)$ for all $x \in A$.

$\Rightarrow$ $f = \chi_B = \gamma(B)$ by Theorem 2.10.

$\Rightarrow$ γ satisfies the condition SRJ.

$\Rightarrow$ Since $\gamma : \mathscr{P}(A) \to 2^A$ is a function, $\gamma : \mathscr{P}(A) \to 2^A$ is surjective.

$\Rightarrow$ $\gamma : \mathscr{P}(A) \to 2^A$ is a bijective function. □

Exercise 2.5

1. Let $A = \{1, 2, 3\}$, $B = \{a, b\}$. Find A^B, B^A, 2^A, and $\mathscr{P}(A)$.

2. Let $\{B_i\}_{i \in I}$ be a family of subclasses of A. Prove that $\prod_{i \in I} B_i \subseteq A^I$.

3. Let $\{A_i\}_{i\in I}$ and $\{B_i\}_{i\in I}$ be families of classes with the same index class I. Show that if $A_i \subseteq B_i$ for all $i \in I$, then $\prod_{i\in I} A_i \subseteq \prod_{i\in I} B_i$.

4. Let $\{A_i\}_{i\in I}$ and $\{B_i\}_{i\in I}$ be families of nonempty classes with the same index class I. Prove that if $\prod_{i\in I} A_i \subseteq \prod_{i\in I} B_i$, then $A_i \subseteq B_i$ for all $i \in I$.

5. Let $\{A_i\}_{i\in I}$ and $\{B_i\}_{i\in I}$ be families of classes with the same index class I. Prove that $(\prod_{i\in I} A_i)\bigcap(\prod_{i\in I} B_i) = \prod_{i\in I}(A_i \bigcap B_i)$.

6. Let $\{A_i\}_{i\in I}$ and $\{B_j\}_{j\in J}$ be families of classes. Prove the following.

   (1) $(\prod_{i\in I} A_i)\bigcap(\prod_{j\in J} B_j) = \prod_{(i,j)\in I\times J} (A_i \bigcap B_j)$.

   (2) $(\prod_{i\in I} A_i)\bigcup(\prod_{j\in J} B_j) = \prod_{(i,j)\in I\times J} (A_i \bigcup B_j)$.

7. Let $\{A_i\}_{i\in I}$ be a family of classes, and for each $i \in I$, let B_i be a subclass of A_i. Let $p_i : \prod_{i\in I} A_i \to A_i$ denote the i-projection for each $i \in I$.

   Prove that $\bigcap_{i\in I} p_i^{-1}(B_i) = \prod_{i\in I} B_i$.

8. Let $\{A_i\}_{i\in I}$ be a family of classes and let $A = \prod_{i\in I} A_i$. For a subset $B \subseteq A$, let $B_i = p_i(B)$ for each $i \in I$. Prove that $B \subseteq \prod_{i\in I} B_i$.

9. For classes A, B and X, prove that $A^X \cup B^X \subseteq (A \cup B)^X$.

10. For classes A, B and X, prove that $(A \cap B)^X = A^X \cup B^X$.

11. For classes A, B and X, prove that $(A \setminus B)^X = A^X \setminus B^X$.

12. Prove that if A and B are sets, then A^B is a set.

2.6 The axiom of replacement

Axioms 3 through 6 are "set" axioms, that is, they are designed for the purpose of establishing the properties of sets. We are now in a position to introduce our last "set" axiom. This axiom is motivated by the following considerations.

We noted earlier that we are to think of a set as a class which is "not too large". Now, if A and B are classes and $f : A \to B$ is a surjective function, then, in an obvious intuitive sense, B has "as many, or fewer elements than A" by the conditions F1 and SRJ. Thus if A is "not too large" and $f : A \to B$ is a surjective function, it stands to reason that B is "not too large". These remarks lead us to state the following as an axiom.

Axiom 7 (Axiom of Replacement). If A is a set and $f : A \to B$ is a surjective function, then B is a set.

Remark 2.36. Axiom 3 of Pair and Axiom 4 of Subset follow from Axiom 7 of Replacement. See Exercises 2.6.1 and 2.6.2. Thus Axiom 3 and Axiom 4 can now be eliminated from our list of axioms. To show that Axiom 7 implies Axiom 3, we need the fact that the empty class $\emptyset$ is a set, which will follow from Axiom 9 of Infinity in Chapter 6.

The statement Axiom 7 is traditionally called the *axiom of replacement.* It has the following consequences.

Theorem 2.37. (1) *If A is a set and A is in one-to-one correspondence with a class B, then B is a set.*

(2) *Every singleton is a set. (under the assumption that the empty class $\emptyset$ is a set which will be followed from Axiom 9 of Infinity in Chapter 6.)*

(3) *If $\{A_i\}_{i\in I}$ is an indexed family of sets and I is a set, then*

(i) $\{A_i \mid i \in I\}$ *is a set, and* (ii) $\bigcup_{i\in I} A_i$ *is a set.*

Proof. (1) A is a set and A is in one-to-one correspondence with a class B.

$\Rightarrow$ There exists a bijective function $f : A \to B$.

$\Rightarrow$ $f : A \to B$ is surjective.

$\Rightarrow$ Since A is a set, B is a set by Axiom 7 of Replacement.

(2) Assume that the empty class $\emptyset$ is a set.

$\Rightarrow$ $\{\emptyset, \emptyset\}$ is a set by Axiom 3 of Pair.

$\Rightarrow$ $\{\emptyset\}$ is a set by Axiom 4 of Subset.

$\Rightarrow$ Since a singleton is in one-to-one correspondence with $\{\emptyset\}$, every singleton is a set by Axiom 7 of Replacement.

(3) (i) Let $\{A_i\}_{i\in I}$ be an indexed family of sets and assume that I is a set. Define a function $\phi : I \to \{A_i \mid i \in I\}$ by $\phi(i) = A_i$ for all $i \in I$.

$\Rightarrow \phi : I \to \{A_i \mid i \in I\}$ is a surjective function.

$\Rightarrow \{A_i \mid i \in I\}$ is a set by Axiom 7 of Replacement.

(ii) Since $\{A_i \mid i \in I\}$ is a set by (i) above and A_i is a set for each $i \in I$, $\bigcup_{i\in I} A_i$ is a set by Axiom 5 of Union. □

Theorem 2.38. *Let $\{A_i\}_{i\in I}$ be an indexed family of sets indexed by a set I and let $A = \bigcup_{i\in I} A_i$. Then*

(1) *A^I is a set and* (2) *$\prod_{i\in I} A_i$ is a set.*

Proof. Let $\{A_i\}_{i\in I}$ be an indexed family of sets indexed by a set I and write $A = \bigcup_{i\in I} A_i$.

(1) Since A_i is a set for each $i \in I$ and I is a set, $\{A_i\}_{i\in I}$ is a set by Theorem 2.37(3).

$\Rightarrow A = \bigcup_{i\in I} A_i$ is a set by Axiom 5 of Union.

$\Rightarrow$ Since I is a set, $I \times A$ is a set by Theorem 1.53.

$\Rightarrow \mathscr{P}(I \times A)$ is a set by Axiom 6 of Power Set.

Note that $A^I \subseteq \mathscr{P}(I \times A)$.

($\because$) $f \in A^I$.

$\Rightarrow f \subseteq I \times A$ by Definition 2.33(2).

$\Rightarrow f \in \mathscr{P}(I \times A)$ by Definition 1.49.

$\Rightarrow A^I$ is a set by Axiom 4 of Subset.

(2) Since $\prod_{i\in I} A_i \subseteq A^I$ and A^I is a set, $\prod_{i\in I} A_i$ is a set by Axiom 4 of Subset. □

The following proposition shows that $\prod_{i\in I} A_i$ in Definition 2.33 is legitimate.

Proposition 2.39. *If each $A_i (i \in I)$ and I are sets and if $f \in \prod_{i \in I} A_i$, then f is a set.*

Proof. Let $\{A_i\}_{i \in I}$ be a family of sets, where I is a set.

Let $A = \bigcup_{i \in I} A_i$ and let $f \in \prod_{i \in I} A_i$.

$\Rightarrow A$ is a set by Axiom 5 of Union and $f : I \to A$ is a function.

$\Rightarrow I \times A$ is a set by Theorem 1.53 and $f \subseteq I \times A$.

$\Rightarrow f$ is a set by Axiom 4 of Subset. □

Exercise 2.6

1. Show that Axiom 3 of Pair follows from Axiom 7 of Replacement. Thus Axiom 3 of Pair can now be eliminated. [Hint: Use the fact that the empty class $\emptyset$ is a set, which will be guaranteed by Axiom 9 of Infinity, and use the Axiom 6 of Power Set.]

2. Show that Axiom 4 of Subset follows from Axiom 7 of Replacement. Thus Axiom 4 of Subset can now be eliminated. [Hint: Use the fact that the empty class $\emptyset$ is a set.]

Chapter 3

Relations

3.1 Introduction

Intuitively, a binary relation in a class A is a statement $R(x, y)$ which is either true or false for each ordered pair (x, y) of elements of A. For instance, the relation ⌜*x divides y*⌟, which we may write $D(x, y)$, is a relation in the class $\mathbb{Z}$ of the integers: $D(x, y)$ is true for every pair (x, y) of integers such that y is a multiple of x. It is false for every other pair of integers.

The *representing graph* of a relation R in A is a graph $G \subseteq A \times A$ which consists of all the pairs (x, y) such that $R(x, y)$ is true. Conversely, if we are given an arbitrary graph $G \subseteq A \times A$, then G defines a relation in A, namely the relation R such that $R(x, y)$ is true if and only if $(x, y) \in G$.

Thus, as we did in the case of functions, we are able to identify relations with their representing graphs. In this way the study of relations is a part of elementary set theory.

3.2 Fundamental concepts and definitions

Definition 3.1. (1) A *relation in* a class A is defined to be any subclass of $A \times A$. Thus a relation G in a class A is a graph G with $\text{dom}(G) \subseteq A$

and $\mathrm{ran}(G) \subseteq A$.

(2) For a relation G in a class A, the *inverse relation* of G, denoted by G^{-1} is the relation defined by $G^{-1} = \{(y, x) \in A \times A \mid (x, y) \in G\}$.

(3) For two relations G and H in a class A, the *composite relation* of G followed by H, denoted by $G \circ H$, is the relation defined by $G \circ H = \{(x, z) \in A \times A \mid \text{there exists } y \in A \text{ such that } (x, y) \in H \text{ and } (y, z) \in G\}$.

(4) The *diagonal graph* or *diagonal relation* 1_A in a class A is defined to be the class $\{(x, x) \mid x \in A\}$. i.e., $1_A = \{(x, x) \mid x \in A\}$.

Definition 3.2. Let G be a relation in a class A.

(1) G is called *reflexive* if $(x, x) \in G$ for all $x \in A$.

(2) G is called *symmetric* if $(x, y) \in G$ implies $(y, x) \in G$.

(3) G is called *antisymmetric* if $(x, y) \in G$ and $(y, x) \in G$ imply $x = y$.

(4) G is called *transitive* if $(x, y) \in G$ and $(y, z) \in G$ imply $(x, z) \in G$.

(5) G is called *irreflexive* if $(x, x) \notin G$ for all $x \in A$.

(6) G is called *asymmetric* if $(x, y) \in G$ implies $(y, x) \notin G$.

(7) G is called *intransitive* if $(x, y) \in G$ and $(y, z) \in G$ imply $(x, z) \notin G$.

(8) G is called an *equivalence relation* if it is reflexive, symmetric, and transitive.

(9) G is called an *order relation* if it is reflexive, antisymmetric, and transitive.

(10) G is called an *strict order relation* if it is irreflexive, asymmetric, and transitive.

There is a variety of interesting and useful alternative ways of defining the above notions. Some are given in the next theorem.

Theorem 3.3. *Let G be a relation in A.*

(1) *G is reflexive if and only if $1_A \subseteq G$.*

(2) *G is symmetric if and only if $G = G^{-1}$.*

(3) *G is antisymmetric if and only if $G \cap G^{-1} \subseteq 1_A$.*

(4) *G is transitive if and only if $G \circ G \subseteq G$.*

Proof. Let G be a relation in A.

(1) G is reflexive.
$\Leftrightarrow$ $(x, x) \in G$ for all $x \in A$ by Definition 3.2(1).
$\Leftrightarrow$ $(x, x) \in G$ for all $(x, x) \in 1_A$ by Definition 3.1(4).
$\Leftrightarrow$ $1_A \subseteq G$ by Definition 1.14(2).
(2) G is symmetric.
$\Leftrightarrow$ $(x, y) \in G$ iff $(y, x) \in G$ for all $x, y \in A$ by Definition 3.2(2).
$\Leftrightarrow$ $(x, y) \in G$ iff $(x, y) \in G^{-1}$ for all $x, y \in A$ by Definition 3.1(2).
$\Leftrightarrow$ $G = G^{-1}$ by Definition 1.14(1).
(3) G is antisymmetric.
$\Leftrightarrow$ If $(x, y) \in G$ and $(y, x) \in G$, then $x = y$ by Definition 3.2(3).
$\Leftrightarrow$ If $(x, y) \in G$ and $(x, y) \in G^{-1}$, then $x = y$ by Definition 3.1(2).
$\Leftrightarrow$ If $(x, y) \in G \cap G^{-1}$, then $x = y$ by Definition 1.17(2).
$\Leftrightarrow$ If $(x, y) \in G \cap G^{-1}$, then $(x, y) \in 1_A$ by Definition 3.1(4).
$\Leftrightarrow$ $G \cap G^{-1} \subseteq 1_A$ by Definition 1.14(2).
(4) G is transitive.
$\Leftrightarrow$ If $(x, y) \in G$ and $(y, z) \in G$, then $(x, z) \in G$ by Definition 3.2(4).
$\Leftrightarrow$ If $(x, z) \in G \circ G$, then $(x, z) \in G$ by Definition 3.1(3).
$\Leftrightarrow$ $G \circ G \subseteq G$ by Definition 1.14(2). □

Example 3.4. Let $\mathbb{Z}$ designate the set of integers.

(1) The equality relation $=$ in $\mathbb{Z}$ is reflexive, symmetric, and transitive. Hence, it is an equivalence relation.

(2) The relation $\leq$ (⌜*less than or equal to*⌟) in $\mathbb{Z}$ is reflexive, antisymmetric, and transitive. Hence it is an order relation.

(3) The relation $<$ (⌜*strictly less than*⌟) in $\mathbb{Z}$ is not an order relation. It is irreflexive, asymmetric, and transitive. Hence it is a strict order relation.

(4) The equality relation $=$ in the class of all sets is reflexive, symmetric, and transitive by Theorem 1.15. Hence, it is an equivalence relation.

(5) The relation $\Leftrightarrow$ on the class of all the sentences is an equivalence relation. Indeed, for all sentences P, Q and R, $\Leftrightarrow$ satisfies the following.

(i) $P \Leftrightarrow P$.

(ii) If $P \Leftrightarrow Q$, then $Q \Leftrightarrow P$.

(iii) If $P \Leftrightarrow Q$ and $Q \Leftrightarrow R$, then $P \Leftrightarrow R$.

Exercise 3.2

1. Each of the following describes a relation in the set $\mathbb{Z}$ of integers. State, for each one, whether it has any of the following properties: reflexive, symmetric, antisymmetric, transitive, irreflexive, asymmetric, and intransitive.

(1) $G = \{(x, y) \mid x + y < 3\}$.

(2) $G = \{(x, y) \mid x \text{ divides } y\}$.

(3) $G = \{(x, y) \mid x \text{ and } y \text{ are relatively prime}\}$.

(4) $G = \{(x, y) \mid x + y \text{ is an even number}\}$.

(5) $G = \{(x, y) \mid x = y \text{ or } x = -y\}$.

(6) $G = \{(x, y) \mid x + y$ is even number and x is a multiple of $y\}$.

(7) $G = \{(x, y) \mid y = x + 1\}$.

2. Let G be a relation in A. Prove each of the following.

   (1) G is irreflexive if and only if $G \cap 1_G = \emptyset$.

   (2) G is asymmetric if and only if $G \cap G^{-1} = \emptyset$.

   (3) G is intransitive if and only if $(G \circ G) \cap G = \emptyset$.

3. Show that if G is an equivalence relation in A, then $G \circ G = G$.

4. Let $\{G_i\}_{i \in I}$ be an indexed family of equivalence relations in A. Show that $\bigcap_{i \in I} G_i$ is an equivalence relation in A.

5. Let $\{G_i\}_{i \in I}$ be an indexed family of order relations in A. Show that $\bigcap_{i \in I} G_i$ is an order relation in A.

6. Let H be a reflexive relation in A. Prove that for any relation G in A, $G \subseteq H \circ G$ and $G \subseteq G \circ H$.

7. Let G be a reflexive relation in A and let H be a reflexive and transitive relation in A. Show that $G \subseteq H$ if and only if $G \circ H = H$. (In particular, this holds if G and H are equivalence relations.)

8. Show that the inverse of an order relation in A is an order relation in A.

9. Let G be a relation in A. Show that G is an order relation if and only if $G \cap G^{-1} = 1_A$ and $G \circ G = G$.

10. Let G and H be equivalence relations in A. Show that $G \circ H$ is an equivalence in A if and only if $G \circ H = H \circ G$.

11. Let G and H be equivalence relations in A. Prove that $G \cup H$ is an equivalence in A if and only if $G \circ H \subseteq G \cup H$ and $H \circ G \subseteq G \cup H$.

12. Let G be an equivalence relation in A and let H and J be arbitrary relations in A. Prove that if $G \subseteq H$ and $G \subseteq J$, then $G \subseteq H \circ J$.

3.3 Equivalence relations and partitions

In the remainder of this chapter we will concern ourselves with equivalence relations in *sets*. The concepts we are about to introduce arise naturally in terms of sets, but cannot be extended to proper classes. To understand why not, the reader should review our discussion in Section 7 of Chapter 1. Briefly, if A is a set and $P(X)$ is a property, then by Theorem 1.52, it is legitimate to form the set of all the subsets $X \subseteq A$ which satisfy $P(X)$. However, if A were an arbitrary class, it would not be permissible to form the "class of all subclasses of A which satisfy $P(X)$". This restriction compels us to confine the following discussion to sets. Intuitively, this should not disturb the reader too much, for a set is almost the same thing as a class. A set is any class except an "excessively large" one.

Definition 3.5. Let A be a set. A *partition* of A is defined to be a family $\{A_i\}_{i \in I}$ of nonempty subsets of A with the following properties:

P1. For all $i, j \in I$, $A_i \cap A_j = \emptyset$ or $A_i = A_j$.

P2. $A = \bigcup_{i \in I} A_i$.

Intuitively, a partition is a family of subsets of A which are disjoint from one another, and whose union is all of A. See Figure 3.1. The subsets are called the *members* of the partition. Note that the subclass $\{A_i\}_{i \in I} \subseteq \mathscr{P}(A)$ may not be a set if A is a proper class.

Remark 3.6. (1) It is customary to allow a given member of the partition to be designated by more than one index. That is, we may have $A_i = A_j$ for some $i \neq j$. Note that $A_1 = A_6$ and $A_2 = A_7$ in Figure 3.1. Hence the condition that two *distinct* members be disjoint is correctly expressed by P1.

(2) Property P1 states that any two members A_i and A_j are either disjoint

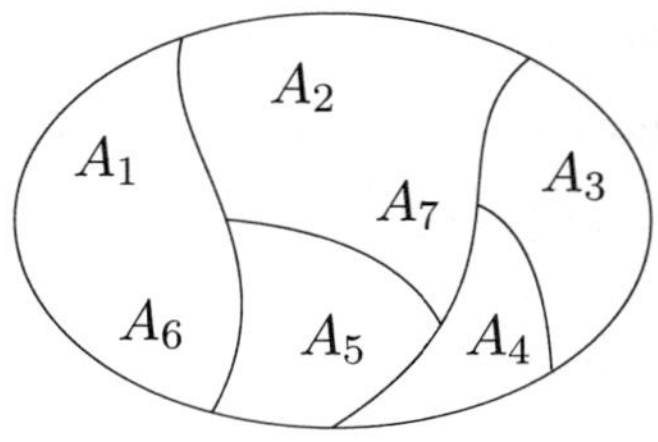

Figure 3.1: Partition $\{A_1, A_2, A_3, A_4, A_5, A_6, A_7\}$ of A

or equal. That is, they have either no elements in common or all their elements in common. In other words, if they have so much as one element in common, they have all their elements in common. Thus, P1 and P2 may also be stated as follows:

P1°. If there exists $x \in A_i \cap A_j$, then $A_i = A_j$.

P2$'$. $A \subseteq \bigcup_{i \in I} A_i$.

($\because$) Independently of Condition P2, we are given that each A_i is a subset of A.

$\Rightarrow \bigcup_{i \in I} A_i \subseteq A$ by Theorem 1.41(1).

$\Rightarrow A \subseteq \bigcup_{i \in I} A_i$ iff $A = \bigcup_{i \in I} A_i$.

It is convenient to write $P2'$ in the form

P2°. If $x \in A$, then $x \in A_i$ for some $i \in I$.

(3) A partition of A is a family $\{A_i\}_{i \in I}$ of nonempty subsets of A satisfying the following conditions:

P1°. If there exists $x \in A_i \cap A_j$, then $A_i = A_j$.

P2°. If $x \in A$, then $x \in A_i$ for some $i \in I$.

Definition 3.7. Let A be a set and let G be an equivalence relation in A.

(1) If $(x, y) \in G$, then we say that x *is equivalent to* y *modulo* G and write sometimes $x \underset{G}{\sim} y$ instead of $(x, y) \in G$. When there is no danger of ambiguity, we will simply write $x \sim y$ in stead of $x \underset{G}{\sim} y$ and say that x *is equivalent to* y.

(2) For an element $x \in A$, the *equivalence class of* x *modulo* G is the set G_x defined by

$$G_x = \{y \in A \mid (y, x) \in G\} = \{y \in A \mid y \underset{G}{\sim} x\}.$$

In other words, G_x is the set of all the elements of A which are equivalent to x. In the mathematical literature, G_x is also denoted by the symbols A_x, $[x]$, $G(x)$ and x/G.

(3) The set of all equivalence classes modulo G is called the *quotient set of* A *by* G, and is customarily denoted by A/G. In other words,

$$A/G = \{G_x \mid x \in A\}.$$

Note. For an equivalence relation $\sim$ in A, Definition 3.2(8) can be written as follows:

(1) $x \sim x$ for all $x \in A$.

(2) If $x \sim y$, then $y \sim x$ for all $x, y \in A$.

(3) If $x \sim y$ and $y \sim z$, then $x \sim z$ for all $x, y, z \in A$.

Lemma 3.8. *Let* G *be an equivalence relation in* A*, denoted by* $\sim$*. Then* $x \sim y$ *if and only if* $G_x = G_y$.

Proof. **(only if)** $x \sim y$ $(x, y \in A)$.

$\Rightarrow$ (i) $G_x \subseteq G_y$.

($\because$) $z \in G_x$.

$\Rightarrow$ $z \sim x$ by Definition 3.7(2).

$\Rightarrow$ Since $x \sim y$, $z \sim y$ by the transitivity of G.

$\Rightarrow$ $z \in G_y$ by Definition 3.7(2).

(ii) Analogously, we can show that $G_y \subseteq G_x$.

$\Rightarrow G_x = G_y$ by Definition 1.14(1).

(if) $G_x = G_y$.

$\Rightarrow$ Since $x \sim x$ by the reflexive property, $x \in G_x$.

$\Rightarrow$ Since $G_x = G_y$ by assumption, $x \in G_y$.

$\Rightarrow x \sim y$ by Definition 3.7(2). □

Theorem 3.9. *Let A be a set and let G be an equivalence relation in A, denoted by $\sim$. Let $\{G_x \mid x \in A\}$ be the family of all the equivalence classes modulo G. Then $\{G_x \mid x \in A\}$ is a partition of A.*

Proof. By Definition 3.7(2), $G_x \subseteq A$ for all $x \in A$.

Since $x \sim x$, $x \in G_x$ and hence each G_x is nonempty.

It remains to prove that the conditions P1° and P2°(P2′) hold.

(P1°) $z \in G_x \cap G_y$.

$\Rightarrow z \in G_x$ and $z \in G_y$ by Definition 1.17(2).

$\Rightarrow z \sim x$ and $z \sim y$ by Definition 3.7(2).

$\Rightarrow x \sim z$ and $z \sim y$ by the symmetry of G.

$\Rightarrow x \sim y$ by the transitivity of G.

$\Rightarrow G_x = G_y$ by Lemma 3.8.

Thus if $z \in G_x \cap G_y$, then $G_x = G_y$.

(P2′) $x \in A$.

$\Rightarrow$ Since $x \sim x$ by the reflexivity of G, $x \in G_x$.

$\Rightarrow x \in \bigcup_{x \in A} G_x$ by Definition 1.38(2).

Thus $A \subseteq \bigcup_{x \in A} G_x$ by Definition 1.14(2). □

Theorem 3.9 has an important converse, which follows.

Theorem 3.10. *Let A be a set and let $\{A_i\}_{i \in I}$ be a partition of A. Let G be the set of all pairs (x, y) of elements of A such that x and y are in the same member of the partition. i.e., $G = \{(x, y) \mid x \in A_i$ and $y \in A_i$ for some $i \in I\}$. Then G is an equivalence relation in A such that $\{A_i\}_{i \in I} = \{G_x \mid x \in A\}$.*

Proof. **(i)** G is reflexive.

(∵) $x \in A$.

$\Rightarrow$ Since $\{A_i\}_{i \in I}$ is a partition of A, $x \in A_i$ for some $i \in I$ by Definition 3.5(P2).

$\Rightarrow$ $x \in A_i$ and $x \in A_i$ by Theorem 1.9(3).

$\Rightarrow$ $(x, x) \in G$ by the definition of G.

(ii) G is symmetric.

(∵) $(x, y) \in G$.

$\Rightarrow$ $x \in A_i$ and $y \in A_i$ for some $i \in I$ by the definition of G.

$\Rightarrow$ $y \in A_i$ and $x \in A_i$ by Theorem 1.9(1).

$\Rightarrow$ $(y, x) \in G$ by the definition of G.

(iii) G is transitive.

(∵) $(x, y) \in G$ and $(y, z) \in G$.

$\Rightarrow$ $x \in A_i$ and $y \in A_i$ for some $i \in I$ and $y \in A_j$ and $z \in A_j$ for some $j \in I$ by the definition of G.

$\Rightarrow$ Since $y \in A_i \cap A_j$ and $\{A_i\}_{i \in I}$ is a partition of A, $A_i = A_j$ by Definition 3.5 and hence $x \in A_i$ and $z \in A_i$.

$\Rightarrow$ $(x, z) \in G$ by the definition of G.

Thus G is an equivalence relation by (i), (ii) and (iii).

(iv) Each A_i is an equivalence class modulo G.

(∵) Since $A_i \neq \emptyset$, there exists $x \in A_i$.

$\Rightarrow$ $y \in A_i$ iff $(y, x) \in G$ iff $y \in G_x$.

$\Rightarrow$ $A_i = G_x$ by Definition 1.14(1). □

Definition 3.11. (1) Let G be an equivalence relation in a class A, and let $\{G_x\}_{x \in A}$ be the family of all the equivalence classes modulo G. Then by Theorem 3.9, $\{G_x\}_{x \in A}$ is a partition of A, which is called the *partition induced by* G, or the *partition corresponding to* G.

(2) Conversely, let $\{A_i\}_{i \in I}$ be a partition of a set A and define

$$G = \{(x, y) \mid x \in A_i \text{ and } y \in A_i \text{ for some } i \in I\}.$$

Then by Theorem 3.10, G is an equivalence relation in A which induces the partition $\{A_i\}_{i \in I}$. G is called the *equivalence relation corresponding to* $\{A_i\}_{i \in I}$.

Remark 3.12. Theorems 3.9 and 3.10 make it clear that every equivalence relation in A corresponds uniquely to a partition of A, and conversely.

(1) For a partition of A, the *corresponding equivalence relation* is the relation which calls elements x and y "equivalent" if they are in the same member of the partition.

(2) For an equivalence relation in A, the *corresponding partition* is the one which puts elements x and y in the same member of the partition if and only if they are equivalent.

(3) The reader should note that G is the equivalence relation corresponding to $\{A_i\}_{i \in I}$ if and only if $\{A_i\}_{i \in I}$ is the partition corresponding to G.

Example 3.13. Let $A = \{a, b, c, d, e\}$ and let $A_1 = \{a, b\}$ and $A_2 = \{c, d\}$ and $A_3 = \{e\}$. Let $G = \{(a,a), (b,b), (c,c), (d,d), (e,e), (a,b), (b,a), (c,d), (d,c)\}$.

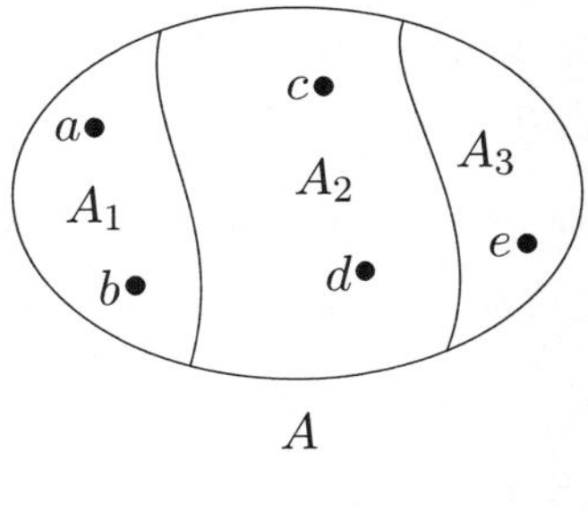

Figure 3.2: partition $\{A_1, A_2, A_3\}$ of A

It is easy to see that $\{A_1, A_2, A_3\}$ is a partition of A (see Figure 3.2), and that G is an equivalence relation in A. G is the equivalence relation corresponding to $\{A_1, A_2, A_3\}$ and $\{A_1, A_2, A_3\}$ is the partition corresponding to G. It should be noted that $A_1 = G_a = G_b$, $A_2 = G_c = G_d$, and $A_3 = G_e$. Thus A/G is the set of three elements $\{G_a, G_c, G_e\}$.

The concept of a quotient set plays a vital role in many parts of advanced mathematics.

Exercise 3.3

1. Let $\mathbb{Z}$ be the set of integers. For each integer n, let $B_n = \{m \in \mathbb{Z} \mid$ there exists $q \in \mathbb{Z}$ such that $m = n + 5q\}$. Prove that $\{B_n\}_{n\in\mathbb{Z}}$ is a partition of $\mathbb{Z}$.

2. Let $\mathbb{R}$ be the set of the real numbers. In each of the following, prove that $\{B_r\}_{r\in\mathbb{R}}$ is a partition of $\mathbb{R} \times \mathbb{R}$. Describe geometrically the members of this partition. Find the equivalence relation corresponding to each partition.

   (1) $B_r = \{(x, y) \mid x \in \mathbb{R}, y \in \mathbb{R}, y = x + r\}$ for each $r \in \mathbb{R}$.

   (2) $B_r = \{(x, y) \mid x \in \mathbb{R}, y \in \mathbb{R}, x^2 + y^2 = r^2\}$ for each $r \in \mathbb{R}$.

   [Hint: $y = x + r$ is the equation of a line and $x^2 + y^2 = r$ is the equation of a circle.]

3. Let $\mathbb{R}$ be the set of the real numbers. Prove that each of the following is an equivalence relation in $\mathbb{R} \times \mathbb{R}$.

   (1) $G = \{((a, b), (c, d)) \mid (a, b) \in \mathbb{R} \times \mathbb{R}, (c, d) \in \mathbb{R} \times \mathbb{R}, a^2 + b^2 = c^2 + d^2\}$.

   (2) $H = \{((a, b), (c, d)) \mid (a, b) \in \mathbb{R} \times \mathbb{R}, (c, d) \in \mathbb{R} \times \mathbb{R}, b - a = d - c\}$.

   (3) $J = \{((a, b), (c, d)) \mid (a, b) \in \mathbb{R} \times \mathbb{R}, (c, d) \in \mathbb{R} \times \mathbb{R}, a + b = c + d\}$.

   Find the partition corresponding to each of these equivalence relations, and describe geometrically the members of this partition.

   [Hint for (2): If $b - a = d - c = k$, note that $((a, b), (c, d)) \in H$ if and only if both (a, b) and (c, d) satisfy the equation $y = x + k$.]

   [Hint for (3): If $a + b = c + d = k$, note that $((a, b), (c, d)) \in J$ if and only if both (a, b) and (c, d) satisfy the equation $y = -x + k$.]

4. If H and J are the equivalence relations of Exercise 3.3.3 above, describe the equivalence relation $H \cap J$. Describe the equivalence classes modulo $H \cap J$.

5. Let H and J be the equivalence relations of Exercise 3.3.3 above. Prove that $H \circ J = J \circ H$ and conclude that $H \circ J$ is an equivalence relation, and describe the equivalence relation $H \circ J$. [Hint: See Exercise 3.2.10.]

6. Let L be the set of all straight lines in the plane. Let G and H be the following relations in L.

$$G = \{(l_1, l_2) \mid l_1 \text{ is parallel to } l_2\}, \quad H = \{(l_1, l_2) \mid l_1 \text{ is perpendicular to } l_2\}.$$

Prove the following (argue informally):

(1) G is an equivalence relation in L.

(2) $H \circ G = H$ and $G \circ H = H$.

(3) $G \cup H$ is an equivalence relation. Describe its equivalence classes.

7. Let A be an arbitrary set. Prove that 1_A and $A \times A$ are equivalence relations in A. Describe the partitions induced, respectively, by 1_A and $A \times A$.

8. Let $\{A_i\}_{i \in I}$ be a partition of A and let $\{B_j\}_{j \in J}$ be a partition of B. Prove that $\{A_i \times B_j\}_{(i,j) \in I \times J}$ is a partition of $A \times B$.

9. Let $f : A \to B$ be a surjective function and let $\{B_i\}_{i \in I}$ be a partition of B. Prove that $\{f^{-1}(B_i)\}_{i \in I}$ is a partition of A.

10. Let $f : A \to B$ be an injective function and let $\{A_i\}_{i \in I}$ be a partition of A. Prove that $\{f(A_i)\}_{i \in I}$ is a partition of $f(A)$.

11. Let G and H be equivalence relations in A. Prove that each equivalence class modulo $G \cap H$ is the intersection of an equivalence class modulo G with an equivalence class modulo H. More exactly, $(G \cap H)_x = G_x \cap H_x$ for all $x \in A$.

12. Let G and H be equivalence relations in A and assume that $G \cup H$ is an equivalence relation in A. Prove that each equivalence class modulo $G \cup H$ is the union of an equivalence class modulo G with an equivalence class modulo H. More exactly, $(G \cup H)_x = G_x \cup H_x$ for all $x \in A$.

3.4 Pre-image, restriction and quotient of equivalence relations

Definition 3.14. (1) Let $f : A \to B$ be a function, and let G be an equivalence relation in B. The *pre-image of G under f*, denoted by $f^{-1}(G)$, is a relation in A defined by

$$f^{-1}(G) = \{(x, y) \mid x \in A, y \in A \text{ and } (f(x), f(y)) \in G\}.$$

It is simple to show that $f^{-1}(G)$ is an equivalence relation in A.

(2) Let G be an equivalence relation in A and let $B \subseteq A$. The *restriction of G to B*, denoted by $G|_B$, is a relation in B defined by

$$G|_B = \{(x, y) \mid x \in B \text{ and } y \in B \text{ and } (x, y) \in G\}.$$

It is simple to show that $G\ |_B$ is an equivalence relation in B.

(3) Let G and H be equivalence relations in A. We call H a *refinement of G* if $H \subseteq G$. In this case, we also say that H is *finer* than G, and that G is *coarser* than H.

Theorem 3.15. *Let G and H be equivalence relations in A. Assume that $H \subseteq G$. Then if $z \in G_x$, then $H_z \subseteq G_x$.*

Proof. Let G and H be equivalence relations in A. Assume that $H \subseteq G$ and let $z \in G_x$. We claim that $H_z \subseteq G_x$.
($\because$) $y \in H_z$. i.e., $(y, z) \in H \subseteq G$.
$\Rightarrow$ Since $(z, x) \in G$, $(y, x) \in G$ by the transitivity of G.
$\Rightarrow y \in G_x$. □

Corollary 3.16. *Let G and H be equivalence relations in A. If $H \subseteq G$, then for each $x \in A, H_x \subseteq G_x$.*

Proof. Since $x \in G_x$, $H_x \subseteq G_x$ by Theorem 3.15. □

Remark 3.17. Let G and H be equivalence relations in A. If H is a refinement of G, then each equivalence class modulo G is a union of equivalence classes modulo H. Thus for all $x \in A$, $\{H_z \mid z \in G_x\}$ is a partition of G_x.

($\because$) P_1. For all $y, z \in G_x$, either $H_y \cap H_z = \emptyset$ or $H_y = H_z$ because H is an equivalence relation in A.

P_2. $G_x = \bigcup_{z \in G_x} H_z$ because $x \in A$ implies that $H_z \subseteq G_x$ for all $z \in G_x$ by Theorem 3.15.

Definition 3.18. Let G and H be equivalence relations in a set A and let H be a refinement of G. The *quotient relation of* G *by* H, which is usually denoted by G/H, is a relation in A/H defined by

$$G/H = \{(H_x, H_y) \mid (x, y) \in G\} (\subseteq A/H \times A/H).$$

Theorem 3.19. *Let G and H be equivalence relations in a set A. Assume that H is a refinement of G. Then G/H is an equivalence relation in A/H.*

Proof. Let G and H be equivalence relations in a set A. Assume that H is a refinement of G.

(1) G/H is reflexive.

($\because$) $H_x \in A/H$. i.e., H_x is an equivalence class of H in A and $x \in A$.

$\Rightarrow (x, x) \in G$ by the reflexivity of G.

$\Rightarrow (H_x, H_x) \in G/H$ by Definition 3.18.

(2) G/H is symmetric.

($\because$) $(H_x, H_y) \in G/H$.

$\Rightarrow (x, y) \in G$ by Definition 3.18.

$\Rightarrow (y, x) \in G$ by the symmetry of G.

$\Rightarrow (H_y, H_x) \in G/H$ by Definition 3.18.

(3) G/H is transitive.

($\because$) $(H_x, H_y) \in G/H$ and $(H_y, H_z) \in G/H$.

$\Rightarrow (x, y) \in G$ and $(y, z) \in G$ by Definition 3.18.

$\Rightarrow (x, z) \in G$ by the transitivity of G.

$\Rightarrow (H_x, H_z) \in G/H$ by Definition 3.18. □

Remark 3.20. Let G and H be equivalence relations in a set A. If H is a refinement of G, then G/H is an equivalence relation in A/H by Theorem 3.19.

(1) We may write $H_x \underset{G/H}{\sim} H_y$ instead of $(H_x, H_y) \in G/H$.
Thus Definition 3.18 may be written in the more suggestive form:

$$H_x \underset{G/H}{\sim} H_y \quad \text{if and only if} \quad x \underset{G}{\sim} y.$$

(2) The reader may easily verify that $H_x \underset{G/H}{\sim} H_y$ if and only if H_x and H_y are subsets of the same equivalence class modulo G.

Example 3.21. Let $A = \{a, b, c, d, e\}$ and let $H = \{(a, a),\ (b, b),\ (c, c),\ (d, d),\ (e, e),\ (a, b),\ (b, a),\ (c, d),\ (d, c)\}$ defined as in Example 3.13. Let $G = \{(a, a),\ (b, b),\ (c, c),\ (d, d),\ (e, e),\ (a, b),\ (b, a),\ (c, d),\ (d, c),\ (c, e),\ (e, c),\ (d, e),\ (e, d)\}$.

(1) G and H are equivalence relations in A and H is a refinement of G.

(2) The partition of A induced by G is $\{G_a, G_c\}$, where $G_a = \{a, b\} = G_b$ and $G_c = \{c, d, e\} = G_d = G_e$. i.e., $A/G = \{G_a, G_c\}$. See Figure 3.3.

(3) The partition of A induced by H is $\{H_a, H_c, H_e\}$, where $H_a = \{a, b\} = H_b$, $H_c = \{c, d\} = H_d$ and $H_e = \{e\}$. i.e., $A/H = \{H_a, H_c, H_e\}$. See Figure 3.3.

(4) Each equivalence class modulo G is a union of equivalence classes modulo H. i.e., $G_a = G_b = H_a$ and $G_c = G_d = G_e = H_c \cup H_e$.

(5) $G/H = \{(H_a, H_a), (H_c, H_c), (H_e, H_e), (H_c, H_e), (H_e, H_c)\}$.

(6) The partition of A/H induced by G/H is $\{(G/H)_{H_a}, (G/H)_{H_c}\}$, where $(G/H)_{H_a} = \{H_a\}$ and $(G/H)_{H_c} = \{H_c, H_e\} = (G/H)_{H_e}$.
i.e., $(A/H)/(G/H) = \{(G/H)_{H_a}, (G/H)_{H_c}\}$. Thus $(A/H)/(G/H)$ is the set $\{\alpha, \beta\}$, where $\alpha = \{H_a\}$ and $\beta = \{H_c, H_e\}$. See Figure 3.4.

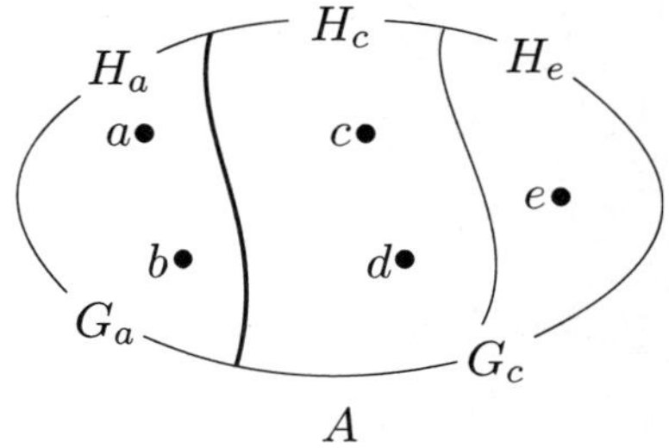

Figure 3.3: partition $A/G = \{G_a, G_c\}$ of A

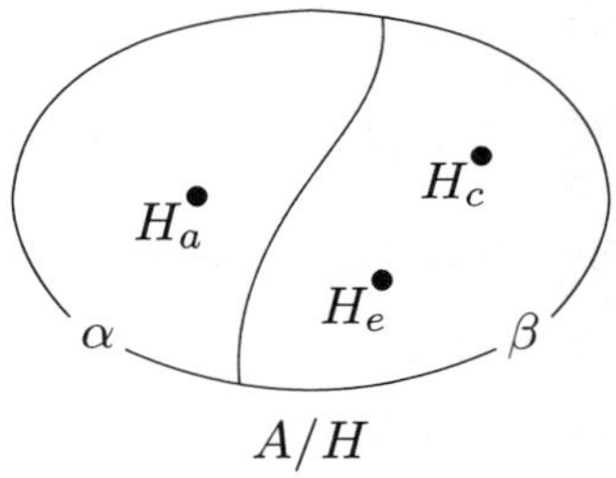

Figure 3.4: partition $(A/H)/(G/H) = \{\alpha, \beta\}$ of A/H

Example 3.22. Consider $H = \{(m,n) \in \mathbb{Z} \times \mathbb{Z} \mid m - n \text{ is a multiple of } 6\}$ and $G = \{(m,n) \in \mathbb{Z} \times \mathbb{Z} \mid m - n \text{ is a multiple of } 3\}$.

(1) G and H are equivalence relations in $\mathbb{Z}$ and H is a refinement of G.

(2) The partition of $\mathbb{Z}$ induced by G is $\{G_i \mid i \in \mathbb{Z}\} = \{G_0, G_1, G_2\}$, where

$G_0 = \{3n \mid n \in \mathbb{Z}\} = G_3 = G_{-3} = G_6 = G_{-6} = \cdots,$

$G_1 = \{3n + 1 \mid n \in \mathbb{Z}\} = G_4 = G_{-2} = G_7 = G_{-5} = \cdots,$

$G_2 = \{3n + 2 \mid n \in \mathbb{Z}\} = G_5 = G_{-1} = G_8 = G_{-4} = \cdots.$

i.e., $\mathbb{Z}/G = \{G_i \mid i \in \mathbb{Z}\} = \{G_0, G_1, G_2\}$.

(3) The partition of $\mathbb{Z}$ induced by H is $\{H_0, H_1, H_2, H_3, H_4, H_5\}$, where

$$H_0 = \{6n \mid n \in \mathbb{Z}\} = H_6 = H_{-6} = H_{12} = H_{-12} = \cdots,$$

$$H_1 = \{6n+1 \mid n \in \mathbb{Z}\} = H_7 = H_{-5} = H_{13} = H_{-11} = \cdots,$$

$$H_2 = \{6n+2 \mid n \in \mathbb{Z}\} = H_8 = H_{-4} = H_{14} = H_{-10} = \cdots,$$

$$H_3 = \{6n+3 \mid n \in \mathbb{Z}\} = H_9 = H_{-3} = H_{15} = H_{-9} = \cdots,$$

$$H_4 = \{6n+4 \mid n \in \mathbb{Z}\} = H_{10} = H_{-2} = H_{16} = H_{-8} = \cdots,$$

$$H_5 = \{6n+5 \mid n \in \mathbb{Z}\} = H_{11} = H_{-1} = H_{17} = H_{-7} = \cdots.$$

i.e., $\mathbb{Z}/H = \{H_i \mid i \in \mathbb{Z}\} = \{H_0, H_1, H_2, H_3, H_4, H_5\}$.

(4) $$G/H = \{(H_0,H_0),(H_1,H_1),(H_2,H_2),(H_3,H_3),(H_4,H_4),(H_5,H_5), (H_0,H_3),(H_3,H_0),(H_1,H_4),(H_4,H_1),(H_2,H_5),(H_4,H_2)\}.$$

(5) The partition of $\mathbb{Z}/H$ induced by G/H is $\{(G/H)_{H_0}, (G/H)_{H_1}, (G/H)_{H_2}\}$, where $(G/H)_{H_0} = \{H_0, H_3\} = (G/H)_{H_3}$, $(G/H)_{H_1} = \{H_1, H_4\} = (G/H)_{H_4}$ and $(G/H)_{H_2} = \{H_2, H_5\} = (G/H)_{H_5}$.

i.e., $$(\mathbb{Z}/H)/(G/H) = \{(G/H)_{H_i} \mid i = 0,1,2,3,4,5\} = \{(G/H)_{H_0}, (G/H)_{H_1}, (G/H)_{H_2}\}.$$

Exercise 3.4

1. Let $A = \{a,b,c,d,e,f\}$ and let G and H be the following equivalence relations in A defined by $G = 1_A \cup \{(a,b),(b,a),(b,c),(c,b),(a,c),(c,a),(d,e),(e,d)\}$ and $H = 1_A \cup \{(b,c),(c,b)\}$. Clearly H is a refinement of G. Exhibit the sets $A/G, A/H, G/H, (A/H)/(G/H)$.

2. Let $\mathbb{R}$ be the set of the real numbers and let G be the equivalence relation in $\mathbb{R} \times \mathbb{R}$ defined by
$$G = \{((a,b),(c,d)) \mid (a,b) \in \mathbb{R}\times\mathbb{R}, (c,d) \in \mathbb{R}\times\mathbb{R}, a^2+b^2 = c^2+d^2\}.$$
Let $f : \mathbb{R} \to \mathbb{R}\times\mathbb{R}$ be the function given by $f(x) = (\sin x, \cos x)$. Describe $f^{-1}(G)$. What are its equivalence classes?

3. Let $f : A \to B$ be a function and let G be an equivalence relation in B. Prove that $f^{-1}(G)$ is an equivalence relation in A.

4. Let $f : A \to B$ be a function and let G be an equivalence relation in B. Prove that each equivalence class modulo $f^{-1}(G)$ is the inverse image of an equivalence class modulo G. More precisely, if $H = f^{-1}(G)$ and $y = f(x)$, prove that $H_x = f^{-1}(G_y)$.

5. Let G be an equivalence relation in A and assume that $B \subseteq A$. Prove that $G|_B$ is an equivalence relation in B.

6. Let G be an equivalence relation in A and assume that $B \subseteq A$. Prove that for each $x \in B$, $(G|_B)_x = G_x \cap B$.

7. Let G, H and J be equivalence relations in A and assume that $G \subseteq H$ and $H \subseteq J$. Prove that H/G is finer than J/G.

8. Let G, H and J be equivalence relations in A and assume that $G \subseteq H$ and $H \subseteq J$. Prove the following.

   (1) $G \subseteq H \circ J$.

   (2) If $H \circ J$ is an equivalence relation in A, then $(H/G) \circ (J/G) = (H \circ J)/G$.

   (3) $(H/G) \circ (J/G)$ is an equivalence relation in A/G.

9. Assume that G and H are equivalence relations in A and that $G \subseteq H$. Prove that $G_x \underset{H/G}{\sim} G_y$ if and only if G_x and G_y are subsets of the same equivalence class modulo H.

10. Let G be an equivalence relation in A and let H be an equivalence relation in B. The *product* of G and H is defined to be the following relation in $A \times B$:

    $$G \cdot H = \{((x, w), (y, z)) \mid (x, y) \in G \text{ and } (w, z) \in H\}.$$

    Prove that $G \cdot H$ is an equivalence relation in $A \times B$.

11. Prove that every equivalence relation in a set A is the pre-image of an equivalence relation in $A \times A$.

    [Hint: Let $f : A \to A \times A$ be the function given by $f(x) = (x, x)$. If G is a relation in A, consider the relation $G \cdot G$ in $A \times A$ defined in Exercise 3.4.10.]

12. Let $f : A \to B$ be a function and let G be an equivalence relation in B. Prove that $f^{-1}(G) = f^{-1} \circ G \circ f$.

3.5 Equivalence relations and functions

Definition 3.23. Let A be a set.

(1) For a function $f : A \to B$, we define a relation G in A by

$$G = \{(x, y) \mid f(x) = f(y)\} \subseteq A \times A.$$

It is easy to see that G is an equivalence relation in A.

G is called the *equivalence relation determined by* f.

(2) Conversely, for an equivalence relation G in A, we define a function $f : A \to A/G$ by $f(x) = G_x$ for all $x \in A$.

It is easy to see that $f : A \to A/G$ is a function.

The function f is called the *canonical function from* A *to* A/G.

Theorem 3.24. *Let G be an equivalence relation in a set A and let f be the canonical function from A to A/G. Then G is the equivalence relation determined by f.*

Proof. Let f be the canonical function from A to A/G, and let H be the equivalence relation determined by f.

$\Rightarrow$ Since $(x, y) \in G$ iff $G_x = G_y$ iff $f(x) = f(y)$ iff $(x, y) \in H$, $G = H$. □

Now we will define three functions $q : A \to A/G$, $h : A/G \to f(A)$ and $i : f(A) \to B$, obtained from a function $f : A \to B$, which play an important role in many mathematical arguments, where G is the equivalence relation determined by f.

Definition 3.25. Let A and B be sets and let $f : A \to B$ be a function. Let G be the equivalence relation determined by $f : A \to B$.

- $q : A \to A/G$ is the function given by $q(x) = G_x$ for all $x \in A$.
- $h : A/G \to f(A)$ is the function given by $h(G_x) = f(x)$ for all $x \in A$.

- $i : f(A) \to B$ is the function given by $i(y) = y$ for all $y \in f(A)$.

Note that $q : A \to A/G$ is the canonical function and that $i : f(A) \to B$ is the inclusion function $i_{f(A)} : f(A) \to B$ of $f(A)$ in B.

Theorem 3.26. *Let A and B be sets. Let $f : A \to B$ be a function, let G be the equivalence relation determined by f, and let q, h, i be the functions defined above. Then the following hold.*

(1) *$q : A \to A/G$ is surjective.* (2) *$h : A/G \to f(A)$ is bijective.*

(3) *$i : f(A) \to B$ is injective.* (4) *$f = i \circ h \circ q$.*

Proof. Clearly, $q : A \to A/G$, $h : A/G \to f(A)$ and $i : f(A) \to B$ are well-defined functions.

(1) $q : A \to A/G$ is surjective because $G_x \in A/G$ implies that $x \in A$ and $q(x) = G_x$.

(2) (i) $h : A/G \to f(A)$ is injective.

($\because$) $h(G_x) = h(G_y)$. $\Rightarrow f(x) = f(y)$. $\Rightarrow (x, y) \in G$. $\Rightarrow G_x = G_y$.

(ii) $h : A/G \to f(A)$ is surjective because $f(x) \in f(A)$ implies that $x \in A$, $G_x \in A/G$ and $f(x) = h(G_x)$.

$\Rightarrow h : A/G \to f(A)$ is bijective by (i) and (ii) above.

(3) $i : f(A) \to B$ is injective because $i(y_1) = i(y_2)$ implies that $y_1 = i(y_1) = i(y_2) = y_2$.

(4) $i \circ h \circ q = f$.

($\because$) Since $(i \circ h \circ q)(x) = i(h(q(x))) = i(h(G_x)) = i(f(x)) = f(x)$ for all $x \in A$, and since $\mathrm{dom}(i \circ h \circ q) = A = \mathrm{dom}(f)$ and $\mathrm{ran}(i \circ h \circ q) = B = \mathrm{ran}(f)$,

$i \circ h \circ q = f : A \to B$ by Theorem 2.10. □

Definition 3.27. By Theorem 3.26, any function $f : A \to B$ can be expressed as a composite of three functions $q : A \to A/G$, $h : A/G \to f(A)$ and $i : f(A) \to B$, which are, respectively, surjective, bijective, and injective, where G is the equivalence relation determined by $f : A \to B$. This is referred to as the *canonical decomposition of* f, and it is customarily exhibited in the following diagram:

$$A \underset{\text{surj}}{\xrightarrow{q}} A/G \underset{\text{bij}}{\xrightarrow{h}} f(A) \underset{\text{inj}}{\xrightarrow{i}} B.$$

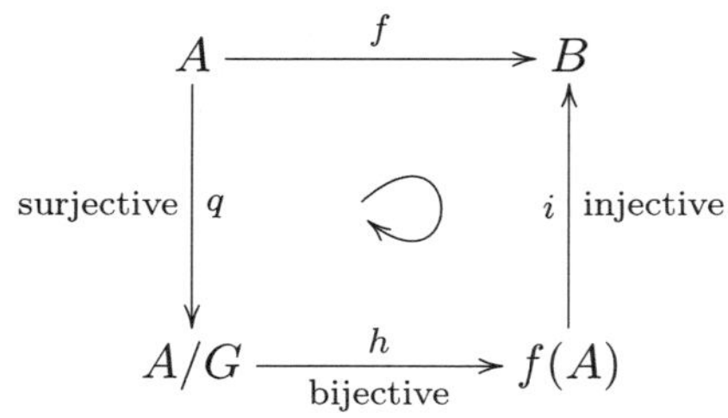

One of the results of Theorem 3.26 is especially useful. Namely, that if $f : A \to B$ is a function and G is the equivalence relation determined by f, then A/G and $f(A)$ are in one-to-one correspondence. This is customarily expressed by writing $A/G \approx f(A)$. In particular,

Corollary 3.28. *If $f : A \to B$ is a surjective function, then $h : A/G \to B$ is a bijective function, where G is the equivalence relation determined by $f : A \to B$.*

Proof. $f : A \to B$ is a surjective function.

$\Rightarrow h : A/G \to B$ is bijective by Theorem 3.26 because $f(A) = B$. $\square$

Definition 3.29. Let A and B be sets, let $f : A \to B$ be a function, and let G be the equivalence relation in A determined by f. Let H be any equivalence relation in A which is finer than G. i.e., $H \subseteq G$. We define a function $f/H : A/H \to B$ by $(f/H)(H_x) = f(x)$ for all $x \in A$.

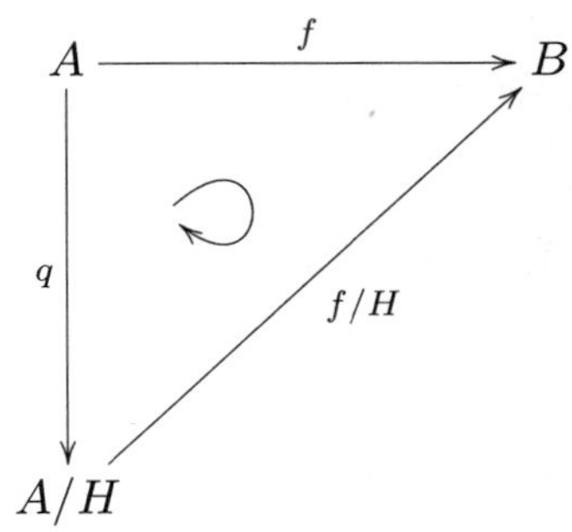

It is easy to see that f/H is a function from A/H to B. The function $f/H : A/H \to B$ is called the *quotient function of f by H*.

Theorem 3.30. *Let $f : A \to B$ be a function. Let G be the equivalence relation determined by f and let H be a refinement of G. Then G/H is the equivalence relation determined by $f/H : A/H \to B$.*

Proof. $(f/H)(H_x) = (f/H)(H_y)$. $\Leftrightarrow f(x) = f(y)$. $\Leftrightarrow (x, y) \in H$.
$\Leftrightarrow (H_x, H_y) \in G/H$. □

Corollary 3.31. *Let G and H be equivalence relations in a set A. Assume that $H \subseteq G$. Then $(A/H)/(G/H)$ and A/G are in one-to-one correspondence.*

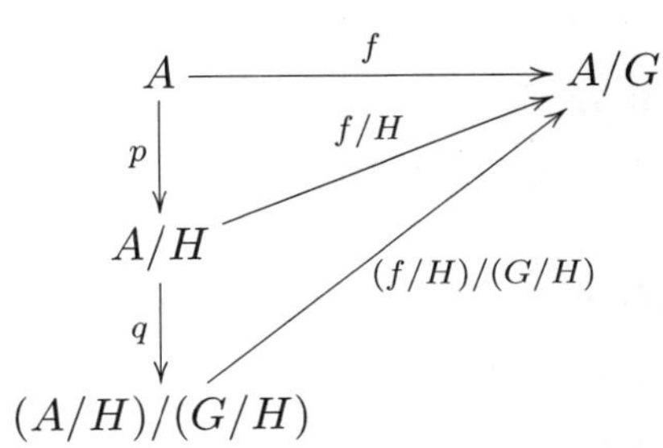

Proof. Let G and H be equivalence relations in A such that $H \subseteq G$ and let $f : A \to A/G$ be the canonical function from A to A/G.
$\Rightarrow G$ is the equivalence relation determined by f by Theorem 3.24.
$\Rightarrow f/H : A/H \to A/G$ is a function by Definition 3.29.

$\Rightarrow$ G/H is the equivalence relation determined by the function $f/H : A/H \to A/G$ by Theorem 3.30.

$\Rightarrow$ Since $f = f/H \circ p$ is surjective, $f/H : A/H \to A/G$ is surjective by Theorem 2.26.

$\Rightarrow$ $(f/H)/(G/H) : (A/H)/(G/H) \to A/G$ is a bijective function by Corollary 3.28. □

Example 3.32. Let $A = \{0, 1, 2, 3, 4, 5, 6\}$ and $B = \{0, 1, 2, 3\}$. Define a function $f : A \to B$ by $f(x) = |x - 3|$ for all $x \in A$. Let G be the equivalence relation determined by f and let $H = \{(0, 0), (1, 1), (2, 2), (3, 3), (4, 4), (5, 5), (6, 6), (2, 4), (4, 2)\} \subset A \times B$.

(1) $G = \{(0, 0), (1, 1), (2, 2), (3, 3), (4, 4), (5, 5), (6, 6), (0, 6), (6, 0), (1, 5), (5, 1), (2, 4), (4, 2)\}$.

(2) $A/G = \{G_i \mid i = 0, 1, 2, 3, 4, 5, 6\} = \{G_0, G_1, G_2, G_3\}$ where $G_0 = G_6 = \{0, 6\}$, $G_1 = G_5 = \{1, 5\}$, $G_2 = G_4 = \{2, 4\}$ and $G_3 = \{3\}$.

(3) $A/H = \{H_i \mid i = 0, 1, 2, 3, 4, 5, 6\} = \{H_0, H_1, H_2, H_3, H_5, H_6\}$, where $H_0 = \{0\}$, $H_1 = \{1\}$, $H_2 = H_4 = \{2, 4\}$, $H_3 = \{3\}$, $H_5 = \{5\}$ and $H_6 = \{6\}$.

(4) H is a refinement of G and $G/H = \{(H_0, H_0), (H_1, H_1), (H_2, H_2), (H_3, H_3), (H_5, H_5), (H_6, H_6), (H_0, H_6), (H_6, H_0), (H_1, H_5), (H_5, H_1)\}$.

(5) The function $f/H : A/H \to B$ is given by $(f/H)(H_0) = f(0) = 3$, $(f/H)(H_1) = f(1) = 2$, $(f/H)(H_2) = f(2) = 1$, $(f/H)(H_3) = f(3) = 0$, $(f/H)(H_5) = f(5) = 2$, $(f/H)(H_6) = f(6) = 3$.

Exercise 3.5

1. Let $f : A \to B$ be a surjective function. Let G be the equivalence relation determined by f and let H be an equivalence relation in A which is coarser than G. Define the *image* of H under f by $f(H) = \{(f(x), f(y)) \mid (x, y) \in H\}$. Prove that $f(H)$ is an equivalence relation in B.

2. Let $f : A \to B$ be a surjective function and let G be the equivalence relation determined by f. Let J be any equivalence relation in B. Prove the following.
   (1) $f^{-1}(J)$ is coarser than G.
   (2) $H = f^{-1}(J)$ if and only if $J = f(H)$. (See Exercise 3.5.1 above.)

3. Let $f : A \to B$ be a function and let G be the equivalence relation determined by f. Prove that $G = f^{-1} \circ f$.

4. Let $f : A \to B$ and $g : B \to C$ be functions and let G be the equivalence relation determined by g. Prove that $f^{-1}(G)$ is the equivalence relation determined by $g \circ f$.

5. Let G and H be equivalence relations in A and assume that $H \subseteq G$. Let $f : A \to A/H$ be the canonical function from A to A/H. Prove that $G/H = f(G)$. (See Exercise 3.5.1 above)

6. Let G and H be equivalence relations in A and assume that $H \subseteq G$. Let $f : A \to A/H$ be the canonical function from A to A/H and let $g : A \to A/G$ be the canonical function from A to A/G. Let $h = g/G$. Prove that $g = h \circ f$.

7. Let G, H and J be equivalence relations in A and assume that $J \subseteq H \subseteq G$. Let $f : A \to A/J$, $g : A \to A/H$ and $h : A \to A/G$ be the canonical functions associated, respectively, with J, H and G. Prove that $h/J = h/H \circ g/J$.

8. Let G and H be equivalence relations in A and assume that $H \subseteq G$. Let $f : A \to A/H$ be the canonical function from A to A/H. Let $h = g/G$. Prove that $f/H : A/H \to A/H$ is surjective.

9. Let G and H be equivalence relations in A and assume that $G \circ H = H \circ G$. Prove the following.
   (1) $A/(G \circ H) \approx (A/G)/(G \circ H/G)$. (2) $A/G \approx (A/G \cap H)/(G/G \cap H)$.

10. Let $f : A \to A$ be a function and let G be the equivalence relation determined by f. Prove that $f \circ f = f$ if and only if $f(z) \in G_x$ for all $z \in G_x$ and for all $x \in A$.

Chapter 4

Partially Ordered Classes

4.1 Fundamental concepts and definitions

Definition 4.1. (1) By a *partially ordered class* we mean a pair[1] of objects $\langle A, G\rangle$, where A is a class and G is an order relation in A.

In this case, we also say that A *is ordered by* G or that G *orders* A.

If it is well understood, in a given discussion, that G is the order relation in A, then we will say loosely that A is a *partially ordered class.*

(2) If A is a set and G is an order relation in A, then we say that $\langle A, G\rangle$ is a *partially ordered set.*

In ordinary mathematical applications, every partially ordered class is a partially ordered set. However, the intuitive idea of an "ordered collection of

[1]For two elements x and y, the *ordered pair* (x, y) was defined formally by $(x, y) = \{\{x\}, \{x, y\}\}$. It can also be defined formally by $(x, y) = \{x \times \{\emptyset\}, y \times \{\{\emptyset\}\}\}$. In general for two classes X_1 and X_2, the *ordered pair* $\langle X_1, X_2\rangle$ can be formally defined by $\langle X_1, X_2\rangle = \{x_1 \times \{\emptyset\}, x_2 \times \{\{\emptyset\}\} \mid x_1 \in X_1, x_2 \in X_2\} = (X_1 \times \{\emptyset\}) \cup (X_2 \times \{\{\emptyset\}\})$. Note that $\langle X_1, X_2\rangle$ may not be a set as mentioned in Remark 2.4(3). We write here customarily $< A, G >$ in stead of (A, G).

elements" is meaningful for any collection A, whether A be a set or proper class; hence it is natural to give the definition in its most general form, letting A be any class. Once again, since every set is a class, everything we have to say about partially ordered classes applies, in particular, to partially ordered sets.

Notation 4.2. Let $\langle A, G\rangle$ be a partially ordered class and let $x, y \in A$. It is customary to use the following notations.

(1) $x \leq y$ denotes the fact that $(x, y) \in G$.

(2) $x \geq y$ has the same meaning as $y \leq x$.

(3) $x \nleq y$ means the fact that $(x, y) \notin G$.

(4) $x < y$ means the fact that $x \leq y$ and $x \neq y$.

Definition 4.3. Let A be a partially ordered class and let $x, y \in A$.

(1) If $x \leq y$, then we say that x is *less than or equal to* y or that y is *greater than or equal to* x.

(2) If $x < y$, then we say that x is *strictly less than* y or that y is *strictly greater than* x.

Remark 4.4. Let $\langle A, G\rangle$ be a partially ordered class.

(1) If B is a subclass of A, then we may consider B to be ordered by the order relation in A. In other words, for all $b_1, b_2 \in B$, $b_1 \leq b_2$ in B if and only if $b_1 \leq b_2$ in A.

i.e., $\langle B, G|_B\rangle$ is a partially ordered class where $G|_B = G \cap (B \times B)$.

(2) Note that the relation $<$ on a partially ordered class A is not an order relation. It is neither reflexive nor antisymmetric. It is irreflexive, asymmetric and transitive.

If A and B are partially ordered classes, there are several possible ways of ordering the class $A \times B$. The two most useful ways of doing so are given in the following definition.

Definition 4.5. Let A and B be partially ordered classes.

(1) By the *lexicographic ordering* of $A \times B$ we mean the following order relation in $A \times B$: If $(a_1, b_1) \in A \times B$ and $(a_2, b_2) \in A \times B$, then

$$(a_1, b_1) \leq (a_2, b_2) \text{ iff (i) } a_1 < a_2 \text{ or } \text{ (ii) } (a_1 = a_2 \text{ and } b_1 \leq b_2).$$

(2) By the *antilexicographic ordering* of $A \times B$ we mean the following order relation in $A \times B$: If $(a_1, b_1) \in A \times B$ and $(a_2, b_2) \in A \times B$,

$$(a_1, b_1) \leq (a_2, b_2) \text{ iff (i) } b_1 < b_2 \text{ or } \text{ (ii) } (b_1 = b_2 \text{ and } a_1 \leq a_2).$$

The lexicographic ordering is so called because it imitates the way we order words in the dictionary. For example, *be* precedes *go* because *b* precedes *g*, and *be* precedes *by* because *e* precedes *y*.

Definition 4.6. Let A be a partially ordered class. Let $x, y \in A$ and $B \subseteq A$.

(1) x and y are said to be *comparable* if either $x \leq y$ or $y \leq x$.

x and y are said to be *incomparable* if x and y are not comparable.

(2) B is called a *fully ordered subclass* of A or a *linearly ordered subclass* of A, or a *chain* of A if x and y are comparable for all $x, y \in B$.

A is called a *fully ordered class* or a *linearly ordered class* if x and y are comparable for all $x, y \in A$.

(3) For $a \in A$, the subclass $S_a = \{x \in A \mid x < a\}$ is called the *initial segment of A determined by a.*

Theorem 4.7. *Let A be a partially ordered class. If P is an initial segment of A and Q is an initial segment of P, then Q is an initial segment of A.*

Proof. P is an initial segment of A and Q is an initial segment of P.

$\Rightarrow$ There exists $a \in A$ such that $P = \{x \in A \mid x < a\}$ and

There exists $b \in P$ such that $Q = \{x \in P \mid x < b\}$ by Definition 4.6.

$\Rightarrow b < a$. Let $Q_1 = \{x \in A \mid x < b\}$.

$\Rightarrow Q_1$ is an initial segment of A by Definition 4.6.

Furthermore $Q = Q_1$ because $x \in Q_1$ iff $x \in A$ and $x < b$ iff $x \in A$ and $x < b < a$ iff $x \in P$ and $x < b$ iff $x \in Q$. □

Theorem 4.7 may be paraphrased as follows: An initial segment of an initial segment of A is an initial segment of A.

Definition 4.8. Let A be a partially ordered class and let $L, U \subseteq A$. Then the pair (L, U) is called a *cut* of A if (L, U) satisfies the following properties:

(1) $L \neq \emptyset$ and $U \neq \emptyset$.

(2) $L \cap U = \emptyset$ and $L \cup U = A$.

(3) If $x \in L$ and $y \leq x$, then $y \in L$.

(4) If $x \in U$ and $y \geq x$, then $y \in U$.

It is convenient to use a graphic device, called a line diagram, to illustrate some properties of a partially ordered class.

Definition 4.9. Given a partially ordered class A, a *Hasse diagram* or *line diagram* of A is a diagram on the plane defined as follows:

(1) The elements of A are represented by points on the diagram,

(2) $x < y$ and there is no z such that $x < z < y$ in A if and only if x and y are connected by a unique line which rises from x to y on the diagram.

In a partially ordered class A, note that $x \leq x$ for all $x \in A$ and that $x \leq y$ for some different $x, y \in A$ if and only if x and y are connected by a union of some rising line segments on the line diagram of A.

Example 4.10. Figure 4.1 represents the line diagram of a partially ordered class $A = \{a, b, c, d, e, f\}$ with 6 elements. Note the following.

(1) $\{a, b, c\}$ and $\{d, e, b, c\}$ are chains of A.

(2) $S_e = \{b, c, f\}$ is the initial segment determined by e.

(3) If $L = \{a, b, c\}, U = \{d, e, f\}$, then (L, U) is a cut of A.

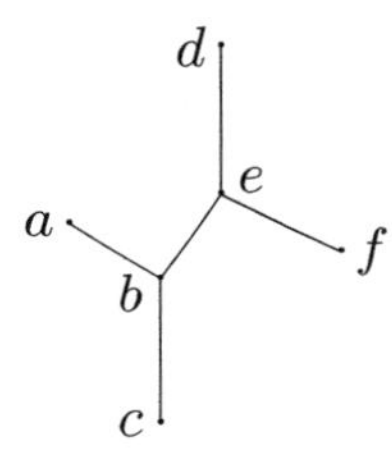

Figure 4.1: the line diagram of $A = \{a, b, c, d, e, f\}$

Example 4.11. The most important order relation in mathematics is the class inclusion relation $\subseteq$. If $\mathscr{A}$ is a class of classes and we consider $\mathscr{A}$ to be ordered by the class inclusion relation, we say that $\mathscr{A}$ is *ordered by inclusion.*

Note that $\subseteq$ is an order relation.

($\because$) (i) $\subseteq$ is reflexive because $A \subseteq A$ for all $A \in \mathscr{A}$.

(ii) $\subseteq$ is antisymmetric because $A \subseteq B$ and $B \subseteq A$ imply $A = B$ for all $A,\ B \in \mathscr{A}$.

(iii) $\subseteq$ is transitive because $A \subseteq B$ and $B \subseteq C$ imply $A \subseteq C$ for all $A,\ B,\ C \in \mathscr{A}$.

Note that a chain $\mathscr{C}$ of $\mathscr{A}$ under the class inclusion means that for any two elements $A, B \in \mathscr{C}$, either $A \subseteq B$ or $B \subseteq A$.

Exercise 4.1

1. Let $A = \{a, b, c, d\}$. If $\mathscr{P}(A)$ is ordered by inclusion draw its line diagram.

2. Let A be the partially ordered class defined by the following diagram (Figure 4.2). List all the chains of A, all the initial segments of A and all the cuts of A.

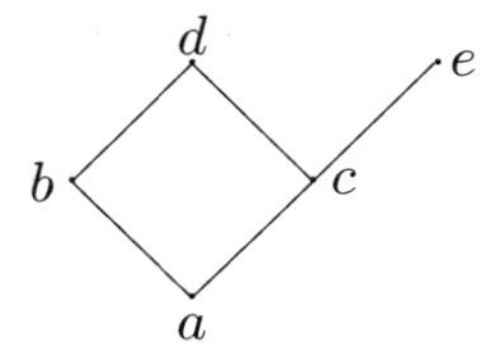

Figure 4.2: partially ordered set A

3. Let A be the partially ordered class of Exercise 4.1.2 above. Draw the line diagram for the following classes:

   (1) The class of all the chains of A (ordered by inclusion).

   (2) The class of all the initial segments of A (ordered by inclusion).

   (3) The class of all the cuts of A (ordered by inclusion on the "left component" L).

4. Let A and B be partially ordered classes and let C be a chain of A and let D be a chain of B. If $A \times B$ is ordered lexicographically (See Definition 4.5(1)), prove that $C \times D$ is a chain of $A \times B$.

5. Let A and B be partially ordered classes and let $A \times B$ be ordered antilexicographically (See Definition 4.5(2)). Prove that if (L, U) is a cut of B, then $(A \times L, A \times U)$ is a cut of $A \times B$.

6. Let A be the partially ordered class and let G be an equivalence relation in A. Assume the following condition: If $x \underset{G}{\sim} z$ and $x \leq y \leq z$, then $y \underset{G}{\sim} z$. Define a relation H in A/G by $H = \{(G_x, G_y) \mid$ for each $w \in G_x$, there exists $z \in G_y$ such that $w \leq z\}$. Prove that H is an order relation in A/G.

4.2 Order preserving functions and isomorphisms

Definition 4.12. Let A and B be partially ordered classes.

(1) A function $f : A \to B$ is said to be *increasing*(or, *order-preserving*) if for all $x, y \in A$, $x \leq y$ implies $f(x) \leq f(y)$.

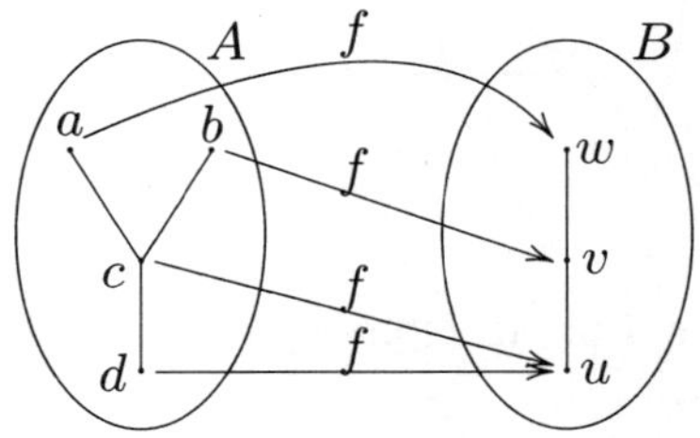

Figure 4.3: an increasing function $f : A \to B$

(2) A function $f : A \to B$ is said to be *strictly increasing* if for all $x, y \in A$, $x < y$ implies $f(x) < f(y)$.

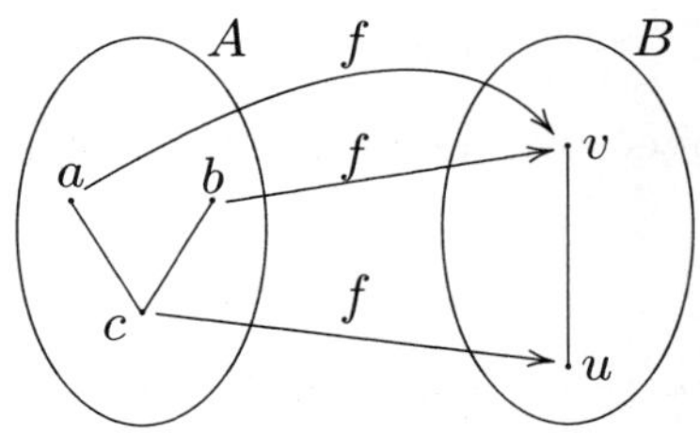

Figure 4.4: a strictly increasing function $f : A \to B$

(3) A function $f : A \to B$ is called an *isomorphism* if it is bijective and $x \leq y$ if and only if $f(x) \leq f(y)$ for all $x, y \in A$.

In this case, we say that A *is isomorphic with* B and write $A \cong B$.

Note that there is a bijective, order-preserving function which are not an isomorphism as the following figure 4.6 shows.

Theorem 4.13. *Let A and B be partially ordered classes and let $f : A \to B$ be an isomorphism. Then $x_1 < x_2$ if and only if $f(x_1) < f(x_2)$ for all $x_1, x_2 \in A$.*

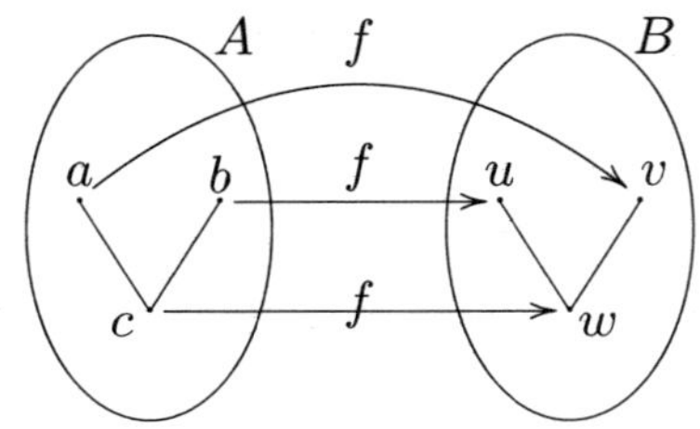

Figure 4.5: an isomorphism $f : A \to B$

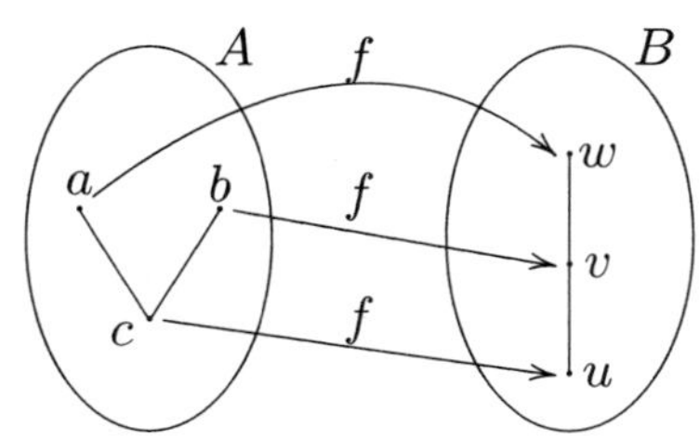

Figure 4.6: a bijective, increasing but not an isomorphism

Proof. $x_1 < x_2$.

$\Leftrightarrow x_1 \le x_2$ and $x_1 \ne x_2$ by Notation 4.2(4).

$\Leftrightarrow f(x_1) \le f(x_2)$ and $f(x_1) \ne f(x_2)$ because f is an isomorphism.

$\Leftrightarrow f(x_1) < f(x_2)$ by Notation 4.2(4). □

Theorem 4.14. *Let A and B be partially ordered classes and let $f : A \to B$ be a bijective function. Then $f : A \to B$ is an isomorphism if and only if $f : A \to B$ and $f^{-1} : B \to A$ are increasing functions.*

Proof. Since $f : A \to B$ is bijective, $f^{-1}(f(x)) = 1_A(x) = x$ for all $x \in A$, by Theorem 2.21.

(if) f and f^{-1} are increasing functions.

$\Rightarrow$ For all $x_1, x_2 \in X$,

(i) $x_1 \le x_2$ implies $f(x_1) \le f(x_2)$ because f is increasing,

(ii) $f(x_1) \le f(x_2)$ implies $x_1 \le x_2$ because f^{-1} is increasing.

$\Rightarrow$ Since $f : A \to B$ is bijective, $f : A \to B$ is an isomorphism.

(only if) $f : A \to B$ is an isomorphism.

$\Rightarrow$ (i) $f : A \to B$ is an increasing function by Definition 4.12.

(ii) $f^{-1} : B \to A$ is an increasing function.

($\because$) $y_1 \leq y_2$ for some $y_1, y_2 \in B$.

$\Rightarrow$ Since f is surjective, there exist $x_1, x_2 \in A$ such that

$y_1 = f(x_1)$, $y_2 = f(x_2)$ and hence $f(x_1) \leq f(x_2)$.

$\Rightarrow$ $x_1 \leq x_2$ because $f : A \to B$ is an isomorphism.

$\Rightarrow$ $f^{-1}(f(x_1)) = x_1 \leq x_2 = f^{-1}(f(x_2))$.

$\Rightarrow$ $f^{-1}(y_1) \leq f^{-1}(y_2)$. □

Theorem 4.15. *Let A, B and C be partially ordered classes.*

(1) *The identity function $1_A : A \to A$ is an isomorphism.*

(2) *If $f : A \to B$ is an isomorphism, then $f^{-1} : B \to A$ is an isomorphism.*

(3) *If $f : A \to B$ and $g : B \to C$ are isomorphisms, then $g \circ f : A \to C$ is an isomorphism.*

Proof. (1) Since $1_A : A \to A$ is a bijective function by Example 2.13(1) and $1_A(x) = x$ for all $x \in A$,

$x_1 \leq x_2$ if and only if $1_A(x_1) \leq 1_A(x_2)$ for all $x_1, x_2 \in A$.

$\Rightarrow$ $1_A : A \to A$ is an isomorphism.

(2) $f : A \to B$ is an isomorphism.

$\Rightarrow$ (i) $f^{-1} : B \to A$ is bijective.

($\because$) Since $f : A \to B$ is bijective by Definition 4.12,

$f^{-1} : B \to A$ is bijective by Theorem 2.19.

(ii) For all $y_1, y_2 \in B$, $y_1 \leq y_2$ if and only if $f^{-1}(y_1) \leq f^{-1}(y_2)$.

($\because$) $y_1, y_2 \in B$.

$\Rightarrow$ Since $f : A \to B$ is bijective, there exist $x_1, x_2 \in A$ such that

$y_1 = f(x_1)$ and $y_2 = f(x_2)$.

i.e., $x_1 = f^{-1}(y_1)$ and $x_2 = f^{-1}(y_2)$.

$\Rightarrow y_1 \leq y_2$ iff $f(x_1) \leq f(x_2)$.

$\Rightarrow y_1 \leq y_2$ iff $x_1 \leq x_2$ because $f : A \to B$ is an isomorphism.

$\Rightarrow y_1 \leq y_2$ iff $f^{-1}(y_1) \leq f^{-1}(y_2)$.

$\Rightarrow f^{-1} : B \to A$ is an isomorphism.

(3) $f : A \to B$ and $g : B \to C$ are isomorphisms.

$\Rightarrow$ (i) $g \circ f : A \to C$ is a bijective function.

($\because$) Since $f : A \to B$ and $g : B \to C$ are bijective

$g \circ f : A \to C$ is bijective by Theorem 2.26.

(ii) For all $x_1, x_2 \in A$, $x_1 \leq x_2$ iff $(g \circ f)(x_1) \leq (g \circ f)(x_2)$.

($\because$) $x_1 \leq x_2$.

$\Leftrightarrow f(x_1) \leq f(x_2)$ because $f : A \to B$ is an isomorphism.

$\Leftrightarrow g(f(x_1)) \leq g(f(x_2))$ because $g : B \to C$ is an isomorphism

$\Leftrightarrow (g \circ f)(x_1) \leq (g \circ f)(x_2)$.

$\Rightarrow g \circ f : A \to C$ is an isomorphism. □

Remark 4.16. (1) If $f : A \to B$ is a bijective function, then $f^{-1} : B \to A$ is bijective by Theorem 2.19 and then $(f^{-1})^{-1} : A \to B$ is bijective by Theorem 2.19 again. Moreover, since $\left(f^{-1}\right)^{-1} \circ f = 1_A$ and $f \circ \left(f^{-1}\right)^{-1} = 1_B$, $(f^{-1})^{-1} = f$ by Theorem 2.21.

(2) By using (1) above, Theorem 4.15(2) can be proved by the following argument:

($\because$) $f : A \to B$ is an isomorphism.

$\Leftrightarrow f$ and f^{-1} are increasing functions by Theorem 4.14.

$\Leftrightarrow (f^{-1})^{-1}$ and f^{-1} are increasing by (1) above.

$\Leftrightarrow f^{-1} : B \to A$ is an isomorphism by Theorem 4.14.

(3) Theorem 4.15 says that the relation "is isomorphic with" is an equivalence relation among partially ordered classes.

($\because$) (i) A is isomorphic with A by Theorem 4.15(1).

(ii) If A is isomorphic with B, then B is isomorphic with A by Theorem 4.15(2).

(iii) If A is isomorphic with B and B is isomorphic with C, then A is isomorphic with C by Theorem 4.15(3).

We will write $A \cong B$ to denote the fact that A is isomorphic with B.

(4) The concept of isomorphism is of great importance in the study of partially ordered classes. Suppose that A and B be partially ordered classes and $f : A \to B$ is an isomorphism.

Let us agree to write x' instead of $f(x)$. Since $f : A \to B$ is bijective, every element x in A corresponds with a unique element x' in B:

$$x \overset{f}{\longmapsto} x', \quad y \overset{f}{\longmapsto} y', \quad z \overset{f}{\longmapsto} z', \quad \text{etc.}$$

(i) The ordering of A is exactly the same as the ordering of B.

i.e., $x \le y (x, y \in A)$ if and only if $x' \le y' (x', y' \in B)$.

(ii) The line diagram of A is the same as the line diagram of B.

Thus essentially, there is no difference between A and B except the letters we use to designate their elements.

Example 4.17. Let $B = \{l, m, n, o, p, q\}$ be ordered as in the following diagram of Figure 4.7.

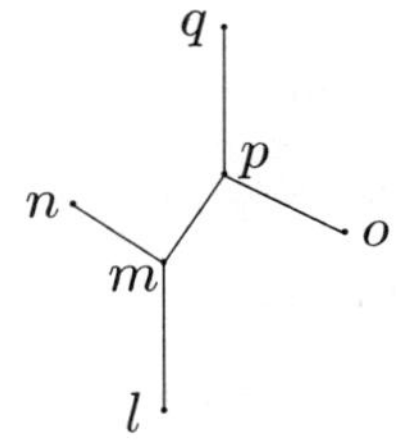

Figure 4.7: the line diagram of B

Then B is isomorphic with the class A of Example 4.10. The isomorphism $f : A \to B$ is given by the following table.

x	$f(x)$
a	n
b	m
c	l
d	q
e	p
f	o

the table of the function $f : A \to B$

In conclusion, if A is isomorphic with B, and if we identify corresponding elements, then A and B are essentially the same partially ordered class.

Exercise 4.2

1. Prove that if $f : A \to B$ is an injective, increasing function, then it is strictly increasing.

2. Let $f : A \to B$ be an increasing function. If C is a chain of A, prove that $f(C)$ is a chain of B.

3. Let A and B be partially ordered classes. Prove the following.

   (1) If $A \times B$ is ordered lexicographically, then the projection function

   $p_1 : A \times B \to A$ is increasing.

   (2) If $A \times B$ is ordered antilexicographically, then the projection function

   $p_2 : A \times B \to B$ is increasing.

   [Note that $p_1(x, y) = x, p_2(x, y) = y$.]

4. A subclass C of a partially ordered class is called *convex* if it satisfies the following condition: If $a \in C$ and $b \in C$ and $a \le x \le b$, then $x \in C$.

   Let A and B be partially ordered classes, let $f : A \to B$ be an increasing function and let C be a convex subclass of B. Prove that $f^{-1}(C)$ is a convex subclass of A.

5. Let A and B be partially ordered classes, let $f : A \to B$ be an increasing function and let H be the equivalence class determined by f. Prove that each equivalence class modulo H is a convex subclass of A.

6. Let A and B be partially ordered classes and let $f : A \to B$ be an increasing function. Assume that $f(A) = B$. Prove that if (L, U) is a cut of B, then $(f^{-1}(L), f^{-1}(U))$ is a cut of A.

7. Prove that the composite of two increasing functions is increasing. Use this result to prove Theorem 4.15(3).

8. Let A and B be partially ordered classes and let $f : A \to B$ be an isomorphism. Prove each of the following.

   (1) If C is a convex subclass of A, then $f(C)$ is a convex subclass of B.

   (2) If (L, U) is a cut of A, then $(f(L), f(U))$ is a cut of B.

   (3) If $[a, b]$ is a closed interval of A, then $f([a, b])$ is a closed interval of B.

   (If $a, b \in A$, then the set $\{x \in A \mid a \leq x \leq b\}$ is called a *closed interval* of A and is denoted by the symbol $[a, b]$.)

9. Let E and F be partially ordered classes and let $q : E \to F$ be an isomorphism. Prove that for arbitrary $x \in E, \overline{q}(S_x) = S_{q(x)}$. Conclude that $S_x \cong S_{q(x)}$.

10. Let A be a partially ordered set. For each $a \in A$, let $I_a = \{x \in A \mid x \leq a\}$ and let $\mathscr{I} = \{I_a \mid a \in A\}$ ordered by the inclusion $\subseteq$. Prove that $\mathscr{I}$ is isomorphic with A.

11. Let A, B and C be mutually disjoint, partially ordered classes. If $A \cong B$, prove that $(A \cup C) \cong (B \cup C)$.

    [Hint: Take the union of the two functions, as in Theorem 2.15.]

12. Let A be a partially ordered set. Define $L_x = \{z \in A \mid z \leq x\}$ and $U_x = \{z \in A \mid z \nleq x\}$. Prove the following.

    (1) For each $x \in A$, (L_x, U_x) is a cut of A.

    (2) The function f defined by $f(x) = (L_x, U_x)$ is an isomorphism between A and the class of all the cuts of the above-described form.

4.3 Distinguished elements and the duality

Certain distinguished elements play an important part in the study of partially ordered classes. We now define them.

Definition 4.18. Let A be a partially ordered class and let $B \subseteq A$.

(1) An element $m \in B$ is called a *maximal element* of B if none of the elements of B are strictly greater than m.
In other words, $m \in B$ is a *maximal element* of B provided that if $x \in B$ and $x \geq m$, then $x = m$.

(2) An element $n \in B$ is called a *minimal element* of B if none of the elements of B are strictly less than n.
In other words, $n \in B$ is a *minimal element* of B provided that if $x \in B$ and $x \leq n$, then $x = n$.

(3) An element $a \in B$ is called the *greatest element* of B if $a \geq x$ for all $x \in B$.

(4) An element $b \in B$ is called the *least element* of B if $b \leq x$ for all $x \in B$.

(5) An element $a \in A$ is called an *upper bound of B in A* if $a \geq x$ for all $x \in B$. The class of all the upper bounds of B in A will be denoted by $\upsilon_A(B)$. An upper bound a of B in A will be called simply an *upper bound of B* and we will write $\upsilon(B)$ for $\upsilon_A(B)$ if there is no risk of ambiguity of A. i.e., $\upsilon_A(B) = \upsilon(B) = \{a \in A \mid a \text{ is an upper bound of } B\}$.

(6) An element $b \in A$ is called a *lower bound of B in A* if $b \leq x$ for all $x \in B$. The class of all the lower bounds of B in A will be denoted by $\lambda_A(B)$. A lower bound b of B in A will be called simply a *lower bound of B* and we will write $\lambda(B)$ for $\lambda_A(B)$ if there is no risk of ambiguity of A. i.e., $\lambda_A(B) = \lambda(B) = \{b \in A \mid b \text{ is a lower bound of } B\}$.

(7) (i) B is said to be *bounded above* or *upper bounded*

if B has an upper bound.

(ii) B is said to be *bounded below* or *lower bounded*

if B has a lower bound.

(iii) B is said to be *bounded* if B is bounded both above and below.

(8) If $\lambda(B)$ has a greatest element, this element is called the *greatest lower bound* or the *infimum* of B in A and is abbreviated by $\inf_A B$ or simply by $\inf B$ if there is no risk of ambiguity of A.

(9) If $\upsilon(B)$ has a least element, this element is called the *least upper bound* or the *supremum* of B in A and is abbreviated by $\sup_A B$ or simply by $\sup B$ if there is no risk of ambiguity of A.

Remark 4.19. Let A be a partially ordered class and let $B \subseteq A$.

(1) If B has a greatest element, then this element is unique.

($\because$) Assume that a and a' are two greatest elements of B.

$\Rightarrow$ Since $a \in B$ and $a' \in B$, $a' \geq a$ and $a \geq a'$.

$\Rightarrow$ $a = a'$ by antisymmetry of $\leq$.

(2) If B has a least element, then this element is unique.

($\because$) Assume that a and a' are two least elements of B.

$\Rightarrow$ Since $a \in B$ and $a' \in B$, $a' \leq a$ and $a \leq a'$.

$\Rightarrow$ $a = a'$ by antisymmetry of $\leq$.

(3) The supremum $\sup B$ and the infimum $\inf B$ of B in A are unique if they exist.

(4) $a = \sup_A B \Leftrightarrow$ (i) a is an upper bound of B in A and

(ii) If a' is an upper bound of B in A, then $a \leq a'$.

(5) $b = \inf_A B \Leftrightarrow$ (i) b is a lower bound of B in A and

(ii) If b' is a lower bound of B in A, then $b \geq b'$.

Example 4.20. (1) Figure 4.8 is the line diagram of a partially ordered class $A = \{a, b, c, d, e, f\}$ that has maximal elements but has no greatest element.

(i) A has maximal elements but no greatest element;
a and d are maximal elements of A,

(ii) A has minimal elements but no least element;
c and f are minimal elements of A.

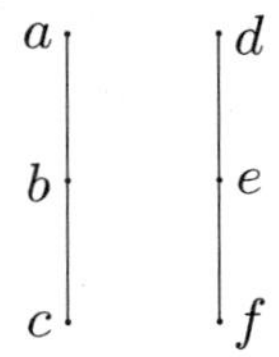

Figure 4.8: the partially ordered set $A = \{a, b, c, d, e, f\}$

(2) In Figure 4.9, let $A = \{a, b, c, d, e, f\}$, $B = \{b, c, e, f\}$, $C = \{a, b, c, e, f\}$ and $D = \{b, c, d, e, f\}$. Then

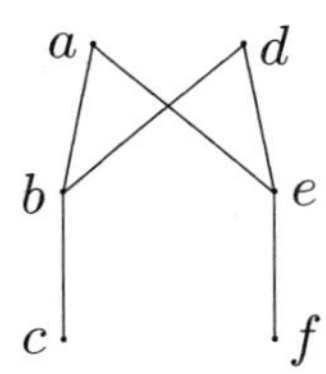

Figure 4.9: the partially ordered set $A = \{a, b, c, d, e, f\}$

(i) B has two upper bounds in A;
a and d are upper bounds of B in A.

(ii) B has no sup in A, although $\sup_C B = a$ and $\sup_D B = d$.

(3) Let $\mathbb{N}$ denote the class of all the positive integers and let $\mathbb{Z}$ denote the class of all the integers. Then

(i) The class $\mathbb{N}$ of all positive integers has a least element but no greatest element and no maximal elements.

(ii) The class $\mathbb{Z}$ of all integers has neither a greatest nor a least element.

(4) Let $\mathscr{A}$ be a class of sets which is ordered by inclusion and let $\mathscr{B} = \{B_i\}_{i\in I}$ be a subclass of $\mathscr{A}$. Assume that $\bigcup_{i\in I} B_i$ and $\bigcap_{i\in I} B_i$ are elements of $\mathscr{A}$. Then we have the following.

① $\sup_{\mathscr{A}}\mathscr{B} = \bigcup_{i\in I} B_i$.

($\because$) (i) Since $B_j \subseteq \bigcup_{i\in I} B_i$ for all $B_j \in \mathscr{B}$, $\bigcup_{i\in I} B_i$ is an upper bound of $\mathscr{B}$.

(ii) If C is any upper bound of $\mathscr{B}$, then $B_i \subseteq C$ for all $i \in I$, and hence $\bigcup_{i\in I} B_i \subseteq C$.

Thus $\bigcup_{i\in I} B_i$ is the least upper bound of $\mathscr{B}$ in $\mathscr{A}$.

② $\inf_{\mathscr{A}}\mathscr{B} = \bigcap_{i\in I} B_i$.

($\because$) (i) Since $\bigcap_{i\in I} B_i \subseteq B_j$ for all $B_j \in \mathscr{B}$, $\bigcap_{i\in I} B_i$ is a lower bound of $\mathscr{B}$.

(ii) If D is any lower bound of $\mathscr{B}$, then $D \subseteq B_i$ for all $i \in I$, and hence $D \subseteq \bigcap_{i\in I} B_i$.

Thus $\bigcap_{i\in I} B_i$ is the greatest lower bound of $\mathscr{B}$ in $\mathscr{A}$.

(5) For any partially ordered class $\langle A, \leq\rangle$, $\emptyset \subseteq A$. Note that the following statements are true.

① $\lambda(\emptyset) = \lambda_A(\emptyset) = A$.

($\because$) Since $\lambda_A(\emptyset) = \{x \mid x \in A, \text{ and } x \leq y \text{ for all } y \in \emptyset\}$, $\lambda_A(\emptyset) \subseteq A$.

Conversely, for any $x \in A$ it is true that $x \leq y$ for all $y \in \emptyset$.

Thus if $x \in A$, then $x \in \lambda_A(\emptyset)$. i.e., $A \subseteq \lambda_A(\emptyset)$.

$\therefore \lambda_A(\emptyset) = A$.

② $\upsilon(\emptyset) = \upsilon_A(\emptyset) = A$.

($\because$) Since $\upsilon_A(\emptyset) = \{x \mid x \in A, \text{ and } x \geq y \text{ for all } y \in \emptyset\}$, $\upsilon_A(\emptyset) \subseteq A$.

Conversely, for any $x \in A$, it is true that $x \geq y$ for all $y \in \emptyset$.

Thus if $x \in A$, then $x \in \upsilon_A(\emptyset)$. i.e., $A \subseteq \upsilon_A(\emptyset)$.

$\therefore \upsilon_A(\emptyset) = A$.

③ If A has a greatest element a, then $a = \inf \emptyset$.

($\because$) Since $A = \lambda_A(\emptyset)$, $a = \sup A = \sup \lambda_A(\emptyset) = \inf_A \emptyset$.

④ If A has a least element b, then $b = \sup \emptyset$.

($\because$) Since $A = \upsilon_A(\emptyset)$, $b = \inf A = \inf \upsilon_A(\emptyset) = \sup_A \emptyset$.

The Duality

When we reason about partially ordered classes, we are led to the interesting notion of *duality*.

If G is an order relation in A, then G^{-1} is also an order relation in A. Let $\langle A, G\rangle$ denote the partially ordered class A ordered by G and let $\langle A, G^{-1}\rangle$ denote the partially ordered class A ordered by G^{-1}. Let $B \subseteq A$. Briefly the *duality* can be explained as follows:

(1) $x \leq y$ in $\langle A, G\rangle \Leftrightarrow x \geq y$ in $\langle A, G^{-1}\rangle$.

(2) a is a maximal element of $\langle A, G\rangle \Leftrightarrow a$ is a minimal element of $\langle A, G^{-1}\rangle$.

(3) a is the greatest element of $\langle A, G\rangle \Leftrightarrow a$ is the least element of $\langle A, G^{-1}\rangle$.

(4) a is an upper bound of B in $\langle A, G\rangle \Leftrightarrow a$ is a lower bound of B in $\langle A, G^{-1}\rangle$. (i.e., $a \in \upsilon_A(B)$ in $\langle A, G\rangle \Leftrightarrow a \in \lambda_A(B)$ in $\langle A, G^{-1}\rangle$.)

(5) $b = \sup B$ in $\langle A, G\rangle \Leftrightarrow b = \inf B$ in $\langle A, G^{-1}\rangle$.

Let $\langle A, G\rangle$ be a partially ordered class and let $B \subseteq A$. Then we have the following "*duality table*" for the partially ordered class $\langle A, G\rangle$.

< Duality Table >

words in $\langle A, G\rangle$	$\overset{\text{duality}}{\longleftrightarrow}$	words in $\langle A, G^{-1}\rangle$
$\leq$	$\overset{\text{duality}}{\longleftrightarrow}$	$\geq$
maximal	$\overset{\text{duality}}{\longleftrightarrow}$	minimal
greatest	$\overset{\text{duality}}{\longleftrightarrow}$	least
upper bound	$\overset{\text{duality}}{\longleftrightarrow}$	lower bound
$\upsilon_A(B)$	$\overset{\text{duality}}{\longleftrightarrow}$	$\lambda_A(B)$
$\sup_A B$	$\overset{\text{duality}}{\longleftrightarrow}$	$\inf_A B$
$\vdots$	$\vdots$	$\vdots$

Example 4.21. For the class $\mathscr{A}$ of all sets ordered by the inclusion, we have the following duality table. Let $A, B \in \mathscr{A}$ and $A_i \in \mathscr{A}$ for $i \in I$.

< Duality Table for Sets >

words in $\langle \mathscr{A}, \subseteq\rangle$	$\overset{\text{duality}}{\longleftrightarrow}$	words in $\langle \mathscr{A}, \supseteq\rangle$
$A \subseteq B$	$\overset{\text{duality}}{\longleftrightarrow}$	$A \supseteq B$
$\bigcup_{i\in I} A_i$	$\overset{\text{duality}}{\longleftrightarrow}$	$\bigcap_{i\in I} A_i$
$\mathscr{U}$	$\overset{\text{duality}}{\longleftrightarrow}$	$\emptyset$
$\lambda(\{A_i \mid A_i \in \mathscr{A}\})$	$\overset{\text{duality}}{\longleftrightarrow}$	$\upsilon(\{A_i \mid A_i \in \mathscr{A}\})$
$\vdots$	$\vdots$	$\vdots$

Definition 4.22. Let S be a statement about partially ordered classes and let S' be the statement obtained from S by replacing each occurrence of $\leq$ by $\geq$ and vice versa, in other words replacing each word by its corresponding word in the duality table for partially ordered class and vice versa. Then S' is a statement which is called the *dual* of the statement S.

We introduce a useful theorem for the duality.

Theorem 4.23 (Principle of Duality). *If a statement S is true for all partially ordered classes, then the dual of S is also true.*

Proof. Assume that a statement S is true for all partially ordered classes and let $\langle A, G\rangle$ be a partially ordered class.
$\Rightarrow$ Since $\langle A, G^{-1}\rangle$ is also a partially ordered class, S is true in $\langle A, G^{-1}\rangle$.
$\Rightarrow$ The dual of S is true in $\langle A, G\rangle$. □

The concept of duality permits a considerable economy in the presentation of theorems about partially ordered classes for every time we prove a theorem, we know that the dual of the theorem is also true.

The following are a few properties of the distinguished elements in a partially ordered class.

Theorem 4.24. *Let A be a partially ordered class and let B and C be subclasses of A. Assume that $B \subseteq C$. If B has a greatest element b, and C has a greatest element c, then $b \leq c$.*

Proof. B and C have the greatest elements b and c, respectively.
$\Rightarrow$ $b \in B$ and $x \leq b$ for all $x \in B$, and $c \in C$ and $x \leq c$ for all $x \in C$.
$\Rightarrow$ Since $B \subseteq C$, $b \in C$ and hence $b \leq c$. □

Theorem 4.25 (Dual to Theorem 4.24). *Let A be a partially ordered class and let B and C be subclasses of A. Assume that $B \subseteq C$. If B has a least element b, and C has a least element c, then $b \geq c$.*

Proof. This is the dual theorem of Theorem 4.24 and hence it directly follows from Theorem 4.23. □

Theorem 4.26. *Let A be a partially ordered class and let B and C be subclasses of A. If $B \subseteq C$, then $\upsilon(C) \subseteq \upsilon(B)$.*

Proof. $x \in \upsilon(C)$.

$\Rightarrow x \geq y$ for all $y \in C$ because x is a upper bound of C.

$\Rightarrow x \geq y$ for all $y \in B$ because $B \subseteq C$.

$\Rightarrow x \in \upsilon(B)$ and hence $\upsilon(C) \subseteq \upsilon(B)$ by Definition 1.14(2) □

Theorem 4.27 (Dual to Theorem 4.26). *Let A be a partially ordered class and let B and C be subclasses of A. If $B \subseteq C$, then $\lambda(C) \subseteq \lambda(B)$.*

Proof. This is the dual theorem of Theorem 4.26. □

Theorem 4.28. *Let A be a partially ordered class and let B and C be subclasses of A. Suppose that B has a* sup *in A and C has a* sup *in A. If $B \subseteq C$, then $\sup B \leq \sup C$.*

Proof. Since $B \subseteq C$, $\upsilon(C) \subseteq \upsilon(B)$ by Theorem 4.26 and hence $\sup B \leq \sup C$ by Theorem 4.25. □

Theorem 4.29 (Dual to Theorem 4.28). *Let A be a partially ordered class and let B and C be subclasses of A. Suppose that B has an* inf *in A and C has an* inf *in A. If $B \subseteq C$, then $\inf B \geq \inf C$.*

Proof. This is the dual theorem of Theorem 4.28. □

Theorem 4.30. *Let A be a partially ordered class and let B be a subclass of A. Then $B \subseteq \upsilon(\lambda(B))$.*

Proof. $x \in B$.

$\Rightarrow x \geq y$ for all $y \in \lambda(B)$ by Definition 4.18(6).

$\Rightarrow x \in \upsilon(\lambda(B))$ by Definition 4.18(5).

Thus $B \subseteq \upsilon(\lambda(B))$ by Definition 1.14(2). □

Theorem 4.31 (Dual to Theorem 4.30). *Let A be a partially ordered class and let B be a subclass of A. Then $B \subseteq \lambda(\upsilon(B))$.*

Proof. This is the dual theorem of Theorem 4.30. □

Lemma 4.32. *Let A be a partially ordered class and let B be a subclass of A. If $\lambda(B)$ has a* sup *in A, then B has an* inf *in A and* $\inf B = \sup \lambda(B)$.

Proof. Assume that $\lambda(B)$ has a sup in A and let $a = \sup \lambda(B)$.

$\Rightarrow$ (i) a is a lower bound of B.

($\because$) $b \in B$.

$\Rightarrow c \leq b$ for all $c \in \lambda(B)$.

$\Rightarrow b$ is an upper bound of $\lambda(B)$ and hence $a \leq b$.

$\therefore$ a is a lower bound of B.

(ii) a is the greatest lower bound of B.

($\because$) d is any lower bound of B.

$\Rightarrow d \in \lambda(B)$.

$\Rightarrow$ Since a is an upper bound of $\lambda(B)$, $a \geq d$.

$\therefore$ a is the greatest lower bound of B.

$\Rightarrow a = \inf B$. □

Lemma 4.33 (Dual to Lemma 4.32). *Let A be a partially ordered class and let B be a subclass of A. If $\upsilon(B)$ has an* inf *in A, then B has a* sup *in A and* $\sup B = \inf \upsilon(B)$.

Proof. This is the dual theorem of Lemma 4.32. □

Definition 4.34. A partially ordered class A is said to be *conditionally complete* if every nonempty subclass of A that is bounded above has a sup.

We have the following alternative definition of *conditionally complete*: A is said to be conditionally complete if every nonempty subclass of A that is bounded below has an inf.

Our next theorem establishes the equivalence of the two definitions.

Theorem 4.35. *Let A be a partially ordered class. Then the following two conditions are equivalent.*

(1) *Every nonempty subclass of A that is bounded above has a* sup.

(2) *Every nonempty subclass of A that is bounded below has an* inf.

Proof. ((1) $\Rightarrow$ (2)) Assume the condition (1).

Let B be a nonempty, lower bounded subclass of A.

$\Rightarrow \lambda(B) \neq \emptyset$ and each element of B is an upper bound of $\lambda(B)$.

$\Rightarrow \lambda(B) \neq \emptyset$ and $\lambda(B)$ is bounded above.

$\Rightarrow \lambda(B)$ has a sup in A by the assumption (1).

$\Rightarrow B$ has an inf by Lemma 4.32.

((2) $\Rightarrow$ (1)) It is the dual of the statement (1) $\Rightarrow$ (2). $\square$

Note that the set $\{q \in \mathbb{Q} \mid x < \sqrt{2}\}$ is bounded above but has no sup in the partially ordered set $\langle \mathbb{Q}, \leq \rangle$. Thus the principal of duality(Theorem 4.23) cannot be applied directly to prove the statement $\lceil(1) \Rightarrow (2)\rfloor$ of Theorem 4.35.

Exercise 4.3

1. Assume that $B \subseteq A$. Prove that if B has a greatest element b, then $b = \sup B$.

2. Assume that $B \subseteq A$. Prove that $\upsilon(B) = \upsilon(\lambda(\upsilon(B)))$. [Hint: Use Theorems 4.26 and 4.30.]

3. Assume that $B \subseteq A$ and $C \subseteq A$. Prove that $\lambda(B \cup C) = \lambda(B) \cap \lambda(C)$.

4. Assume that $B \subseteq A$. Prove that if B has a sup b, then $\lambda(\upsilon(B)) \cap \upsilon(B) = \{b\}$.

5. Assume that $C \subseteq B$ and $B \subseteq A$. Prove that $\sup_A C \leq \sup_B C$.

6. Let B and C be subclasses of a partially ordered class A. Prove that if $\sup B = \sup C$, then $\upsilon(B) = \upsilon(C)$.

7. Let A and B be partially ordered classes and let $f : A \to B$ be a strictly increasing function. Prove that if b is a maximal element of B, then each element of $f^{-1}(b)$ is a maximal element of A.

8. Let A and B be partially ordered classes and let $f : A \to B$ be an increasing function. Prove that if a is the greatest element of A, then $f(a)$ is the greatest element of $f(A)$.

9. Let A and B be partially ordered classes and let $f : A \to B$ be an increasing function. Assume that $C \subseteq A$. Prove that if c is an upper bound of C, then $f(c)$ is an upper bound of $f(A)$.

10. Let A and B be partially ordered classes and let $f : A \to B$ be an isomorphism. Prove each of the following.

    (1) a is a maximal element of A iff $f(a)$ is a maximal element of B.

    (2) a is the greatest element of A iff $f(a)$ is the greatest element of B.

    (3) Suppose that $C \subseteq A$. Then x is an upper bound of C iff $f(x)$ is an upper bound of $f(C)$.

    (4) Suppose that $C \subseteq A$. Then $b = \sup C$ iff $f(b) = \sup f(C)$.

11. Let A be a partially ordered class. Prove the following.

    (1) If every subclass of A has a sup and an inf, then A has a least element and a greatest element. [Hint: Use Examples 4.20(5).]

    (2) The following two statements are equivalent:

    (i) Every subclass of A has a sup. (ii) Every subclass of A has an inf.

12. Let A and B be partially ordered classes. Prove the following.

    (1) Suppose that $A \times B$ is ordered lexicographically. If (a, b) is a maximal element of $A \times B$, then a is a maximal element of A.

    (2) Suppose that $A \times B$ is ordered antilexicographically. If (a, b) is a maximal element of $A \times B$, then b is a maximal element of B.

4.4 Lattices

Definition 4.36. (1) Let A be a partially ordered class. Then A is called a *lattice* if every doubleton $\{x, y\}$ in A has a sup and an inf in A.

In a lattice A, it is customary to write $x \vee y = \sup\{x, y\}$ and $x \wedge y = \inf\{x, y\}$ for $x, y \in A$. We call $x \vee y$ the *join* of x and y and the expression $x \vee y$ is read "x join y". Also we call $x \wedge y$ the *meet* of x and y and the expression $x \wedge y$ is read "x meet y".

(2) Let A be a lattice and let $B \subseteq A$. Then B is called a *sublattice* of A if $x \vee y \in B$ and $x \wedge y \in B$ for all $x, y \in B$.

Remark 4.37. The duality theorem(Theorem 4.23) says that a statement is true for all partially ordered classes if and only if its dual statement is true for all partially ordered classes.

Since there is a partially ordered class in which some doubleton $\{x, y\}$ in A has no sup in A but has an inf in A, it is not true that every doubleton $\{x, y\}$ in A has a sup in A if and only if every doubleton $\{x, y\}$ in A has an inf in A.

Proposition 4.38. *Let A be a lattice and let $a, b, c \in A$. Then the following hold.*

(1) $a \leq a \vee b$ *and* $b \leq a \vee b$.

(2) *If* $a \leq c$ *and* $b \leq c$, *then* $a \vee b \leq c$.

(3) $a \wedge b \leq a$ *and* $a \wedge b \leq b$.

(4) *If* $c \leq a$ *and* $c \leq b$, *then* $c \leq a \wedge b$.

Proof. It is obvious and left for the reader. □

Theorem 4.39. *Let A be a lattice and let $x, y, z \in A$. Then the join $\vee$ and the meet $\wedge$ have the following properties.*

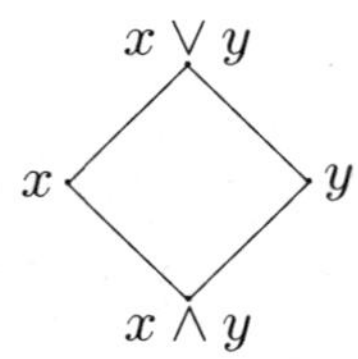

Figure 4.10: $\sup\{x, y\}$ and $\inf\{x, y\}$

L1. *$x \vee x = x$ and $x \wedge x = x$.*

L2. *$x \vee y = y \vee x$ and $x \wedge y = y \wedge x$.*

L3. *$(x \vee y) \vee z = x \vee (y \vee z)$ and $(x \wedge y) \wedge z = x \wedge (y \wedge z)$.*

L4. *$(x \vee y) \wedge x = x$ and $(x \wedge y) \vee x = x$.*

Proof. L1. $x \vee x = \sup\{x, x\} = x$ and $x \wedge x = \inf\{x, x\} = x$.

L2. $x \vee y = \sup\{x, y\} = \sup\{y, x\} = y \vee x$ and

$x \wedge y = \inf\{x, y\} = \inf\{y, x\} = y \wedge x$.

L3. $x \vee (y \vee z) = (x \vee y) \vee z$.

($\because$) $x \leq x \vee (y \vee z)$ and $y \leq y \vee z \leq x \vee (y \vee z)$ by Proposition 4.38(1).

$\Rightarrow x \vee y \leq x \vee (y \vee z)$ by Proposition 4.38(2).

$\Rightarrow$ Since $z \leq y \vee z \leq x \vee (y \vee z)$ by Proposition 4.38(1),

$(x \vee y) \vee z \leq x \vee (y \vee z)$ by Proposition 4.38(2).

Similarly, we can show that $x \vee (y \vee z) \leq (x \vee y) \vee z$.

$\Rightarrow x \vee (y \vee z) = (x \vee y) \vee z$ by the antisymmetricity of $\leq$.

The equality $x \wedge (y \wedge z) = (x \wedge y) \wedge z$ is the dual of $x \vee (y \vee z) = (x \vee y) \vee z$.

L4. $x = (x \vee y) \wedge x$.

($\because$) $x \leq x \vee y$ and $x \leq x \vee x = x$ by Proposition 4.38(1) and L1 above.

$\Rightarrow x \leq (x \vee y) \wedge x$ by Proposition 4.38(4).

$\Rightarrow x$ is a lower bound of $\{x \vee y, x\}$.

$\Rightarrow$ Since $z \leq x$ for any lower bound z of $\{x \vee y, x\}$,

x is the greatest lower bound of $\{x \vee y, x\}$.

$\Rightarrow x = \inf\{x \vee y, x\} = (x \vee y) \wedge x$.

The equality $(x \wedge y) \vee x = x$ is the dual of $x = (x \vee y) \wedge x$. □

A lattice may alternatively be defined as an algebraic system with two operations[2] $\vee$ and $\wedge$ which have properties L1 through L4. This fact, which is of great importance in the study of lattices, is a consequence of the following theorem.

Theorem 4.40. *Let A be a class in which two operations, denoted $\vee$ and $\wedge$, are given and have properties L1 through L4. Define a relation $\leq$ in A by*

$$x \leq y \text{ if and only if } x \vee y = y.$$

Then $\leq$ is an order relation in A, and A is a lattice.

Proof. (i) $\leq$ is reflexive.

(∵) For any $x \in A$, since $x \vee x = x$ by L1, $x \leq x$.

(ii) $\leq$ is antisymmetric.

(∵) $x \leq y$ and $y \leq x$ for some $x, y \in A$.

$\Rightarrow x \vee y = y$ and $y \vee x = x$.

$\Rightarrow$ Since $x \vee y = y \vee x$ by L2, $x = y$.

(iii) $\leq$ is transitive.

(∵) $x \leq y$ and $y \leq z$ for some $x, y, z \in A$.

$\Rightarrow x \vee y = y$ and $y \vee z = z$.

$\Rightarrow$ Since $x \vee (y \vee z) = (x \vee y) \vee z$ by L3,

$x \vee z = x \vee (y \vee z) = (x \vee y) \vee z = y \vee z = z$.

$\Rightarrow x \leq z$.

$\Rightarrow \leq$ is an order relation in A by (i), (ii) and (iii) above.

(iv) $\sup\{x, y\} \in A$ and $\inf\{x, y\} \in A$ for any $x, y \in A$.

---

[2]For a class $\mathscr{A}$ and an integer $n = 0, 1, 2, \cdots$, by an *n-ary operation* of $\mathscr{A}$ we mean a function from $\mathscr{A}^n$ to $\mathscr{A}$. In particular, the 0-ary operations are called the *nullary operations*, the 1-ary operations are called the *unary operations*, the 2-ary operations are called the *binary operations* and the 3-ary operations are called the *ternary operations*. Also n-ary operations are simply called operations.

($\because$) $x, y \in A$.

$\Rightarrow$ Since $\vee$ and $\wedge$ are operations in A, there exist $x \vee y$ and $x \wedge y$ in A for any $x, y \in A$.

$\Rightarrow$ $x \vee y$ is a least upper bound of $\{x, y\}$. i.e., $x \vee y = \sup\{x, y\}$.

($\because$) By applying L1, L2 and L3, $x \vee (x \vee y) = (x \vee x) \vee y = x \vee y$ and $y \vee (x \vee y) = y \vee (y \vee x) = (y \vee y) \vee x = y \vee x = x \vee y$.

$\Rightarrow$ $x \leq x \vee y$ and $y \leq x \vee y$.

$\Rightarrow$ $x \vee y$ is an upper bound of $\{x, y\}$.

Next let z be any upper bound of $\{x, y\}$.

$\Rightarrow$ $x \leq z$ and $y \leq z$.

$\Rightarrow$ $x \vee z = z$ and $y \vee z = z$.

$\Rightarrow$ $(x \vee y) \vee z = x \vee (y \vee z) = x \vee z = z$ by applying L3.

$\Rightarrow$ $x \vee y \leq z$. and hence $x \vee y = \sup\{x, y\}$.

Analogously, we can also show that $x \wedge y = \inf\{x, y\}$.

$\Rightarrow$ $\sup\{x, y\}$ and $\inf\{x, y\}$ exist in A for any $x, y \in A$ and hence A is a lattice by (i), (ii), (iii) and (iv) above.. □

Remark 4.41. By Theorems 4.39 and 4.40, a lattice may be defined in two different ways:

(1) as a partially ordered class $\langle A, \leq \rangle$ in which every pair of elements has a sup and an inf, or

(2) as an algebraic system $\langle A, \vee, \wedge \rangle$ with two operations $\vee$ and $\wedge$ satisfying the rules L1 through L4.

Definition 4.42. A *Boolean algebra* is defined to be a lattice $(A, \vee, \wedge)$ with the following three additional properties.

L5. There exist elements $0, 1 \in A$ such that $x \vee 0 = x$ and $x \wedge 1 = x$ for all $x \in A$.

L6. For each $x \in A$, there exists $x' \in A$ such that $x \wedge x' = 0$ and $x \vee x' = 1$.

L7. $x \vee (y \wedge z) = (x \vee y) \wedge (x \vee z)$ and $x \wedge (y \vee z) = (x \wedge y) \vee (x \wedge z)$ for all x, y and z in A.

Note that a Boolean algebra A has three operations $\vee$, $\wedge$ and $'$ satisfying L1 through L7 and that the symbols 0 and 1 in L5 do not represent the real numbers 0 and 1.

Example 4.43. (1) The universal class $\mathscr{U}$ of all sets with the operations $\cup$ and $\cap$ defined in Definition 1.17, is a lattice. Indeed, the operations $\cup$ and $\cap$ satisfy the conditions L1 through L4 by Theorem 1.20 and hence $\mathscr{U}$ with the operations $\cup$ and $\cap$ is a lattice by Theorem 4.40.

But the universal class $\mathscr{U}$ is not a Boolean algebra. Also note that since $A \cup B = \sup\{A, B\}$, $A \cap B = \inf\{A, B\}$ by Example 4.20(4) and since $A \subseteq B$ if and only if $A \cup B = B$ by Theorem 1.20(2), the order relation in the lattice $\mathscr{U}$ is the inclusion $\subseteq$ by Theorem 4.40.

(2) Let $\mathscr{C}$ denote the class of all sentences with the operations $\vee$ and $\wedge$ defined in Definition 1.3. The operations $\vee$ and $\wedge$ satisfy L_1 through L_4 by Theorem 1.9 and hence $\mathscr{C}$ with the operations $\vee$ and $\wedge$ is a lattice by Theorem 4.40. Note that the logical equivalence $\Leftrightarrow$ is an equivalence relation on $\mathscr{C}$. Let $\mathscr{S}$ denote the quotient class $\mathscr{C}/_{\Leftrightarrow}$. Then the operations $\vee$ and $\wedge$ of $\mathscr{C}$ induce well-defined operations $\vee$ and $\wedge$ of $\mathscr{S}$. In other words, $[P] \vee [Q] = [P \vee Q]$ and $[P] \wedge [Q] = [P \wedge Q]$ for all $[P], [Q] \in \mathscr{S}$, where $[P]$ denotes the equivalence class of $P \in \mathscr{C}$ under $\Leftrightarrow$. Since $\mathscr{S}$ with $\vee$ and $\wedge$ satisfies L_1 through L_4 by applying Theorem 1.9, the algebraic system $(\mathscr{S}, \vee, \wedge)$ is a lattice by Theorem 4.40. Now let T be a sentence which is always true. Such a sentence T is called a *tautology*. For example, $P \vee \neg P$ is such a sentence for any sentence $P \in \mathscr{C}$. Let F be a sentence which is always false. For example, $P \wedge \neg P$ is such a sentence for any sentence $P \in \mathscr{C}$. Then $[T], [F] \in \mathscr{S}$, and for all $[P], [Q], [R] \in \mathscr{S}$,

L_5: $[P] \vee [F] = [P]$ and $[P] \wedge [T] = [P]$ $\big(1 = [T], 0 = [F]\big)$,

L_6: $[P] \wedge [\neg P] = [F]$ and $[P] \vee [\neg P] = [T]$ $\big([P]' = [\neg P]\big)$, and

L_7: $[P] \vee ([Q] \wedge [R]) = ([P] \vee [Q]) \wedge ([P] \vee [R])$ and

$[P] \wedge ([Q] \vee [R]) = ([P] \wedge [Q]) \vee ([P] \wedge [R])$.

Thus $\mathscr{S}$ with $\vee$ and $\wedge$ is a Boolean algebra by Definition 4.42.

Note that the operation $\Rightarrow$ of $\mathscr{C}$ induces the operation $\Rightarrow$ of $\mathscr{S}$. Since $(P \Rightarrow Q) \Leftrightarrow ((P \vee Q) \Leftrightarrow Q)$ for any sentences $P, Q \in \mathscr{C}$ by Remark 1.6(2), $[P] \Rightarrow [Q]$ if and only if $[P] \vee [Q] = [Q]$ in $\mathscr{S}$. Hence the operation $\Rightarrow$ of $\mathscr{S}$ is the order relation of $\mathscr{S}$ by Theorem 4.40.

(3) For any set A, the power set $\mathscr{P}(A)$ of A is a Boolean algebra under the operations $\cup$, $\cap$ and $'$ defined by $S' = A \setminus S$ for any $S \in \mathscr{P}(A)$, where $0 = \emptyset$ and $1 = A$ in the Boolean algebra $\mathscr{P}(A)$.

(4) Consider the set $B = \{0, 1\}$ consisting of two distinct elements 0 and 1 and define the operations $\vee$ and $\wedge$ and $'$ in B by setting

$0 \vee 0 = 0$, $0 \vee 1 = 1 \vee 0 = 1 \vee 1 = 1$,

$1 \wedge 1 = 1$, $0 \wedge 1 = 1 \wedge 0 = 0 \wedge 0 = 0$, and $0' = 1$ and $1' = 0$.

Then B with $\vee$, $\wedge$ and $'$ satisfy the conditions $L1$ through $L7$ and hence $(B, \vee, \wedge, ')$ is a Boolean algebra.

Definition 4.44. Let A be a partially ordered class. A is called a *complete lattice* if every subclass of A has a sup. Alternatively, A is called a complete lattice if every subclass of A has an inf.

The purpose of our next theorem is to show that these two alternative definitions are equivalent. It will follow that if A is a complete lattice (in the sense of either definition), then every subclass of A has a sup and an inf in A. In particular, every doubleton in A has a sup and an inf in A, hence we are justified in calling A a lattice.

Theorem 4.45. *Let A be a partially ordered class. Then the following are equivalent.*

(1) *Every subclass of A has a sup in A.*

(2) *Every subclass of A has an* inf *in A.*

Proof. **((1) ⇒ (2))** Assume that every subclass of A has a sup in A.

⇒ $\sup_A A, \sup_A \emptyset \in A$ and since $\inf_A \emptyset = \sup_A A$ and $\inf A = \sup_A \emptyset$ by Examples 4.20(5) and $\sup_A \emptyset, \sup_A A \in A$ by assumption, A has both a greatest element $\sup A$ and a least element $\inf A = \sup \emptyset$ in A.

⇒ $\inf_A \emptyset = \sup A \in A$ for $\emptyset \subset A$, and every nonempty subclass of A is upper bounded by $\sup A$ and lower bounded by $\inf A$.

⇒ $\inf_A \emptyset \in A$ and every nonempty (and hence upper bounded) subclass of A has a sup in A by assumption.

⇒ $\inf_A \emptyset \in A$ and every nonempty (and hence lower bounded) subclass of A has an inf in A by Theorem 4.35.

⇒ Every subclass B of A has an $\inf_A B$ in A.

((2) ⇒ (1)) The statement ⌈(2) ⇒ (1)⌋ is the dual of the statement ⌈(1) ⇒ (2)⌋ and the statement ⌈(1) ⇒ (2)⌋ is true for all partially ordered classes.

⇒ The statement ⌈(2) ⇒ (1)⌋ is true by the duality theorem(Theorem 4.23). □

Example 4.46. Let A be a set and let $\mathscr{G}$ be the set of all the equivalence relations in A, ordered by inclusion.

The diagonal graph 1_A is the least element of $\mathscr{G}$.

The relation $A \times A$ is the greatest element of $\mathscr{G}$.

Let $\{G_i\}_{i\in I}$ be a subset of $\mathscr{G}$. Then $\bigcup_{i\in I} G_i$ is not an element of $\mathscr{G}$ in general, but $\bigcap_{i\in I} G_i$ is an element of $\mathscr{G}$ which is an inf of $\{G_i\}_{i\in I}$ in $\mathscr{G}$. Hence $\mathscr{G}$ is a complete lattice, by Theorem 4.45.

Exercise 4.4

1. Let A be a class with two operations $\vee$ and $\wedge$ which satisfy L1 through L4. Prove that $x \vee y = y$ if and only if $x \wedge y = x$.

2. In Theorem 4.40, prove that $x \wedge y = \inf\{x, y\}$.

3. Let A be a lattice. Prove that the following statements are true.

   (1) If $a \leq b$, then for all $x \in A$, $a \vee x \leq b \vee x$ and $a \wedge x \leq b \wedge x$.

   (2) If $a \leq b$ and $c \leq d$, then $a \vee c \leq b \vee d$ and $a \wedge c \leq b \wedge d$.

4. Let A be a lattice. If $[a, b]$ and $[c, d]$ are closed intervals of A, prove that $[a, b] \cap [c, d] = [a \vee c, b \wedge d]$. (See Exercise 4.2.8 for the definition of closed intervals.)

5. Let A be a lattice. Prove that every closed interval $[a, b]$ of A is a sublattice.

6. Let A be a lattice. For $a \in A$, let $I_a = \{x \in A \mid x \leq a\}$. Prove that I_a is a sublattice of A.

7. By a *distributive lattice*, we mean a class with two operations $\vee$ and $\wedge$ satisfying L1, L2, L3, L4 and L7. If A is a distributive lattice, prove that

   $c \vee x = c \vee y$ and $c \wedge x = c \wedge y$ imply $x = y$.

8. In an arbitrary lattice A, prove the so-called distributive inequality.

   $x \wedge (y \vee z) \geq (x \wedge y) \vee (x \wedge z)$ and $x \vee (y \wedge z) \leq (x \vee y) \wedge (x \vee z)$.

9. Let A be a lattice and let $x, y, z \in A$. Prove that if $x \leq z$, then $x \vee (y \wedge z) \leq (x \vee y) \wedge z$.

10. Draw the line diagram of a lattice A and a subclass $B \subseteq A$ such that B is not a sublattice of A.

11. Draw the line diagram of two lattices whose intersection is not a lattice.

12. Let A be a partially ordered set. Prove that the class of all the convex subsets of A is a complete lattice.

13. Draw the line diagram of a partially ordered class which is conditionally complete, but is not a complete lattice.

14. State what is wrong if we prove $\lceil(1) \Rightarrow (2)\rfloor$ of Theorem 4.45 as follows. Assume the statement (1) holds. Then since the statement (2) is the dual statement of (1), (2) holds.

4.5 Fully ordered classes and well-ordered classes

Recall that a partially ordered class A is said to be *fully ordered* if every two elements of A are comparable.

Example 4.47 (Fully Ordered Classes)**.** The following are fully ordered classes.

(1) The class $\mathbb{Z}$ of the integers, with the usual order.

(2) The class $\mathbb{Q}$ of the rational numbers, with the usual order.

(3) The class $\mathbb{R}$ of the real numbers, with the usual order.

Remark 4.48. For a fully ordered class, the following hold.

(1) Every subclass of a fully ordered class is fully ordered.

(2) Every fully ordered class is a lattice.

Theorem 4.49. *Let A be a fully ordered class and let B be a partially ordered class. Let $f : A \to B$ be a function. If $f : A \to B$ is bijective and increasing, then f is an isomorphism.*

Proof. Assume that $f : A \to B$ is bijective and increasing. Let $x, y \in A$.
$\Rightarrow$ (i) $x \leq y$ implies $f(x) \leq f(y)$ because f is an increasing function.
(ii) $f(x) \leq f(y)$ implies $x \leq y$.
($\because$) $f(x) \leq f(y)$.
$\Rightarrow$ Since A is fully ordered, either $x \leq y$ or $x > y$.
$\Rightarrow x \leq y$.
($\because$) Suppose that $x > y$.
$\Rightarrow$ Since $f : A \to B$ is increasing, $f(x) \geq f(y)$.
$\Rightarrow$ Since $f(x) \leq f(y)$ by assumption, $f(x) = f(y)$.
$\Rightarrow$ Since $f : A \to B$ is bijective, $x = y$.
This is contrary to our assumption that $x > y$.
$\Rightarrow$ Since f is bijective, f is an isomorphism. □

Definition 4.50. A partially ordered class A is said to be *well-ordered* if every nonempty subclass of A has a least element.

Proposition 4.51. (1) *Every well ordered class is fully ordered.*

(2) *Every well ordered class is conditionally complete.*

Proof. Assume that A is well ordered class.

(1) Let $x, y \in A$.

$\Rightarrow \{x, y\}$ has a least element which is either x or y.

$\Rightarrow \inf\{x, y\} = x$ or $\inf\{x, y\} = y$.

$\Rightarrow x \leq y$ or $y \leq x$ and hence A is fully ordered.

(2) Let B be a nonempty upper bounded subclass of A.

$\Rightarrow v(B) \neq \emptyset$.

$\Rightarrow$ Since A is well-ordered, $v(B)$ has a least element which is $\sup B$ by Definition 4.18(9).

$\Rightarrow B$ has a sup and hence A is conditionally complete. □

Definition 4.52. Let A be a partially ordered class.

(1) An element $b \in A$ is called the *immediate successor* of $a \in A$ if $a < b$ and there exists no element $c \in A$ such that $a < c < b$.

(2) An element $a \in A$ is called the *immediate predecessor* of $b \in A$ if b is an immediate successor of a.

(3) A subset $B \subseteq A$ is called a *section* of A if B satisfies the following condition:

If $x \in A$ and there exists $y \in B$ such that $x \leq y$, then $x \in B$.

Remark 4.53. In a well-ordered class A, every element of A (with the exception of the greatest element of A, if it exists) has an immediate successor.

Proof. Assume that $x \in A$ is not the greatest element of A.

Let $T = \{y \in A \mid y > x\}$.

$\Rightarrow T \neq \emptyset$.

$\Rightarrow$ Since A is well-ordered, T has a least element, say x_0.

$\Rightarrow x < x_0$ and there exists no $y \in A$ such that $x < y < x_0$.

$\Rightarrow x_0$ is the immediate successor of x. □

Let A be a nonempty well-ordered class. Then A has the least element x_1 by Remark 4.53. If x_1 is not the greatest element of A, then by Remark 4.53, x_1 has an immediate successor, say x_2. Again if x_2 is not the greatest element of A, then by Remark 4.53, x_2 has an immediate successor, say x_3 and so on.

Example 4.54. (1) The class $\{1, 2, 3, 4, 5\}$, ordered in the usual way, is a well-ordered class.

(2) The class $\mathbb{N} = \{1, 2, 3, 4, \cdots\}$ of all the positive integers, ordered in the usual way, is a well-ordered class.

(3) Let $W = \left\{0, \frac{1}{2}, \frac{3}{4}, \frac{7}{8}, \cdots, 1, 1 + \frac{1}{2}, 1 + \frac{3}{4}, 1 + \frac{7}{8}, \cdots, 2, 2 + \frac{1}{2}, 2 + \frac{3}{4}, 2 + \frac{7}{8}, \cdots, \right\}$. Then W is a well-ordered subclass of the real numbers.

Remark 4.55. (1) For (1) and (2) of Example 4.54, only the least element of the class, namely 1, does not have an immediate predecessor.

(2) In Example 4.54(3), there are three elements, namely $0, 1, 2 \in W$ which do not have an immediate predecessor in W.

(3) $\mathbb{Z}, \mathbb{Q}$ and $\mathbb{R}$, in the usual order, are not well-ordered.

Indeed, $V = \{\cdots, -3, -2, -1, 0\}$ is a subclass of each of these classes, and V does not have a least element in each of $\mathbb{Z}, \mathbb{Q}$ and $\mathbb{R}$.

Theorem 4.56. *Let A be a well-ordered class and let B be a subclass of A. Then B is a section of A if and only if $B = A$ or B is an initial segment of A.*

Proof. **(if)** If $B = A$ or B is an initial segment of A, then obviously, B is a section of A.

(only if) Assume that B is a section of A. If $B = A$, we are done.

Assume that $B \neq A$. i.e., $A \setminus B \neq \emptyset$.

$\Rightarrow$ Since A is a well-ordered class, $A \setminus B$ has the least element, say m.

i.e., $m = \inf_A(A \setminus B)$ and $m \in A \setminus B$.

Consider the initial segment S_m of A determined by m.

$\Rightarrow$ (i) $S_m \subseteq B$.

($\because$) $x \in S_m$.

$\Rightarrow x < m$ and hence $x \notin A \setminus B$.

$\Rightarrow x \in B$ and hence $S_m \subseteq B$ by Definition 1.14(2).

(ii) $B \subseteq S_m$.

($\because$) Let $x \in B$. To show that $x \in S_m$, suppose that $x \notin S_m$.

$\Rightarrow$ Since A is well-ordered, A is fully ordered by Proposition 4.51 and hence $x \leq m$ or $m \leq x$.

$\Rightarrow$ Since $x \notin S_m$, $m \leq x$.

$\Rightarrow$ Since B is a section and $x \in B$, $m \in B$.

This contradicts to our choice of $m \in A \setminus B$.

$\Rightarrow x \in S_m$ and hence $B \subseteq S_m$ by Definition 1.14(2).

$\Rightarrow B = S_m$ is the initial segment of A determined by m. □

Theorem 4.57 (Principle of Transfinite Mathematical Induction).
Let A be a well-ordered class and let $P(x)$ be a statement which is either true or false for each element $x \in A$. Suppose that the following condition holds:

Ind. *If $P(y)$ is true for every $y < x$, then $P(x)$ is true.*

Then $P(x)$ is true for every element $x \in A$.

Proof. Suppose that $P(x_0)$ is not true for some $x_0 \in A$ and let $C = \{y \in A \mid P(y) \text{ is false}\}$.

⇒ $C \neq \emptyset$ because $x_0 \in C$.

⇒ Since A is well-ordered, C has a least element m.

⇒ Since m is a least element of C, $P(y)$ is true for every $y < m$.

⇒ By the condition Ind, $P(m)$ is true.

But, since m is the least element of C, $P(m)$ is false.

This is a contradiction.

⇒ $P(x)$ must be true for every $x \in A$. □

Remark 4.58. (1) Usually we add the condition that for the least element x_0 of A, $P(x_0)$ is true as an assumption for Theorem 4.57, but this condition follows from the assumption Ind (which is called the *induction hypothesis*) logically. Indeed, the statement that ⌈if $y < x_0$, then $P(y)$ is true⌋ is true. Thus by the assumption Ind, $P(x_0)$ is true.

(2) The assumption Ind in Theorem 4.57 is perfect. But in the practical application of Theorem 4.57, the assumption Ind should be checked in the following steps; first the condition that $P(x_0)$ is true for the least element x_0 of A and then the condition Ind holds for $x > x_0$.

The following two corollaries are special cases of Theorem 4.57.

Corollary 4.59 (Finite Mathematical Induction).
Let $P(x)$ be a statement which is either true or false for each element $x \in \mathbb{N}$. Suppose the following condition holds:

Induction Hypothesis: *If $P(n)$ is true for every $n < x$, then $P(x)$ is true.*

Then $P(x)$ is true for every element $x \in \mathbb{N}$.

Corollary 4.60 (Finite Mathematical Induction).
Let X be an infinite subset of $\mathbb{N}$ and let $P(x)$ be a statement which is either true or false for each element $x \in X$. Suppose that for each $y \in X$, the following condition holds:

Induction Hypothesis: *If $P(n)$ is true for every $n \in X$ with $n < y$, then $P(y)$ is true.*

Then $P(x)$ is true for every element $x \in X$.

Exercise 4.5

1. Let A be a fully ordered set. Prove that the set of all sections of A (ordered by inclusion) is fully ordered.

2. Let A be a fully ordered class and let B be a partially ordered class. Let $f : A \to B$ be an increasing function. Prove that f is injective if and only if f is strictly increasing.

3. Let A be a fully ordered class and let B be a partially ordered class. Let $f : A \to B$ be a bijective function. Prove that if f is increasing, then f is an isomorphism.

4. Let A be a fully ordered class and let $\{L, U\}$ be a partition of A. Prove that $\{L, U\}$ is a cut of A if and only if $x \leq y$ for all $x \in L$ and for all $y \in U$.

5. Let A be a fully ordered class. Prove that if B and C are convex subclasses of A and $B \cap C \neq \emptyset$ then $B \cup C$ is convex.

6. Let A be a fully ordered class. Let B and C be convex subclasses of A and suppose that $B \cap C \neq \emptyset$. Prove that every upper bound of $B \cap C$ is an upper bound of B or an upper bound of C. Conclude that $v(B \cap C) = v(B) \cup v(C)$.

7. Let A be a well-ordered class. If $x \in A$, prove that the immediate successor of x and the immediate predecessor of x (if it exists) are unique.

8. Let A be a partially ordered class. Prove that B is a section of A if and only if $(B, A \setminus B)$ is a cut of A.

9. Let A be a well-ordered class. Prove the following.

   (1) The intersection of any family of sections of A is a section of A.

   (2) The union of any family of sections of A is a section of A.

10. Let A be a well-ordered class and let B and C be initial segments of A. Prove that if $B \subseteq C$, then B is an initial segment of C.

11. Let A be a fully ordered class. Let $B \subseteq A$ and $m \in B$. Prove that B has a least element if and only if $S_m \cap B$ has a least element, where S_m denotes the initial segment of A determined by m.

12. Let A be a fully ordered class. Prove that A is well ordered if and only if every initial segment of A is well ordered.

    [Hint: Use the result of Exercise 4.5.11 above.]

13. Let A be a well-ordered class. If $a \in A$, let a' designate the immediate successor of a and let a'' designate the immediate predecessor of a (if it exists). Prove the following.

    (1) $a \leq b$ if and only if $a' \leq b'$.
    (2) $a = b$ if and only if $a' = b'$.

    (3) $a < b$ if and only if $a' < b'$.
    (4) $a = b$ if and only if $a'' = b''$.

    (5) $a = b'$ if and only if $a'' = b$.

14. Let A be a well-ordered class. For $a \in A$, let a' designate the immediate successor of a. An element q in A will be called a *limit element* of A if q is not the least element of A and q does not have an immediate predecessor. Prove the following.

    (1) q is a limit element of A if and only if $a < q$ implies $a' < q$.

    (2) q is a limit element of A if and only if $q = \sup\{x \in A \mid x < q\}$.

4.6 Isomorphism between well-ordered classes

The purpose of this section is to prove a remarkable property of well-ordered classes: if A and B are any two well-ordered classes, either A is isomorphic with B, or else one of the two is isomorphic with an initial segment of the other. What this means, roughly speaking, is that *well-ordered classes do not differ from one another except in their size.* This fact has many applications in mathematics, and will be essential to our later discussion of infinite sets and cardinal and ordinal numbers. We begin by proving four preparatory lemmas.

Lemma 4.61. *Let A be a well-ordered class and let B be a subclass of A. If $f : A \to B$ is an isomorphism, then $x \leq f(x)$ for all $x \in A$.*

Proof. Assume that $f : A \to B$ is an isomorphism and let
$P = \{x \in A \mid f(x) < x\}$. To show that $P = \emptyset$, suppose that $P \neq \emptyset$.
$\Rightarrow$ Since A is well-ordered, there exists the least element a of P.
$\Rightarrow$ $f(a) < a$.
$\Rightarrow$ Since f is an isomorphism, $f(f(a)) < f(a) < a$.
$\Rightarrow$ $f(a) \in P$ and $f(a) < a$.
This is impossible because a is the least element of P.
$\Rightarrow$ $P = \emptyset$ and the lemma is proved. □

Lemma 4.62. *Let A be a well-ordered class and let B be a subclass of an initial segment of A. Then there is no isomorphism $f : A \to B$ from A onto B.*

Proof. Let S_a be an initial segment of A and let $B \subseteq S_a$.
Suppose that there exists an isomorphism $f : A \to B$.
$\Rightarrow$ $a \leq f(a)$ by Lemma 4.61.
$\Rightarrow$ $f(a) \notin S_a$. This is impossible because $\mathrm{ran}(f) = B \subseteq S_a$.
$\Rightarrow$ There is no isomorphism $f : A \to S_a$. □

Corollary 4.63. *No well-ordered class is isomorphic with an initial segment of itself.*

Proof. This is immediate from Lemma 4.62. □

Lemma 4.64. *Let A and B be well-ordered classes. If A is isomorphic with an initial segment of B, then B is not isomorphic with any subclass of A.*

Proof. Let $f : A \to S_b$ be an isomorphism, for an initial segment S_b of B and let C be a subclass of A.
Suppose that there exists an isomorphism $g : B \to C$.
$\Rightarrow$ Since $\mathrm{ran}(g) \subseteq C \subseteq A$, we can regard g as a function from B to A,

where both $g : B \to A$ and $f : A \to S_b$ are injective and increasing.

$\Rightarrow$ $f \circ g : B \to S_b$ is injective and increasing.

$\Rightarrow$ Since B is fully ordered by Proposition 4.51, $f \circ g : B \to (f \circ g)(B)$ as a function from B onto $(f \circ g)(B)$ is an isomorphism by Theorem 4.49 and $(f \circ g)(B) \subseteq S_b$, this is impossible by Lemma 4.62.

$\Rightarrow$ There is no isomorphism $g : B \to C$. □

Lemma 4.65. *Let X be a well-ordered class and let S_x and S_y be any initial segments of X. If $x < y$, then S_x is an initial segment of S_y.*

Proof. Assume that $x < y$.

$\Rightarrow$ $S_x \subseteq S_y$ and S_x is a section of S_y.

$\Rightarrow$ Since $S_x \neq S_y$, S_x is an initial segment of S_y by Theorem 4.56. □

Theorem 4.66. *Let A and B be well-ordered classes. Exactly one of the following three cases must hold :*

(1) *A is isomorphic with B.*

(2) *A is isomorphic with an initial segment of B.*

(3) *B is isomorphic with an initial segment of A.*

Proof. Let A and B be well-ordered classes and let C be the following subclass of A:

$C = \{x \in A \mid S_x \cong S_r \text{ for some } r \in B\}$ $(\subseteq A)$ and

$D = \{r \in B \mid S_r \cong S_x \text{ for some } x \in C\}$ $(\subseteq B)$.

Note that $C \neq \emptyset$ and $D \neq \emptyset$ because A and B are well-ordered.

(Step 1) We will define a function $f : C \to D$.

For each $x \in C$, the element $r \in B$ with $S_x \cong S_r$ is unique.

($\because$) Suppose that $S_x \cong S_r$ and $S_x \cong S_t$, for $x \in C$.

To show that $r = t$, suppose that $r \neq t$, say $r < t$.

$\Rightarrow$ S_r is an initial segment of S_t by Lemma 4.65.

$\Rightarrow$ $S_r \ncong S_t$ by Corollary 4.63. But $S_r \cong S_x \cong S_t$.

This is a contradiction and hence $r = t$.

Define $f : C \to D$ by $f(x) = r$ where $S_r \cong S_x$.

$\Rightarrow f : C \to D$ is a function.

(Step 2) The function $f : C \to D$ is an isomorphism.

($\because$) (i) $f : C \to D$ is injective.

($\because$) Suppose that $f(u) = f(v) = r$.

$\Rightarrow S_u \cong S_r \cong S_v$.

$\Rightarrow$ If $u \neq v$, say $u < v$, then S_u is an initial segment of S_v by Lemma 4.65 and hence $S_u \ncong S_v$ by Corollary 4.63. This contradicts to our assumption that $S_u \cong S_v$.

$\Rightarrow u = v$ and hence f is injective.

(ii) $f : C \to D$ is surjective.

($\because$) For each $r \in D$, $S_r \cong S_x$ for some $x \in C$ and hence $f(x) = r$. Thus $f : C \to D$ is surjective.

(iii) $f : C \to D$ is increasing.

($\because$) Suppose that $u \leq v$ and let $f(u) = r$ and $f(v) = t$.

$\Rightarrow S_u \subseteq S_v, S_u \cong S_r$ and $S_v \cong S_t$.

To show that $r \leq t$, suppose that $t < r$.

$\Rightarrow S_t$ is an initial segment of S_r by Lemma 4.65.

$\Rightarrow S_v$ is isomorphic with an initial segment of S_r and S_r is isomorphic with a subclass of S_v.

This is impossible by Lemma 4.64.

$\Rightarrow r \leq t$. i.e., $f(u) \leq f(v)$ and hence $f : C \to D$ is increasing.

$\Rightarrow f : C \to D$ is an isomorphism by Theorem 4.49 by (i), (ii) and (iii) above.

(Step 3) C is a section of A and D is a section of B.

($\because$) (i) C is a section of A. (i.e., $c \in C$ and $x < c$ imply $x \in C$.)

($\because$) Suppose that $c \in C$ and $x < c$ and let $f(c) = r$.

$\Rightarrow S_c \cong S_r$.

$\Rightarrow$ There exists an isomorphism $g : S_c \to S_r$.

$\Rightarrow g|_{S_x} : S_x \to S_{g(x)}$ is an isomorphism because $g(S_x) = S_{g(x)}$.

$\Rightarrow S_x \cong S_{g(x)}$ and hence $x \in C$.

$\Rightarrow C$ is a section of A.

(ii) Analogously, we can show that D is a section of B.

(Step 4) Either C is not an initial segment of A or D is not an initial segment of B.

($\because$) Suppose that C is an initial segment of A, say $C = S_x$,

and that D is an initial segment of B, say $D = S_r$.

$\Rightarrow$ Since $f : C \to D$ is an isomorphism, $C \cong D$. i.e., $S_x \cong S_r$.

$\Rightarrow$ Since $x \in C$ by definition of C, $x \in S_x$. This is a contradiction.

$\Rightarrow C$ or D is not an initial segment.

(Step 5) One of the assertions (1), (2) and (3) of the theorem must hold.

($\because$) Assume that A is not isomorphic with any initial segment of B

and that B is not isomorphic with any initial segment of A.

$\Rightarrow$ Since either C is not an initial segment of A

or D is not an initial segment of B by Step 4, and

since C is a section of A and D is a section of B by Step 3,

$C = A$ or $D = B$ by Theorem 4.56.

$\Rightarrow$ Since $f : C \cong D$, $A \cong D(\subseteq B)$ or $B \cong C(\subseteq A)$.

$\Rightarrow C$ and D are not initial segments.

($\because$) Suppose that C were an initial segment of A.

$\Rightarrow D$ is not an initial segment of B by Step 4.

$\Rightarrow D = B$ and hence $B = D \cong C$.

$\Rightarrow B$ is isomorphic with the initial segment C of A.

This contradicts to our assumption.

$\Rightarrow C$ is not an initial segment of A.

Similarly, D is not an initial segment of B.

$\Rightarrow$ Since C and D are sections of A and B respectively by Step 3,

$C = A$ and $D = B$ by Theorem 4.56.

$\Rightarrow$ Since $C \cong D$ by Step 2, $A \cong B$.

(Step 6) No two of the assertions (1), (2) and (3) of the theorem hold simultaneously.

($\because$) The statements (1) and (2) cannot hold simultaneously and the statements (1) and (3) cannot hold simultaneously by Corollary 4.63. The statements (2) and (3) cannot hold simultaneously by Lemma 4.64. $\square$

Corollary 4.67. *Let A be a well-ordered class and let $B \subseteq A$. Then B is isomorphic with A or an initial segment of A.*

Proof. Let A be a well-ordered class and let $B \subseteq A$.

$\Rightarrow$ B is well-ordered.

$\Rightarrow$ By Theorem 4.66, one of the following three cases must hold.

(1) B is isomorphic with A.

(2) B is isomorphic with S_a for some initial segment S_a of A.

(3) A is isomorphic with S_b for some initial segment S_b of B.

$\Rightarrow$ Since $B \subseteq A$, the case (3) cannot hold.

($\because$) Suppose that the case (3) hold.

$\Rightarrow$ B is not isomorphic with any subclass of A by Lemma 4.64.

But $B \cong B$ and B is a subclass of A.

This is a contradiction.

$\Rightarrow$ B is isomorphic with A or an initial segment of A. $\square$

Exercise 4.6

1. In the proof of Theorem 4.66, prove that $g|_{S_x} : S_x \to S_{g(x)}$ is an isomorphism.

2. In the proof of Theorem 4.66, prove that D is a section of B.

3. Let A be a well-ordered class. Prove that the identity function $1_A : A \to A$ is the only isomorphism from A to A.

4. Let A and B be well-ordered classes. Prove that if $f : A \to B$ and $g : B \to A$ are isomorphisms, then $g = f^{-1}$.

5. Let A and B be well-ordered classes. Prove that there exists at most one isomorphism $f : A \to B$.

6. Let A and B be well-ordered classes. Prove that if A is isomorphic with a subclass of B and if B is isomorphic with a subclass of A, then A is isomorphic with B.

7. Let A and B be well-ordered classes. Prove that if A is isomorphic with a class containing B and if B is isomorphic with a class containing A, then A is isomorphic with B.

8. Let A and B be well-ordered classes. Suppose that A has no greatest element and that every element of B (except the least element) has an immediate predecessor. Prove that B is isomorphic with a section of A.

Chapter 5

The Axiom of Choice and Related Principles

5.1 Introduction

From here on, throughout the remainder of this book, we will be concerned mainly with sets.

In Section 7 of Chapter 1, we explained that the notion of *class* is appealing because of its intuitive simplicity and generality.

(1) Classes suffer from a serious deficiency: Given an arbitrary property of classes, we cannot form the class of all classes having that property.

(2) But sets do not suffer from this deficiency: We have already proven the following.

- *Remark* 1.6(1): If we are given some property of sets, we may form the class of all the sets which have that property.

- **Theorem** 1.52: If A is a set, then the class of all the subsets of A which satisfy any given property is again a set.

We shall need to use these facts frequently in the following pages.

In this Chapter we will discuss a concept which is one of the most important, and at the same time one of the most controversial principals of mathematics.

In 1904, Zermelo brought attention to an assumption which is used implicitly in a variety of mathematical arguments. This assumption does not follow from any previously known postulates of mathematics or logic, hence it must be taken as a new axiom. Zermelo called it the *Axiom of Choice.* The Axiom of Choice has significant implications in many branches of mathematics, and consequences so powerful as, sometimes, to defy credibility. The controversy over this principle continues in our day. We will present some of its aspects in this chapter.

In order to illustrate where the Axiom of Choice intrudes in common mathematical arguments, let us examine the following statement.

Theorem 5.1. *Let A be a nonempty, partially ordered set and suppose that there are no maximal elements in A. Then there exists a nonterminating, increasing sequence $x_1 < x_2 < x_3 < \cdots$ of elements of A.*

Proof. Let A be a nonempty, partially ordered set and suppose that there are no maximal elements in A.

$\Rightarrow$ We may select an arbitrary element of A and call it x_1.

By induction, suppose that we are given $x_1 < x_2 < \cdots < x_n$ and define $A_n = \{x \in A \mid x > x_n\}$.

$\Rightarrow A_n \neq \emptyset$.

($\because$) Suppose that $A_n = \emptyset$.

$\Rightarrow x_n$ is maximal which contradicts to our assumption.

$\Rightarrow A_n \neq \emptyset$.

Select an arbitrary element of A_n and call it x_{n+1}.

$\Rightarrow$ We have $x_1 < x_2 < x_3 < \cdots < x_n < x_{n+1}$ and the inductive process defines an increasing sequence $S_n = \{x_1, x_2, \cdots, x_n\}$ for each $n \in \mathbb{N}$, where $\mathbb{N}$ is the set of all positive integers.

i.e., $S_1 = \{x_1\}, S_2 = \{x_1, x_2\}, S_3 = \{x_1, x_2, x_3\}$, and so on.

Let $S = \bigcup_{n \in \mathbb{N}} S_n$.

$\Rightarrow$ The set S is the nonterminating sequence $x_1 < x_2 < x_3 < \cdots$ that we are seeking. □

Remark 5.2. A careful examination of the above argument will reveal that we have used an assumption which is by no means self-evident or undisputably plausible.

- What we have, in fact, done is to assume that we can make an *infinite succession of arbitrary choices.* It is common enough, in mathematics, to make one arbitrary choice (we do this every time we say "let x be an arbitrary element of A"), and experience confirms that we can make a finite succession of choices. But to make an infinite succession of choices is to carry an argument through an infinite number of steps - and nothing in our experience or in the logic we habitually use, justifies such a process.

- In the proof of Theorem 5.1, it was necessary to choose the elements x_1, x_2, x_3, etc, *in succession*, for each choice depended on the preceding ones.

  The fact that the choices are successive may appear to be the most disturbing element in the whole proof, for this involves a time factor (an infinite *succession* of acts, each one requiring a certain amount of time, would take infinitely long).

- However, the argument can be altered in such a way that all the choices are made simultaneously and independently of one another. We proceed as follows.

  Let us admit that from each nonempty subset $B \subseteq A$ it is possible to choose an arbitrary element r_B, to be called the "*representative*" of B.

Note that in this case, each choice is independent of the others. Hence, in a manner of speaking, all the choices can be made simultaneously.

Returning to the proof of Theorem 5.1, if x_n and A_n are given, we may define x_{n+1} to be the representative of A_n. In other words, instead of choosing representatives for A_1, A_2, A_3, etc., in succession, we have chosen representatives for all the nonempty subsets of A in advance. (Of course, this requires that we make many more choices than are needed for our original argument, but this is the price we must pay to substitute simultaneous choices for successive ones.)

- The preceding paragraph makes it clear that the successive nature of the choices is not the crux of the problem. The problem is: *Can we make infinitely many choices* - be they successive or simultaneous ?

  It is worth noting that in certain particular cases the answer to this question is an obvious "yes."

  For example, if A is a well-ordered set, we may define the "representative" of each nonempty subset $B \subseteq A$ to be the least element of B.

  ($\because$) Since A is well-ordered, we have a law at our disposal which provides us with a representative for each nonempty subset of A.

  The situation in Theorem 5.1, however, is completely different.

  ($\because$) We do not have *any ready-made rule* which is able to furnish us with representatives.

  It is only the latter case - as in Theorem 5.1 - which is of interest to us here.

- In the proof of Theorem 5.1 we speak of "selecting" an element of A_n; clearly, we do not wish to introduce the notion of "selecting" as a new undefined concept of set theory, so we avoid the use of this word by letting an appropriate function "select" representatives.

Definition 5.3. Let A be a set and write $\mathscr{P}'(A)$ for $\mathscr{P}(A) \setminus \{\emptyset\}$.
By a *choice function for* A we mean a function $r : \mathscr{P}'(A) \to A$ such that

$$\text{for all } B \in \mathscr{P}'(A),\ r(B) \in B.$$

We will sometimes write r_B for $r(B)$ and call r_B the *representative* of B.

Example 5.4. let $A = \{a, b, c\}$; an example of a choice function for A is the function $r : \mathscr{P}'(A) \to A$ given in the following table.

B	r(B)
{ a, b, c }	a
{ a, b }	a
{ a, c }	c
{ b, c }	c
{ a }	a
{ b }	b
{ c }	c

Remark 5.5. In the light of Definition 5.3, the question we have been asking can be expressed as follows:

If A is a set, does there exist a choice function for A?

A crucial comment needs to be made at this point:

The proof of Theorem 5.1 does not require that we construct a choice function, it requires merely that a choice function exist!

($\because$) Assume that we have a choice function r for A

$\Rightarrow$ In the controversial step of the proof we just let $x_{n+1} = r(A_n)$.

$\Rightarrow$ If we are assured that a choice function r exists, there is no further difficulty.

- The Axiom of Choice asserts that every set has a choice function. Its intent is to state the existence of a choice function for every set, even where, admittedly, none can be actually constructed.

  It must be emphasized here that to state the existence of a choice function is quite a different thing from producing one, or even claiming that one can be produced.

  ($\because$) We are merely asserting that among all possible functions from $\mathscr{P}'(A)$ to A, there is one at least which maps every B onto an element $x \in B$.

- The essence of the Axiom of Choice is that it is an existential statement rather than a constructive one.

  Once again, it states that among all possible functions from $\mathscr{P}'(A)$ to A, there is at least one which satisfies the condition of Definition 5.3. This does not seem grossly unreasonable.

  The Axiom of Choice makes no claim that a choice function can be constructed. Hence, it does not assert that the sequence of choices described in the proof of Theorem 5.1 can be effectively carried out. And indeed this is not necessary in order for the proof to work.

- Before the reader decides whether or not the Axiom of Choice seems plausible to him, he should examine some of the equivalent propositions which are developed in the next few sections of this chapter.

  Some of them are very powerful indeed, and a rejection of any one of them would entail a rejection of all of them (including of course, the Axiom of Choice).

  It is important to note that what all of these principles have in common is the fact that they are non-constructive. They assert the existence of mathematical objects which cannot be explicitly produced. It is precisely this respect of the Axiom of Choice and related principles which makes them unacceptable to the intuitionists (See Section 0.5), who claim that mathematical existence and constructibility are one.

- Without reentering into the arguments for and against admitting nonconstructive propositions into mathematics, we can say this: intuitively, the Axiom of Choice cannot be rejected outright, nor can we feel truly certain of its validity.

  A more important consideration, however, is the fact that it has been proven that the Axiom of Choice does not contradict the other axioms of set theory, nor is it a consequence of them.

  Thus it has the same status as another famous axiom in mathematics, namely Euclid's "fifth Postulate." We can have a "standard" set theory in which we postulate the Axiom of Choice, and "nonstandard" set theories in which we postulate alternatives to the Axiom of Choice.

- In conclusion, since the Axiom of Choice is neither a consequence of the other axioms of set theory nor in conflict with them, it is impossible to make a decision *pro* or *con* on purely logical grounds.

  Since the Axiom of Choice involves an area of mathematics (namely, infinite sets) which is outside the realm of our experience, it will never be possible to confirm it or deny it by "observation".

  In the final analysis, the decision must be a purely personal one for each individual mathematician to make. It is a matter of personal taste.

5.2 The axiom of choice

In the preceding section, we have given the background for our next axiom:

Axiom 8 (Axiom of Choice). Every set has a choice function.

Remark 5.6. The empty set $\emptyset$ has a choice function. The function $\emptyset : \emptyset \to \emptyset$ is called the *empty function.*

($\because$) It is true that for all $x \in \emptyset$, there exists a unique element $y \in \emptyset$ such that $(x, y) \in \emptyset$.

$\Rightarrow$ There exists a function from $\mathscr{P}'(\emptyset)(= \emptyset)$ to $\emptyset$, which we denote by $\emptyset : \emptyset \to \emptyset$.

In the literature there are several other ways of stating the Axiom of Choice which are equivalent to our Axiom 8. We shall present two of these statements here and several more in the exercises following this section.

Ch 1. Let $\mathscr{A}$ be a set whose elements are mutually disjoint, nonempty sets. There exists a set C which consists of exactly one element from each $A \in \mathscr{A}$.

Definition 5.7. For a family $\mathscr{A}$ of mutually disjoint, nonempty sets, the set C described in Ch 1 is called a *choice set* for $\mathscr{A}$.

Thus Ch 1 asserts that every set of mutually disjoint, nonempty sets has a choice set.

Ch 2. Let $\{A_i\}_{i \in I}$ be a set of sets. If I is nonempty and each A_i is nonempty, then $\prod_{i \in I} A_i$ is nonempty.

Theorem 5.8. Axiom 8, Ch 1 *and* Ch 2 *are equivalent.*

Proof. For a set A, recall the notation $\mathscr{P}'(A) = \mathscr{P}(A) \setminus \{\emptyset\}$ and note that $\mathscr{P}(A)$ is a set by Axiom 6 of Power Set and that $\mathscr{P}'(A)$ is a set by Axiom 4 of Subset.

(Axiom 8 $\Rightarrow$ Ch 1) Assume Axiom 8 and let $\mathscr{A}$ be a set whose elements are mutually disjoint, nonempty sets, and let $A = \bigcup_{X \in \mathscr{A}} X$.

$\Rightarrow$ A is a set by Axiom 5 of Union and $\mathscr{A} \subseteq \mathscr{P}'(A)$.

$\Rightarrow$ By Axiom 8, there exists a choice function $r : \mathscr{P}'(A) \to A$.

$\Rightarrow$ The choice function $r : \mathscr{P}'(A) \to A$ induces a function $\overline{r} : \mathscr{P}\big(\mathscr{P}'(A)\big) \to \mathscr{P}(A)$ defined by $\overline{r}(\mathscr{B}) = \{r(B) \mid B \in \mathscr{B}\} \in \mathscr{P}(A)$ for all $\mathscr{B} \in \mathscr{P}\big(\mathscr{P}'(A)\big)$.

$\Rightarrow$ Since $\mathscr{A} \in \mathscr{P}\big(\mathscr{P}'(A)\big)$, $r(\mathscr{A}) \in \mathscr{P}(A)$ and hence $r(\mathscr{A}) \subset A$.

$\Rightarrow$ $r(\mathscr{A})$ is a set by Axiom 4 of Subset and since $\mathscr{A}$ is a set of mutually disjoint, nonempty sets and $r(B) \in B$ for each $B \in \mathscr{A}$, $r(\mathscr{A})$ is a choice set C for $\mathscr{A}$.

(Ch 1 $\Rightarrow$ Axiom 8) Assume Ch 1 and let A be a nonempty set.

Let $Q_B = \{(B, x) \mid x \in B\}$ for each $B \in \mathscr{P}'(A)$.

$\Rightarrow$ (i) For $B, D \in \mathscr{P}'(A)$, if $B \neq D$ then $Q_B \cap Q_D = \emptyset$.

($\because$) Suppose that $Q_B \cap Q_D \neq \emptyset$.

$\Rightarrow$ There exists $(B, x) = (D, y) \in Q_B \cap Q_D$.

$\Rightarrow$ $B = D$ by Theorem 1.25(3).

(ii) $\{Q_B \mid B \in \mathscr{P}'(A)\}$ is a set of sets.

($\because$) Since A is a set, $\mathscr{P}(A)$ is a set by Axiom 6 of Power Set.

$\Rightarrow$ $\mathscr{P}(A) \times A$ is a set by Theorem 1.53.

$\Rightarrow$ $\mathscr{P}(\mathscr{P}(A) \times A)$ is a set by Axiom 6 of Power Set.

Now since (B, x) for $x \in B \in \mathscr{P}'(A)$ is a set by Axiom 3, $Q_B = \bigcup_{x \in B} \{(B, x)\}$ for $B \in \mathscr{P}'(A)$ is a set by Axiom 5.

$\Rightarrow$ Since $\{Q_B \mid B \in \mathscr{P}'(A)\} \subseteq \mathscr{P}(\mathscr{P}(A) \times A)$, $\{Q_B \mid B \in \mathscr{P}'(A)\}$ is a set of sets by Axiom 4 of Subset.

$\Rightarrow$ The family $\{Q_B\}_{B \in \mathscr{P}'(A)}$ is a set of mutually disjoint, nonempty sets.

$\Rightarrow$ By Ch 1, there exists a set C which consists of exactly one element (B, x) from each Q_B. For each $(B, x) \in C$, write $x = r(B)$.

$\Rightarrow$ $r : \mathscr{P}'(A) \to A$ is a choice function for A.

(Axiom 8 $\Rightarrow$ Ch 2) Assume Axiom 8 and let $\{A_i\}_{i\in I}$ be a set of nonempty sets and assume that I is a nonempty set. Let $A = \bigcup_{i\in I} A_i$.

$\Rightarrow$ A is a nonempty set by Axiom 5 of Union and $\{A_i \mid i \in I\} \subseteq \mathscr{P}'(A)$.

$\Rightarrow$ By Axiom 8 of Choice, there exists a function $r : \mathscr{P}'(A) \to A$ such that $r(B) \in B$ for each $B \in \mathscr{P}'(A)$.

$\Rightarrow$ $r(A_i) \in A_i$ for each $i \in I$. Define $\mathbf{a} : I \to A$ by $\mathbf{a}(i) = r(A_i)$.

$\Rightarrow$ $\mathbf{a} : I \to A$ is a function and $\mathbf{a}(i) \in A_i$ for each $i \in I$. i.e., $\mathbf{a} \in \prod_{i\in I} A_i$.

$\Rightarrow$ $\prod_{i\in I} A_i$ is nonempty.

(Ch 2 $\Rightarrow$ Axiom 8) Assume Ch 2 and let A be a nonempty set.

Let $Q_B = \{(B, x) \mid x \in B\}$ for each $B \in \mathscr{P}'(A)$.

$\Rightarrow$ Q_B is a nonempty set for each $B \in \mathscr{P}'(A)$ by the argument in the proof of (Ch 1 $\Rightarrow$ Axiom 8) and $\mathscr{P}'(A)$ is a set.

$\Rightarrow$ $\{Q_B \mid B \in \mathscr{P}'(A)\}$ is a set of nonempty sets indexed by the set $\mathscr{P}'(A)$.

$\Rightarrow$ $\prod_{B\in\mathscr{P}'(A)} Q_B \neq \emptyset$ by Ch 2.

$\Rightarrow$ There exists an element $r \in \prod_{B\in\mathscr{P}'(A)} Q_B$.

$\Rightarrow$ There exists a function $r : \mathscr{P}'(A) \to \bigcup_{B\in\mathscr{P}'(A)} Q_B$ such that $r(B) \in Q_B$ for each $B \in \mathscr{P}'(A)$.

$\Rightarrow$ For each $B \in \mathscr{P}'(A)$, $r(B) = (B, x) \in Q_B$ with $x \in B$.

Define $c(B) = x$, where $r(B) = (B, x)$.

$\Rightarrow$ $c : \mathscr{P}'(A) \to A$ is a choice function for A. □

In Chapter 2 we promised to give a characterization of surjective functions whose proof depends on the Axiom of Choice.

Theorem 5.9. *Let A and B be sets and let $f : A \to B$ be a function. Then $f : A \to B$ is surjective if and only if there exists a function $g : B \to A$ such that $f \circ g = 1_B$.*

Proof. **(if)** There exists a function $g : B \to A$ such that $f \circ g = 1_B$.

$\Rightarrow f : A \to B$ is surjective.

(∵) Let $y \in B$ and let $x = g(y) \in A$.

$\Rightarrow y = 1_B(y) = (f \circ g)(y) = f(g(y)) = f(x)$. i.e., $(x, y) \in f$.

$\Rightarrow y \in \text{ran}(f)$ and hence $B = \text{ran}(f)$.

(only if) $f : A \to B$ is a surjective function.

$\Rightarrow$ (i) There exists a choice function r for A by Axiom 8 of Choice.

(ii) For each $y \in B$, $f^{-1}(y)$ is a nonempty subset of A and hence $r(f^{-1}(y)) \in f^{-1}(y)$.

Define $g : B \to A$ by $g(y) = r(f^{-1}(y)) \in f^{-1}(y) \subseteq A$ for all $y \in B$.

$\Rightarrow f \circ g : B \to B$ is a function.

$\Rightarrow$ Since $(f \circ g)(y) = f(g(y)) = f(r(f^{-1}(y))) = y = 1_B(y)$ for all $y \in B$, $f \circ g = 1_B$ by Theorem 2.10. □

Note. For the sake of simplicity we have proven Theorem 5.9 in the case where A is a set. Using a slightly stronger form of the Axiom of choice we can prove Theorem 5.9 in the more general case where A is any class. We omit the details.

Exercise 5.2

1. Let A be a set and let $f : A \to B$ be a surjective function. Prove that there exists a subset $C \subseteq A$ such that C is in one-to-one correspondence with B. [Hint: Use Theorem 5.9.]

2. Let A be a set and let $f : B \to C$ and $g : A \to C$ be functions. Suppose that $\text{ran}(f) \subseteq \text{ran}(g)$. Prove that there exists a function $h : B \to A$ such that $g \circ h = f$. [Hint: Use Axiom 8 of Choice.]

3. Let $\{A_i\}_{i \in I}$ be an indexed family of classes, where I is a set. Prove that there exists $J \subseteq I$ such that

   (i) $\{A_i \mid i \in I\} = \{A_j \mid j \in J\}$ and

   (ii) in $\{A_j\}_{j \in J}$, each A_j is indexed only once.

i.e., $A_i = A_j (i, j \in J)$ implies $i = j$.

[Hint: Use Theorem 2.37(3) and Axiom 8 of Choice.]

4. Prove that the statement of Theorem 5.9 implies Axiom 8 of Choice.

In each of the following problems a proposition is stated. Prove that this proposition is equivalent to Axiom 8 of Choice.

5. Let $\mathscr{A}$ be a set of disjoint, nonempty set. There exists a function f, whose domain is $\mathscr{A}$, such that for all $A \in \mathscr{A}, f(A) \in A$.

6. Let E be a set and suppose that $G \subseteq E \times E$. Let $A = \text{dom}(G)$ and $B = \text{ran}(G)$. Then there exists a function $f : A \to B$ such that $f \subseteq G$.

7. Let $\mathscr{A}$ be a set whose elements are nonempty sets, and let $A = \bigcup_{X \in \mathscr{A}} X$. Then, corresponding to every function $g : \mathscr{A} \to \mathscr{A}$, there exists a function $g^* : \mathscr{A} \to A$ such that $g^*(B) \in g(B)$.

8. Let B be a set and let $f : A \to B$ be a function. Then there exist subsets $C \subseteq A$ and $g \subseteq f$ such that $g : C \to B$ is an injective function and $\text{ran}(g) = \text{ran}(f)$.

5.3 An application of the axiom of choice

The purpose of this section is to develop a consequence of the Axiom of Choice. The result we are about to prove is valuable as a stepping stone which will enable us to prove the important maximal principles that follow in the next section.

Let A be a partially ordered set such that every chain of A has a sup in A. Assume that A has a least element p. We intend to show that there exists an element $a \in A$ such that a has no immediate successor. See Theorem 5.17.

In order to show this, we will suppose that every element $a \in A$ has an immediate successor; this assumption will lead to a contradiction.

If every element of A has an immediate successor, then we can define a function $f : A \to A$ such that $f(x)$ is an immediate successor of x for each $x \in A$.

($\because$) For each $x \in A$, let $T_x = \{y \in A \mid y \text{ is an immediate successor of } x\}$.

$\Rightarrow$ (i) Since A is a set, there exists a choice function $g : \mathscr{P}'(A) \to A$ by Axiom 8 of Choice.

(ii) Since x has an immediate successor, $T_x \neq \emptyset$ and hence $T_x \in \mathscr{P}'(A)$ and $g(T_x) \in T_x$ for all $x \in A$.

Define $f : A \to A$ by $f(x) = g(T_x) \in T_x \subseteq A$ for all $x \in A$.

$\Rightarrow$ $f : A \to A$ is a required function.

Definition 5.10. Let A be a partially ordered set. Assume that A has a least element p and that every chain of A has a sup in A. Furthermore assume that every element of A has an immediate successor. Let $f : A \to B$ be a function such that $f(x)$ is an immediate successor of x for each $x \in A$.

(1) A subset B of A is called a *p-sequence* if the following conditions are satisfied.

(α) $p \in B$.

(β) If $x \in B$, then $f(x) \in B$.

(γ) If C is a chain of B, then $\sup C \in B$.

(2) There are p-sequences. For an example, A itself is a p-sequence.

Let P denote the intersection of all the p-sequences. Then P is a p-sequence by the following Lemma 5.11 and that $P \neq \emptyset$ because $p \in P$.

An element $x \in P$ is said to be *select* if it is comparable with every element of P.

In the following Lemmas 5.11, 5.12, 5.13, 5.15, and Corollaries 5.14, 5.16, we assume that (i) A is a partially ordered set with a least element p, (ii) every chain of A has a sup in A, (iii) every element of A has an immediate successor. Let $f : A \to A$ be a function such that $f(x)$ is an immediate successor of x for each $x \in A$.

Lemma 5.11. *Any intersection of p-sequences is a p-sequence.*

Proof. The proof is left as an exercise for the reader. □

Lemma 5.12. *Assume that x is select, $y \in P$, and $y < x$. Then $f(y) \leq x$.*

Proof. x is select, $y \in P$, and $y < x$.

$\Rightarrow$ Since $y \in P$ and P is a p-sequence, $f(y) \in P$ by the condition (β).

$\Rightarrow$ Since x is select, either $f(y) \leq x$ or $x < f(y)$.

$\Rightarrow$ $f(y) \leq x$.

($\because$) Suppose that $x < f(y)$.

$\Rightarrow$ Since $y < x$ by hypothesis, $y < x < f(y)$.

This contradicts the assertion that $f(y)$ is an immediate successor of y and hence $f(y) \leq x$. □

Lemma 5.13. *Assume that $x \in P$ is select and let*

$$B_x = \{y \in P \mid y \leq x \text{ or } y \geq f(x)\}.$$

Then B_x is a p-sequence.

Proof. We will show that B_x satisfies the three conditions which define a p-sequence.

(α) Since p is the least element of A, $p \leq x$ and hence $p \in B_x$.

(β) $y \in B_x$.

$\Rightarrow$ $y \leq x$ or $y \geq f(x)$.

$\Rightarrow$ If $y < x$, then $f(y) \leq x$ by Lemma 5.12 and hence $f(y) \in B_x$.

If $y = x$, then $f(y) = f(x)$ and hence $f(y) \in B_x$.

If $y \geq f(x)$, since $f(y) > y$, $f(y) > f(x)$ and hence $f(y) \in B_x$.

$\Rightarrow$ $f(y) \in B_x$.

(γ) C is a chain of B_x and $m = \sup C \in A$.

$\Rightarrow$ For each $y \in B_x$, $y \leq x$ or $y \geq f(x)$.

$\Rightarrow$ If $y \geq f(x)$ for some $y \in C$, then $m \in B_x$ because $m \geq y \geq f(x)$.

If $y < f(x)$ for all $y \in C$, then $m \in B_x$.

($\because$) $y < f(x)$ for all $y \in C$.

$\Rightarrow$ Since $f(x)$ is an immediate successor of x, $y \leq x$ for all $y \in C$.

$\Rightarrow$ x is an upper bound of C.

$\Rightarrow$ $m \leq x$ and hence $m \in B_x$.

$\Rightarrow$ $\sup C = m \in B_x$. □

Corollary 5.14. *If x is select, then $y \leq x$ or $y \geq f(x)$ for all $y \in P$.*

Proof. x is select.

$\Rightarrow$ Since B_x is a p-sequence by Lemma 5.13 and since P is the intersection of all the p-sequence, $P \subseteq B_x$.

$\Rightarrow$ Since $B_x \subseteq P$ by definition, $P = B_x$.

$\Rightarrow$ $y \leq x$ or $y \geq f(x)$ for all $y \in P$ by the definition of B_x in Lemma 5.13. □

Lemma 5.15. *The set of all select elements is a p-sequence.*

Proof. Let $S = \{x \in P \mid x \text{ is select}\}$.

(α) Since $p \leq y$ for each $y \in P$, p is select. i.e., $p \in S$.

(β) $x \in S$.

$\Rightarrow$ By Corollary 5.14, either $y \leq x$ or $y \geq f(x)$ for all $y \in P$.

$\Rightarrow$ Since $x < f(x)$, either $y < f(x)$ or $y \geq f(x)$ for all $y \in P$.

$\Rightarrow$ $f(x)$ is select. i.e., $f(x) \in S$.

(γ) C is a chain of select elements, $m = \sup C$ and $y \in P$.

$\Rightarrow$ (i) Since P is a p-sequence, $m \in P$.

(ii) If $y \leq x$ for some $x \in C$, then $y \leq x \leq m$ and hence $y \leq m$.

(iii) If $x \leq y$ for all $x \in C$, then $y \in \upsilon(C)$ and hence $m \leq y$.

$\Rightarrow$ m is select and hence $m \in S$. □

Corollary 5.16. *P is fully ordered.*

Proof. Let S denote the set of all the select elements.

$\Rightarrow$ S is a p-sequence by Lemma 5.15.

$\Rightarrow$ Since P is the intersection of all the p-sequences, $P \subseteq S$.

$\Rightarrow$ Since $S \subseteq P$ by Definition 5.10(2), $P = S$.

$\Rightarrow$ Each element of P is select.

$\Rightarrow$ Any two elements of P are comparable.

$\Rightarrow$ P is fully ordered. □

The main purpose of this section is to prove the following theorem.

Theorem 5.17. *Let A be a partially ordered set. Assume that A has a least element p and that every chain of A has a sup in A. Then there is an element $x \in A$ which has no immediate successor.*

Proof. Suppose that every element of A has an immediate successor.

$\Rightarrow$ There exists a function $f : A \to A$ such that $f(x)$ is an immediate successor of x for each $x \in A$.

Let P denote the intersection of all the p-sequences and let $m = \sup P$.

$\Rightarrow$ P is a p-sequence by Lemma 5.11.

$\Rightarrow$ Since P is a chain of P by Corollary 5.16, $m \in P$ by the condition (γ).

$\Rightarrow$ By the condition (β), $f(m) \in P$ and hence $f(m) \leq m$.

This contradicts the assertion that $f(m)$ is an immediate successor of m. □

Exercise 5.3

1. Show that any intersection of p-sequences is a p-sequence.

2. Determine whether every element of each of the following has an immediate successor or not.

   (1) $\mathbb{N}$ with the usual partial order.

   (2) $\mathbb{Z}$ with the usual partial order.

   (3) $\mathbb{Q}$ with the usual partial order.

3. Find a $(1, 1)$-sequence for the partially ordered set $\mathbb{N} \times \mathbb{N}$ with the lexicographic order, where $(1, 1)$ is the least element of $\mathbb{N} \times \mathbb{N}$.

5.4 Maximal principles

The propositions we are about to develop are widely used in mathematics to prove theorems by "nonconstructive" methods. They assert the existence of mathematical objects which cannot be constructed. For example, in linear algebra, a maximal principle can be used to prove that every vector space has a basis, although, in general, it is impossible to exhibit such a basis.

The maximal principles are consequences of Axiom 8 of Choice. Furthermore, as we shall verify in Section 6, they are equivalent to Axiom 8 of Choice.

Theorem 5.18 (Hausdorff's Maximal Principle). *Every partially ordered set has a maximal chain.*

Proof. Let A be a partially ordered set, and let $\mathscr{S}$ be the set of all the chains of A, ordered by inclusion.

$\Rightarrow$ $\mathscr{S}$ is a partially ordered set and has a least element, namely the empty set. Now let $\mathscr{C}$ be a chain of $\mathscr{S}$ and let $K = \bigcup_{C \in \mathscr{C}} C$

$\Rightarrow$ By Example 4.20(4), $K = \sup \mathscr{C}$.

$\Rightarrow$ $K \in \mathscr{S}$.

($\because$) Let $x, y \in K$.

$\Rightarrow$ $x \in D$ and $y \in E$ for some elements $D, E \in \mathscr{C}$.

$\Rightarrow$ Since $\mathscr{C}$ is a chain of $\mathscr{S}$, $E \subseteq D$ or $D \subseteq E$. Say $E \subseteq D$.

$\Rightarrow x, y \in D$.

$\Rightarrow$ Since D is a chain of A, x and y are comparable.

$\Rightarrow K$ is a chain of A and hence $K \in \mathscr{S}$.

$\Rightarrow \mathscr{S}$ satisfies the conditions of Theorem 5.17.

$\Rightarrow$ There exists $C \in \mathscr{S}$ which has no immediate successor by Theorem 5.17.

$\Rightarrow$ There exists no $x \in A \setminus C$ such that $C \cup \{x\}$ is a chain of A.

$\Rightarrow C$ is a maximal chain. □

Definition 5.19. A partially ordered set A is said to be *inductive* if every chain of A has an upper bound in A

Theorem 5.20 (Zorn's Lemma)**.** *Every inductive set has at least one maximal element.*

Proof. A is an inductive set.

$\Rightarrow$ Since A is a partially ordered set, A has a maximal chain C by Hausdorff's maximal principle(Theorem 5.18).

$\Rightarrow$ Since A is inductive, C has an upper bound $m \in A$.

$\Rightarrow m$ is a maximal element of A.

($\because$) Suppose that there exists an element $x \in A$ such that $x > m$.

$\Rightarrow x \notin C$ and $x \geq y$ for all $y \in C$.

$\Rightarrow x$ is comparable with every element of C.

$\Rightarrow C \cup \{x\}$ is a chain which contradicts to the assertion that C is a maximal chain.

$\Rightarrow$ There exists no element $x \in A$ such that $x > m$.

$\Rightarrow m$ is a maximal element of A. □

Exercise 5.4

1. Let A be a partially ordered set and let $\mathscr{A}$ be the set of all the well-ordered subsets of A. For $C \in \mathscr{A}$ and $D \in \mathscr{A}$, define $C \lessdot D$ if and only if C is a section of D.

   (1) Prove that $\lessdot$ is a partial order relation in $\mathscr{A}$.

   (2) Prove that $\mathscr{A}$, ordered by $\lessdot$, is inductive.

2. Derive Hausdorff's Maximal Principle from Zorn's Lemma.

3. Prove that Zorn's Lemma is equivalent to the following:

   Let A an inductive set and let $a \in A$, then A has at least one maximal element b such that $b \geq a$.

4. Prove that Hausdorff's Maximal Principle is equivalent to the following:

   If A is a partially ordered set and B is a chain of A, then A has a maximal chain C such that $B \subseteq C$.

5. Let A be any set with more than one element. Prove that there exists a bijective function $f : A \to A$ such that $f(x) \neq x$ for all $x \in A$.

6. A set $\mathscr{A}$ of sets is said to be *disjointed* if for all $C, D \in \mathscr{A}$, $C \cap D = \emptyset$. Let $\mathscr{F}$ be a set of sets. Prove that $\mathscr{F}$ has a maximal disjointed subset.

7. Let A be a set and let $\mathscr{A}$ be a set of subsets of A. $\mathscr{A}$ is said to be of *finite character* if $\mathscr{A}$ satisfies the following: $B \in \mathscr{A}$ if and only if every finite subset of B belongs to $\mathscr{A}$.

   Let $\mathscr{A}$ be ordered by inclusion and suppose that $\mathscr{A}$ is of finite character.

   (1) Prove that $\mathscr{A}$ is an inductive set.

   (2) Prove that $\mathscr{A}$ has a maximal element.

8. Prove that every vector space V has a basis.

   [Hint: Consider the set $\mathscr{A}$ of all the linearly independent subsets of V. Use Zorn's Lemma; it is easily verified that any maximal linearly independent subset of V is a basis of V.]

9. Let G be a group and let A be an arbitrary subset of G such that A includes the identity element of G. Prove that among the subgroups of G which are subsets of A, there is a maximal one.

10. Show that it is not true in general that every partially ordered set has a maximal well-ordered subset. [Hint: Consider the partially ordered set $\mathbb{Z}$ of integers. This problem asserts that Theorem 5.17 of the text [6] or [7] is not true.]

5.5 The well-ordering theorem

The well-ordering theorem, which will be presented in this section, is one of the most important consequences of Axiom of Choice and is an outstanding example of a nonconstructive proposition.

It asserts that any set can be well-ordered. That is, if A is any set, there exists an order relation G such that A, ordered by G, is a well-ordered set. The proof of the well-ordering theorem gives no indication how such a well-ordering of the elements of A is to be accomplished. It asserts merely that a well-ordering exists.

A finite set A can obviously be well-ordered. for example, if $A = \{a, b, c\}$, then $a < b < c$, $b < a < c$, $c < b < a$ are different well-orderings of A. However, no method has yet been discovered to well-order sets such as the set $\mathbb{R}$ of the real numbers. In fact, in the opinion of most mathematicians, it is impossible to construct a well-ordering of $\mathbb{R}$. (This would mean giving a rule for rearranging the elements of $\mathbb{R}$ in such a way that $\mathbb{R}$ would then become a well-ordered set.)

Once again, the well-ordering theorem does not assert that every set can be *effectively* well-ordered. It merely states that, among all possible graphs $G \subseteq A \times A$, there exists one at least which is an order relation which well-orders A.

Let us now prove the well-ordering theorem. Note that the proof relies heavily on Axiom of Choice.

Definition 5.21. Let A be an arbitrary set and let $\mathscr{A} = \{(B, G) \mid B \subseteq A$ and G is an order relation in B which well-orders $B\}$. We introduce the symbol $\prec$ and define $(B, G) \prec (B', G')$ if and only if

(1) $B \subseteq B'$, (2) $G \subseteq G'$, and

(3) $x \in B$ and $y \in B' \setminus B$ imply $(x, y) \in G'$.

Note that the last condition (3) asserts, roughly, that all the elements of B precede all the elements of $B' \setminus B$. It is easy to verify that $\prec$ is an order relation in $\mathscr{A}$. The details are left for the readers.

Lemma 5.22. *Let A be an arbitrary set and let* $\mathscr{A} = \{(B, G) \mid B \subseteq A,\ G \text{ is an order relation in } B \text{ which well-orders } B\}$. *Let $\mathscr{C} = \{(B_i, G_i)\}_{i \in I}$ be a chain of $\mathscr{A}$ and let $B = \bigcup_{i \in I} B_i$ and $G = \bigcup_{i \in I} G_i$. Then $(B, G) \in \mathscr{A}$.*

Proof. $\mathscr{C} = \{(B_i, G_i)\}_{i \in I}$ is a chain of $\mathscr{A}$, $B = \bigcup_{i \in I} B_i$ and $G = \bigcup_{i \in I} G_i$.

$\Rightarrow$ $B \subseteq A$ by Theorem 1.41(1). We will show that G well-orders B.

First we verify that G is an order relation in B.

(i) G is reflexive.

($\because$) $x \in B$.

$\Rightarrow$ $x \in B_i$ for some $i \in I$.

$\Rightarrow$ $(x, x) \in G_i \subseteq G$ and hence G is reflexive.

(ii) G is antisymmetric.

($\because$) $(x, y) \in G$ and $(y, x) \in G$.

$\Rightarrow$ $(x, y) \in G_i$ and $(y, x) \in G_j$ for some $i, j \in I$.

$\Rightarrow$ Since $\mathscr{C}$ is a chain of $\mathscr{A}$, $G_i \subseteq G_j$ or $G_j \subseteq G_i$, say $G_i \subseteq G_j$.

$\Rightarrow$ $(x, y) \in G_j$ and $(y, x) \in G_j$.

$\Rightarrow$ Since G_j is an order relation, $x = y$ and hence G is antisymmetric.

(iii) G is transitive.

($\because$) $(x, y) \in G$ and $(y, z) \in G$.

$\Rightarrow$ $(x, y) \in G_i$ and $(y, z) \in G_j$ for some $i, j \in I$.

$\Rightarrow$ Since $\mathscr{C}$ is a chain, $G_i \subseteq G_j$ or $G_j \subseteq G_i$, Say $G_i \subseteq G_j$.

$\Rightarrow (x, y) \in G_j$ and $(y, z) \in G_j$.

$\Rightarrow (x, z) \in G_j \subseteq G$ and hence G is transitive.

Thus G is an order relation in B by (i), (ii) and (iii) above.

(iv) B is well-ordered by G.

($\because$) D is a nonempty subset of B.

$\Rightarrow D \cap B_{i_0} \neq \emptyset$ for some $i_0 \in I$ and $D \cap B_{i_0} \subseteq B_{i_0}$.

$\Rightarrow$ Since G_{i_0} well-orders B_{i_0}, $D \cap B_{i_0}$ has a least element $b \in B_{i_0}$ in (B_{i_0}, G_{i_0}). i.e., $(b, y) \in G_{i_0}$ for all $y \in D \cap B_{i_0}$.

$\Rightarrow (b, x) \in G$ for all $x \in D$. i.e., b is the least element of D in (B, G).

($\because$) $x \in D \subseteq B = \bigcup_{i \in I} B_i$.

$\Rightarrow$ Either $x \in B_{i_0}$ or $x \in B_j \setminus B_{i_0}$ for some $j \in I$ with $j \neq i_0$ and hence $(b, x) \in G_{i_0} \subseteq G$ if $x \in D \cap B_{i_0}$.

Now assume that $x \in D \cap \left(B_j \setminus B_{i_0}\right)$ for some $j \in I (j \neq i_0)$.

$\Rightarrow$ Since $x \in B_j \setminus B_{i_0}$, $B_j \nsubseteq B_{i_0}$ and $(B_{i_0}, G_{i_0}), (B_j, G_j) \in \mathscr{C}$.

$\Rightarrow (B_j, G_j) \nprec (B_{i_0}, G_{i_0})$ and hence $(B_{i_0}, G_{i_0}) \prec (B_j, G_j)$ because $\mathscr{C}$ is a chain.

$\Rightarrow$ Since $b \in B_{i_0}$ and $x \in B_j \setminus B_{i_0}$, $(b, x) \in G_j \subseteq G$ by Definition 5.21(3).

$\Rightarrow B$ is well-ordered by G.

Therefore $(B, G) \in \mathscr{A}$ by (i) - (iv). □

Lemma 5.23. *If $\mathscr{C}$, B and G are defined as above, then (B, G) is an upper bound of $\mathscr{C}$.*

Proof. Let $(B_i, G_i) \in \mathscr{C}$. Then clearly $B_i \subseteq B$ and $G_i \subseteq G$.

To show that $(B_i, G_i) \prec (B, G)$, let $x \in B_i$ and $y \in B \setminus B_i$.

$\Rightarrow y \in B_j$ for some $j \in I$.

$\Rightarrow$ Since $y \in B_j$ and $y \notin B_i$, $B_j \nsubseteq B_i$.

$\Rightarrow (B_j, G_j) \nprec (B_i, G_i)$,

$\Rightarrow$ Since $\mathscr{C}$ is a chain, $(B_i, G_i) \prec (B_j, G_j)$.

$\Rightarrow$ Since $x \in B_i$ and $y \in B_j \setminus B_i$, $(x, y) \in G_j \subseteq G$

by Definition 5.21(3).

$\Rightarrow (B_i, G_i) \prec (B, G)$ by Definition 5.21.

Thus (B, G) is an upper bound of $\mathscr{C}$. □

Theorem 5.24 (Well-ordering Theorem). *Any set can be well-ordered.*

Proof. Let A be an arbitrary set and let $\mathscr{A} = \{(B, G) \mid B \subseteq A$ and G is an order relation in B which well-orders $B\}$.

$\Rightarrow$ Every chain of $\mathscr{A}$ has an upper bound in $\mathscr{A}$ by lemmas 5.22 and 5.23.

$\Rightarrow$ By Zorn's Lemma(Theorem 5.20), $\mathscr{A}$ has a maximal element (B, G).

$\Rightarrow A = B$.

($\because$) Suppose that $A \neq B$.

$\Rightarrow$ There exists $x \in A \setminus B$. Define $x \geq b$ for all $b \in B$ and let $G^* = G \cup \{(b, x) \mid b \in B\}$ and $B^* = B \cup \{x\}$.

$\Rightarrow G^*$ well-orders B^*, $G \subseteq G^*$ and $B \subseteq B^*$.

$\Rightarrow (B^*, G^*) \in \mathscr{A}$, $(B, G) \prec (B^*, G^*)$ and $(B, G) \neq (B^*, G^*)$.

This contradicts to the maximality of (B, G) and hence $A = B$.

$\Rightarrow$ Since B is well ordered, A is well-ordered. □

5.6 Conclusion

It is clear that the Axiom of Choice can be derived from the well-ordering theorem.

Lemma 5.25. *Well-ordering Theorem implies Axiom of Choice.*

Proof. Assume the well-ordering Theorem(Theorem 5.24) and let A be any set.

$\Rightarrow$ By Theorem 5.24, A is well-ordered by some order $\leq$.

$\Rightarrow$ For every nonempty subset B of A, B has the least element and define $r(B) =$ the least element of B.

$\Rightarrow r : \mathscr{P}'(A) \to A$ is a function and $r(B) \in B$.

$\Rightarrow r : \mathscr{P}'(A) \to A$ is a choice function for A.

$\Rightarrow A$ has a choice function $r : \mathscr{P}'(A) \to A$. □

Thus we have proved the following

Theorem 5.26. *Under the assumption of Axioms 1 through 7, the following are equivalent.*

(1) *The axiom of choice(Axiom 8).*

(2) *The Hausdorff's maximal principle(Theorem 5.18).*

(3) *The Zorn's lemma(Theorem 5.20).*

(4) *The well-ordering theorem(Theorem 5.24).*

Proof. (1) ⇒ (2) by Theorem 5.18.

(2) ⇒ (3) by Theorem 5.20.

(3) ⇒ (4) by Theorem 5.24.

(4) ⇒ (1) by Lemma 5.25. □

Thus we have established the complete equivalence of the four propositions listed in Theorem 5.26. In the remainder of this book we will accept the Axiom of Choice as one of the axioms of set theory. Thus we will feel free to use the Axiom of Choice in all of our arguments, except if we make an explicit statement to the contrary.

Chapter 6

The Natural Numbers

6.1 Introduction

- Probably the most fundamental - and the most primitive - of all mathematical concepts is that of natural numbers. The natural, or "counting", numbers are the first mathematical abstraction which we learn as children. Every human society, even the most backward and remote, possesses a system of some kind for counting objects.

- We all have a clear intuitive understanding of what the natural numbers are. They are $0, 1, 2, 3$, and so forth. But this intuitive perception, no matter how clear and immediate it may be, is not sufficient for the purposes of mathematics. In order to do mathematics with numbers, we must articulate this vague perception and transform it into a precise definition. The definition must, of course, faithfully reflect the intuitive notion from which it sprang.

- It is our aim in this section to define the natural numbers. To be explicit, we will construct a set of objects to be called "natural numbers", and these "natural numbers" will be endowed with all of the properties which are associated with the natural numbers in our mind.

- We should carefully note two important requirements of our definition. In the first place, we would like to present the natural numbers without introducing any new undefined notions. We were faced with a similar problem when we were about to introduce the notion of function. We solved it by defining a function to be a certain kind of class (specifically, a class of ordered pairs). Following this example, it is clear that we ought to define natural numbers to be classes (more specifically, *sets*) of some kind. In fact, for each n we will define "n" to be a set which (intuitively) has n elements.

  Secondly, each natural number must be uniquely defined; that is, for each n there must be just one, *unique* object which can be recognized as the "natural number n". Thus we must devise a means of fixing exactly one set, among all the sets with n elements, and calling this set "n".

- The numbers "0", "1", "2", etc., which we are about to define will serve as standards in much the same way that the standard yard in Washington, D.C., serves as a norm for measuring length. It matters little whether the standard yard is made of platinum or stainless steel, or whether it is decorated with figures of dancing mermaids. Its only use is as a standard of reference, so anything having the same length is, by definition, one yard long. Analogously, it does not matter too much which specific sets we define "0", "1", "2", etc., to be. They will be used as standards of reference, so a set will be said to "have n elements" if it is one-to-one correspondence with the natural number "n".

We will proceed as follows. First we define

$$0 = \emptyset.$$

In order to define "1", we must fix a set with exactly one element. Thus

$$1 = \{0\} = \{\emptyset\}.$$

Continuing in this fashion, we define

$$2 = \{0, 1\} = \{\emptyset, \{\emptyset\}\},$$
$$3 = \{0, 1, 2\} = \{\emptyset, \{\emptyset\}, \{\emptyset, \{\emptyset\}\}\},$$
$$4 = \{0, 1, 2, 3\} = \{\emptyset, \{\emptyset\}, \{\emptyset, \{\emptyset\}\}, \{\emptyset, \{\emptyset\}, \{\emptyset, \{\emptyset\}\}\}\},$$
$$\vdots$$

Our natural numbers are constructions beginning with the empty set. The preceding definitions can be restated, a little more precisely, as follows.

Definition 6.1. Let A be a set. We define the *successor* of A to be the set A^+, given by $A^+ = A \cup \{A\}$.

Thus, A^+ is obtained by adjoining to A exactly one new element, namely the element A.

Definition 6.2 (Definition of Natural Numbers)**.** Now we define

$$0 = \emptyset,$$
$$1 = 0^+ \ (= 0 \cup \{0\} = \{0\} = \{\emptyset\}),$$
$$2 = 1^+ \ (= 1 \cup \{1\} = \{0, 1\} = \{\emptyset, \{\emptyset\}\}),$$
$$3 = 2^+ \ (= 2 \cup \{2\} = \{0, 1, 2\} = \{\emptyset, \{\emptyset\}, \{\emptyset, \{\emptyset\}\}\}),$$
$$4 = 3^+ \ (= 3 \cup \{3\} = \{0, 1, 2, 3\} = \{\emptyset, \{\emptyset\}, \{\emptyset, \{\emptyset\}\}, \{\emptyset, \{\emptyset\}, \{\emptyset, \{\emptyset\}\}\}\}),$$
$$\vdots$$

We have just outlined a method for producing sets which we call "natural numbers". Beginning with the empty set, we have given directions for constructing, successively, $0 = \emptyset, 1 = 0^+, 2 = 1^+$, and so forth. An important question now is the following: Is there such a thing as the *set of all the natural numbers*(or even the *class of all the natural numbers*)? That is, is there a set (or a class) which contains $\emptyset$, and which contains X^+ whenever it contains X? Certainly, our method does not enable us to construct it. We are merely given instructions for producing numbers $0, 1, 2, \ldots, n$ *up to any* n. Thus we cannot yet speak of the set (or class) of all the natural numbers.

Definition 6.3. A set A is called a *successor set* if the following hold for A.

(i) $\emptyset \in A$.

(ii) If $X \in A$, then $X^+ \in A$.

Remark 6.4. If A is a successor set, then A contains all the natural numbers.

Motivated by this observation, we introduce the following important axiom.

Axiom 9 (Axiom of Infinity). There exists a successor set.

Now the empty class $\emptyset$ is a set because $\emptyset$ is an element of a successor set whose existence was guaranteed by Axiom 9 of Infinity.

Remark 6.5. (1) As we have noted, every successor set includes all the natural numbers. Thus it would make sense to define the "set of the natural numbers" to be the smallest successor set.

(2) Now it is easy to verify that any intersection of successor sets is a successor set. In particular, the intersection of all the successor sets is a successor set (It is obviously the smallest successor set).

Thus, we are led naturally to the following definition.

Definition 6.6. By the *set of the natural numbers* we mean the intersection of all the successor sets. The set of the natural numbers is designated by the symbol ω. And every element of ω is called a *natural number.* Note that $0 \in \omega$ and the set $\omega \setminus \{0\}$ is designated by the symbol $\mathbb{N}$.

6.2 Elementary properties of natural numbers

In this section we will show that the natural numbers, as we have just defined them, satisfy the five conditions commonly known as the Peano axioms. This set of conditions, it is well known, is another way of defining and characterizing the natural numbers.

Theorem 6.7. *For each $n \in \omega$, $n^+ \neq 0$.*

Proof. Let $n \in \omega$.

$\Rightarrow$ By definition, $n^+ = n \cup \{n\}$.

$\Rightarrow n \in n^+$, but since $0 = \emptyset$, $n \notin 0$.

$\Rightarrow n^+ \neq 0$. □

Theorem 6.8 (Finite Mathematical Induction). *Let X be a subset of ω. Assume that X has the following properties(called induction hypothesis):*

(1) $0 \in X$.

(2) *If $n \in X$, then $n^+ \in X$.*

Then $X = \omega$.

Proof. Let $X \subseteq \omega$ and assume the conditions (1) and (2).

$\Rightarrow X$ is a successor set.

$\Rightarrow$ Since ω is the intersection of all the successor sets, $\omega \subseteq X$.

$\Rightarrow$ Since $X \subseteq \omega$, $X = \omega$. □

Lemma 6.9. *Let m and n be natural numbers.*

If $m \in n^+$, then $m \in n$ or $m = n$.

Proof. Let m and n be natural numbers and assume that $m \in n^+$.

$\Rightarrow$ Since $n^+ = n \cup \{n\}$, $m \in n$ or $m \in \{n\}$.

$\Rightarrow$ Since $\{n\}$ is a singleton, $m \in n$ or $m = n$. □

Definition 6.10. A set A is called *transitive* if for each $x \in A$, $x \subseteq A$.

For example, the number 3 is a transitive set. Indeed, its elements are $0, 1, 2$, that is, $\emptyset, \{\emptyset\}, \{\emptyset, \{\emptyset\}\}$. It is clear that each of these elements is a subset of 3. The same is true of every natural number, as we shall prove next.

Lemma 6.11. *Every natural number is a transitive set.*

Proof. Let $X = \{n \in \omega \mid n \text{ is a transitive set}\}$. We will show that $X = \omega$ by using the mathematical induction(Theorem 6.8).

(i) $0 \in X$.

(∵) Suppose that 0 is not a transitive set.

⇒ There exists $y \in 0$ such that y is not a subset of 0.

This is absurd, since $0 = \emptyset$.

(ii) $n \in X$ implies $n^+ \in X$.

(∵) Suppose that $n \in X$.

⇒ $n \in \omega$ and n is a transitive set.

To show that n^+ is transitive, let $m \in n^+$.

⇒ $m \in n$ or $m = n$ by Lemma 6.9.

⇒ If $m \in n$, then since n is transitive, $m \subseteq n \subseteq n^+$.

If $m = n$, then since $n \subseteq n^+$, $m \subseteq n^+$.

⇒ $m \subseteq n^+$ and hence $n^+ \in X$.

⇒ $X = \omega$ by mathematical induction(Theorem 6.8).

⇒ Every natural number is a transitive set. □

Theorem 6.12. *Let m and n be natural numbers. If $m^+ = n^+$, then $m = n$.*

Proof. Assume that $m^+ = n^+$.

⇒ Since $m \in m^+$ and $n \in n^+$, $m \in n^+$ and $n \in m^+$.

⇒ ($m \in n$ or $m = n$) and ($n \in m$ or $n = m$) by Lemma 6.9.

⇒ ($m \in n$ and $n \in m$) or $m = n$ by Theorem 1.9.

⇒ Since m and n are transitive sets by Lemma 6.11,

($m \subseteq n$ and $n \subseteq m$) or $m = n$.

⇒ $m = n$ or $m = n$ by Definition 1.14(1).

⇒ $m = n$ by Theorem 1.9(3). □

Definition (Peano Axioms). The Peano axioms for the set ω are the following.

P1 $0 \in \omega$.

P2 If $n \in \omega$, then $n^+ \in \omega$.

P3 For each $n \in \omega$, $n^+ \neq 0$.

P4 If X is a subset of ω such that

(i) $0 \in X$, and (ii) if $n \in X$, then $n^+ \in X$, then $X = \omega$.

P5 If $m, n \in \omega$ and $m^+ = n^+$, then $m = n$.

Theorem 6.13 (Peano Axioms for Natural Numbers). *The set ω of natural numbers satisfies the Peano axioms.*

Proof. P1 and P2 follow immediately from our definition of ω.

P3 is given by Theorem 6.7,

P4 is given by Theorem 6.8 and

P5 is given by Theorem 6.12.

Thus our set ω satisfies the Peano axioms. □

Exercise 6.2

1. Prove that A is a transitive set if and only if the following holds: If $B \in C$ and $C \in A$, then $B \in A$.

2. Prove that if A and B are transitive sets, then $A \cup B$ and $A \cap B$ are transitive sets.

3. Let A and B be sets. Prove that if $A = B$, then $A^+ = B^+$.

4. Use Theorem 6.8 to prove that for every natural number n, $n \notin n$.

5. Let $m, n, p \in \omega$. Prove the following.

   (a) $n \neq n^+$.

   (b) If $m \in n$, then $n \notin m$.

   (c) If $n \in m$ and $m \in p$, then $n \in p$.

   (d) If $m \in n$, then $m^+ \subseteq n$.

6. (a) Prove by mathematical induction that if $A \in n$ and $n \in \omega$, then $A \in \omega$.

   Conclude that ω is a transitive set.

   (b) Prove that if $A^+ \in \omega$, then $A \in \omega$.

7. Prove that no natural number is a successor set.

8. Prove that no natural number is a subset of any of its element.

9. Prove by mathematical induction that if $n \in \omega$, then either $n = 0$ or $n = m^+$ for some $m \in \omega$.

10. Let $n \in \omega$. Prove the following.

    (a) $\cup n^+ = n$. (b) $\cup \omega = \omega$. (See Notation 1.39.)

11. Let A be a nonempty subset of ω. Prove that if $\cup A = A$, then $A = \omega$.

6.3 Finite recursion

(1) Induction is commonly used not only as a method of proof but also as a method of definition. For example, a familiar way of introducing exponents in arithmetic is by means of the "inductive definition"

I. $a^0 = 1$,

II. $a^{n+1} = a^n a$ for all $n \in \omega$.

The pair of Conditions I and II is meant to be interpreted as a rule which specifies the meaning of a^n for each natural number n. Thus,

$a^0 = 1$ by I,

$a^1 = a^{0+1} = a^0 a = 1a = a$ by Conditions I and II,

$a^2 = a^{1+1} = a^1 a = aa$ by Condition II again.

Continuing in this fashion using Condition II repeatedly, the numbers $a^0, a^1, a^2, \ldots, a^n$ are defined in succession up to any chosen n.

(2) Inductive definitions such as the one we have just seen abound in mathematics. The situation, in almost every case, is the following:

We have a set A, a function $f : A \to A$, and a fixed element $c \in A$. Then we define a function $\gamma : \omega \to A$ by means of the two Conditions

I. $\gamma(0) = c$,

II. $\gamma(n^+) = f(\gamma(n))$ for all $n \in \omega$.

The reader should recognize that exactly this situation prevails in our preceding example. In that example, A is the set of the real numbers, c is 1, f is the function defined by $f(x) = xa$, and $\gamma(n)$ is denoted by a^n.

(3) For another example, let $\mathbb{R}$ be the set of the real numbers and let $f : \mathbb{R} \to \mathbb{R}$ be the function defined by $f(x) = x^2$.

We define a function $\gamma : \omega \to \mathbb{R}$ by

I. $\gamma(0) = 2$,

II. $\gamma(n^+) = f(\gamma(n)) = (\gamma(n))^2$.

The reader will recognize this as an inductive definition of the function $\gamma(n) = 2^{2^n}$.

In each of the foregoing examples, it is reasonable to believe that if γ exists, then Conditions I and II determine the values $\gamma(n)$ for every $n \in \omega$. However, Conditions I and II are insufficient in themselves to guarantee that a function such as γ exists. If we are to accept definition by induction as a legitimate way of constructing mathematical objects, then we must first establish the fact that γ, the function which we purport to be defining, actually exists and is uniquely determined. The purpose of the following theorem is to perform this important task.

Theorem 6.14 (Finite Recursion Theorem). *Let A be a set, c a fixed element of A, and $f : A \to A$ a function from A to A. Then there exists a unique function $\gamma : \omega \to A$ such that*

I. $\gamma(0) = c$, II. $\gamma(n^+) = f(\gamma(n))$ *for all* $n \in \omega$.

Proof. Let A be a set, c a fixed element of A, and $f : A \to A$ a function.

(Existence) We will establish the existence of γ.

It should be carefully noted that γ is a set of ordered pairs which is a function from ω to A and satisfies Conditions I and II.

More specifically, γ is a subset of $\omega \times A$ with the following four properties:

(1) For each $n \in \omega$, there exists $x \in A$ such that $(n, x) \in \gamma$.

(2) If $(n, x_1) \in \gamma$ and $(n, x_2) \in \gamma$, then $x_1 = x_2$.

(3) $(0, c) \in \gamma$. (i.e., $\gamma(0) = c$.)

(4) If $(n, x) \in \gamma$, then $(n^+, f(x)) \in \gamma$.

(i.e., if $x = \gamma(n)$, then $\gamma(n^+) = f(x)$.)

We will define a graph γ with these four properties.

Let $\mathscr{A} = \{G \mid G \subseteq \omega \times A \text{ and } G \text{ satisfies the conditions (3) and (4)}\}$.

$\Rightarrow$ (i) $\mathscr{A}$ is nonempty, because $\omega \times A \in \mathscr{A}$.

(ii) Any intersection of elements of $\mathscr{A}$ is an element of $\mathscr{A}$.

Define $\gamma = \bigcap_{G \in \mathscr{A}} G \in \mathscr{A}$.

$\Rightarrow$ • $\gamma \subseteq \omega \times A$ and γ satisfies the conditions (3) and (4) by construction.

• γ satisfies the condition (1).

It will be shown by induction that $\mathrm{dom}(\gamma) = \omega$, which clearly implies the condition (1).

(i) Since $(0, c) \in \gamma$ by the condition (3), $0 \in \mathrm{dom}(\gamma)$.

(ii) Suppose that $n \in \mathrm{dom}(\gamma)$.

$\Rightarrow$ There exists $x \in A$ such that $(n, x) \in \gamma$.

$\Rightarrow$ $(n^+, f(x)) \in \gamma$ by the condition (4), and hence $n^+ \in \mathrm{dom}(\gamma)$.

$\Rightarrow$ $\mathrm{dom}(\gamma) = \omega$ by mathematical induction(Theorem 6.8).

• γ satisfies the condition (2).

Let $X = \{n \in \omega \mid (n, x) \in \gamma \text{ for no more than one } x \in A\}$.

It will be shown by induction(Theorem 6.8) that $X = \omega$.

(i) To show that $0 \in X$, suppose that $0 \notin X$.

$\Rightarrow$ There exist $c, d \in A$ such that $(0, c) \in \gamma, (0, d) \in \gamma$, and $c \neq d$. Let $\gamma^* = \gamma \setminus \{(0, d)\}$.

$\Rightarrow \gamma^*$ satisfies the condition (3) obviously.

Moreover, γ^* satisfies the condition (4).

($\because$) To show that γ^* satisfies (4), let $(n, x) \in \gamma^*$.

$\Rightarrow$ Since $\gamma^* \subseteq \gamma$, $(n, x) \in \gamma$.

$\Rightarrow$ Since γ satisfies the condition (4), $(n^+, f(x)) \in \gamma$.

$\Rightarrow$ Since $n^+ \neq 0$ by Theorem 6.7, $(n^+, f(x)) \neq (0, d)$.

$\Rightarrow (n^+, f(x)) \in \gamma^*$.

$\Rightarrow \gamma^* \in \mathscr{A}$.

$\Rightarrow$ Since $\gamma = \bigcap_{G \in \mathscr{A}} G \in \mathscr{A}$, $\gamma \subseteq \gamma^*$.

This is impossible and hence $0 \in X$.

(ii) Next we will show that $n \in X$ implies $n^+ \in X$.

Assume that $n \in X$ and suppose that $n^+ \notin X$.

$\Rightarrow$ There exists $x \in A$ such that $(n, x) \in \gamma, (n^+, f(x)) \in \gamma$, and $(n^+, u) \in \gamma$, where $u \neq f(x)$. Let $\gamma^0 = \gamma \setminus \{(n^+, u)\}$.

$\Rightarrow \gamma^0$ satisfies the condition (3).

($\because$) Since $n^+ \neq 0$ by Theorem 6.7, $(n^+, u) \neq (0, c)$.

To show that γ^0 satisfies the condition (4), let $(m, v) \in \gamma^0$.

$\Rightarrow (m, v) \in \gamma$, so $(m^+, f(v)) \in \gamma$.

$\Rightarrow$ We will consider two cases according as $m^+ \neq n^+$ or $m^+ = n^+$.

If $m^+ \neq n^+$, then $(m^+, f(v)) \in \gamma^0$

($\because$) $(m^+, f(v)) \neq (n^+, u)$.

If $m^+ = n^+$, then $(m^+, f(v)) \in \gamma^o$.

($\because$) Assume that $m^+ = n^+$.

$\Rightarrow m = n$ by Theorem 6.12.

$\Rightarrow (m, v) = (n, v)$.

$\Rightarrow$ Since $n \in X$ and $(n, x) \in \gamma$, $v = x$.

$\Rightarrow (m^+, f(v)) = (n^+, f(x)) \in \gamma^o$.

$\Rightarrow (m^+, f(v)) \in \gamma^0$.

$\Rightarrow \gamma^0$ satisfies the condition (4) and hence $\gamma^0 \in \mathscr{A}$.

$\Rightarrow$ Since $\gamma = \bigcap_{G \in \mathscr{A}} G, \gamma \subseteq \gamma^0$.

This is impossible and hence $n^+ \in X$.

$\Rightarrow X = \omega$ by mathematical induction(Theorem 6.8) and

γ satisfies the condition (2) obviously.

$\Rightarrow \gamma : \omega \to A$ is the required function.

(Uniqueness) We will prove that such a γ is unique.

Let γ and γ' be two functions, from ω to A which satisfy the conditions I and II and let $Y = \{n \in \omega \mid \gamma(n) = \gamma'(n)\}$.

$\Rightarrow$ (i) $0 \in Y$ because $\gamma(0) = c = \gamma'(0)$.

(ii) $n \in Y$ implies $n^+ \in Y$ because $\gamma(n^+) = f(\gamma(n)) = f(\gamma'(n))$

$= \gamma'(n^+)$.

$\Rightarrow Y = \omega$ by mathematical induction(Theorem 6.8).

$\Rightarrow \gamma = \gamma'$. □

Corollary 6.15. *Let f, c, and γ be as in Theorem 6.14. If $f : A \to A$ is an injective function and $c \notin$ ran(f), then $\gamma : \omega \to A$ is an injective function.*

Proof. Let $f : A \to A$ be an injective function and let $c \notin$ ran(f).

Let $X = \{m \in \omega \mid \gamma(m) = \gamma(n)$ for some $n \in \omega$ implies $m = n\}$.

The proof is by mathematical induction on m.

$\Rightarrow$ (i) $0 \in X$.

($\because$) Assume that $\gamma(0) = \gamma(n)$ for some $n \in \omega$.

To show that $n = 0$, suppose that $n \neq 0$.

$\Rightarrow n = k^+$ for some $k \in \omega$ by Definition 6.2.

$\Rightarrow c = \gamma(0) = \gamma(n) = \gamma(k^+) = f(\gamma(k))$ which is impossible because $c \notin$ ran(f).

$\Rightarrow 0 = n$ and hence $0 \in X$.

(ii) $m \in X$ implies $m^+ \in X$.

($\because$) Assume that $m \in X$. To show that $m^+ \in X$,

assume that $\gamma(m^+) = \gamma(n)$ for some $n \in \omega$.

$\Rightarrow$ If $n = 0$, then $c = \gamma(0) = \gamma(n) = \gamma(m^+) = f(\gamma(m))$

which is impossible because $c \notin \text{ran}(f)$ and hence $n \neq 0$.

$\Rightarrow n = k^+$ for some $k \in \omega$ by Definition 6.2.

$\Rightarrow f(\gamma(m)) = \gamma(m^+) = \gamma(n) = \gamma(k^+) = f(\gamma(k))$.

$\Rightarrow$ Since f is injective, $\gamma(m) = \gamma(k)$.

$\Rightarrow$ Since $m \in X$ by the induction hypothesis, $m = k$.

$\Rightarrow m^+ = k^+ = n$ and hence $m^+ \in X$.

$\Rightarrow X = \omega$ by mathematical induction(Theorem 6.8).

$\Rightarrow \gamma : \omega \to A$ is an injective function. □

Exercise 6.3

1. Let $a \in \mathbb{R}$, where $\mathbb{R}$ is the set of the real numbers. Define a^n by the following two conditions.

   $$a^0 = 1, \qquad a^{n+1} = a^n a \text{ for all } n \in \omega.$$

   Prove that for each $n \in \omega, a^n$ is a uniquely defined real number. (In other words, prove that $\gamma(n) = a^n$ is a uniquely determined function from ω to $\mathbb{R}$. Hint: Use Theorem 6.14.)

2. Let A be a set and let $f : A \to A$ be a function. Define f^n by the following two conditions.

   $$f^0 = 1_A, \qquad f^{n+1} = f^n f \text{ for all } n \in \omega.$$

   Prove that for each $n \in \omega, f^n$ is a uniquely determined element of A^A.

3. Let A be a set and let $f : A \to B$ be an injective function, where $B \subsetneq A$. Prove that A has a subset which is in one-to-one correspondence with ω. [Hint: Use Corollary 6.15 to prove that there is an injective function $\gamma : \omega \to D$ where $D \subseteq A$.]

4. By a *strictly increasing sequence* in a partially ordered set A, we mean a function $\gamma : \omega \to A$ such that $\gamma(0) < \gamma(1) < \gamma(2) < \cdots$. Let A be a partially ordered set which has no maximal elements. Prove that there is a strictly increasing sequence in A.

6.4 Arithmetic of natural numbers

One of the most important applications of the recursion theorem is its use in defining addition and multiplication of natural numbers.

Definition 6.16 (Addition in ω). Let m be a natural number and consider the function $f : \omega \to \omega$ defined by $f(n) = n^+$ for all $n \in \omega$. Then by the recursion theorem(Theorem 6.14), there exists a unique function $\gamma_m : \omega \to \omega$ defined by the following two Conditions

$$\text{I. } \gamma_m(0) = m, \qquad \text{II. } \gamma_m(n^+) = (\gamma_m(n))^+ \text{ for all } n \in \omega.$$

The *addition of natural numbers* is now defined as follows:

$$m + n = \gamma_m(n) \text{ for all } m, n \in \omega.$$

Conditions I and II above immediately can be rewritten as follows.

$$m + 0 = m, \tag{6.1}$$

$$m + n^+ = (m + n)^+. \tag{6.2}$$

We proceed to derive a few simple properties of addition.

Lemma 6.17. *$n^+ = 1 + n$ for all $n \in \omega$, where 1 is defined to be 0^+.*

Proof. Let $X = \{n \in \omega \mid n^+ = 1 + n\}$.

$\Rightarrow$ (i) $0 \in X$ because $0^+ = 1 = 1 + 0$ by Equation 6.1.

(ii) $n \in X$ implies $n^+ \in X$ because $1 + n^+ = (1 + n)^+ = (n^+)^+$ by Equation 6.2 and the hypothesis that $n \in X$.

$\Rightarrow X = \omega$ by mathematical induction(Theorem 6.8).

$\Rightarrow n^+ = 1 + n$ for all $n \in \omega$. □

For an example, $2 + 3$ can be calculated as follows:

$$\begin{aligned} 2 + 3 &= \gamma_2(3) = \gamma_2(2^+) = \gamma_2(2)^+ = 1 + \gamma_2(2) = 1 + \gamma_2(1^+) \\ &= 1 + \gamma_2(1)^+ = 1 + (1 + \gamma_2(1)) = 1 + (1 + \gamma_2(0^+)) \\ &= 1 + (1 + \gamma_2(0)^+) = 1 + (1 + 2^+) = 1 + (1 + 3) = 1 + 4 = 5. \end{aligned}$$

Lemma 6.18 (Identity Element for Addition). *$0+n=n$ for all $n \in \omega$.*

Proof. Let $X=\{n \in \omega \mid 0+n=n\}$.

$\Rightarrow$ (i) $0 \in X$ because $0+0=0$ by Equation 6.1.

(ii) $n \in X$ implies $n^+ \in X$ because $0+n^+=(0+n)^+=n^+$ by Equation 6.2 and the hypothesis that $n \in X$.

$\Rightarrow X=\omega$ by mathematical induction(Theorem 6.8).

$\Rightarrow 0+n=n$ for all $n \in \omega$. □

Theorem 6.19 (Associative Law for Addition).
$(m+n)+k=m+(n+k)$ for all $m, n, k \in \omega$.

Proof. For arbitrary natural numbers m, n, let

$$L_{mn}=\{k \in \omega \mid (m+n)+k=m+(n+k)\}.$$

$\Rightarrow$ (i) $0 \in L_{mn}$ because $(m+n)+0=m+n=m+(n+0)$ by Equation 6.1.

(ii) $k \in L_{mn}$ implies $k^+ \in L_{mn}$.

($\because$) Suppose that $k \in L_{mn}$.

$$\begin{aligned}\Rightarrow (m+n)+k^+ &= ((m+n)+k)^+ \text{ (by Equation 6.2)}\\ &= (m+(n+k))^+ \text{ (because } k \in L_{mn})\\ &= m+(n+k)^+ \text{ (by Equation 6.2)}\\ &= m+(n+k^+) \text{ (by Equation 6.2).}\end{aligned}$$

$\Rightarrow k^+ \in L_{mn}$.

$\Rightarrow L_{mn}=\omega$ by mathematical induction(Theorem 6.8).

$\Rightarrow (m+n)+k=m+(n+k)$ for all $m, n, k \in \omega$. □

Theorem 6.20 (Commutative Law for Addition).
$m+n=n+m$ for all $m, n \in \omega$.

Proof. For an arbitrary natural number m, let

$$L_m=\{n \in \omega \mid m+n=n+m\}.$$

$\Rightarrow$ (i) $0 \in L_m$ because $m+0=m=0+m$ by Equation 6.1 and Lemma 6.18.

(ii) $n \in L_m$ implies $n^+ \in L_m$.

($\because$) Suppose that $n \in L_m$.

$\Rightarrow m + n^+ = (m+n)^+$ (by Equation 6.2)

$= (n+m)^+$ (by the hypothesis $n \in L_m$)

$= 1 + (n+m)$ (by Lemma 6.17)

$= (1+n) + m$ (by Theorem 6.19)

$= n^+ + m$ (by Lemma 6.17)

$\Rightarrow n^+ \in L_m$.

$\Rightarrow L_m = \omega$ by mathematical induction(Theorem 6.8).

$\Rightarrow m + n = n + m$ for all $m, n \in \omega$. □

Definition 6.21 (Multiplication in ω). Let m be a natural number and consider the function $g : \omega \to \omega$, defined by $g(n) = n + m$ for all $n \in \omega$. Then by the recursion theorem(Theorem 6.14), there exists a unique function $\beta_m : \omega \to \omega$ defined by the following two conditions

$$\text{I. } \beta_m(0) = 0, \qquad \text{II. } \beta_m(n^+) = \beta_m(n) + m \text{ for all } n \in \omega.$$

The *multiplication of natural numbers* is now defined as follows:

$$mn = \beta_m(n) \text{ for all } m, n \in \omega.$$

Conditions I and II above immediately can be rewritten as follows:

$$m0 = 0, \tag{6.3}$$

$$mn^+ = mn + m. \tag{6.4}$$

The following are a few simple properties of multiplication.

Lemma 6.22. $0n = 0$ *for all* $n \in \omega$.

Proof. Let $X = \{n \in \omega \mid 0n = 0\}$.

$\Rightarrow$ (i) $0 \in X$ because $00 = 0$ by Equation 6.3.

(ii) $n \in X$ implies $n^+ \in X$ because $0n^+ = 0n + 0 = 0n = 0$

by Equations 6.1 and 6.4.

$\Rightarrow X = \omega$ by mathematical induction(Theorem 6.8).

$\Rightarrow 0n = 0$ for all $n \in \omega$. □

Lemma 6.23 (Identity Element for Multiplication).

$1n = n$ *for all* $n \in \omega$.

Proof. The proof (by induction) is left as an exercise for the reader. □

For all $m \in \mathbb{N}$, note that $m \times 1 = m \times 0^+ = (m \times 0) + m = 0 + m = m$, and that 3×2 can be calculated as follows:

$$\begin{aligned}3 \times 2 &= 3 \times 1^+ = (3 \times 1) + 3 = (3 \times 0^+) + 3 = (3 \times 0 + 3) + 3 \\ &= (0+3)+3 = 3+3 = \gamma_3(3) = \gamma_3(2^+) = 1 + \gamma_3(2) = 1 + \gamma_3(1^+) \\ &= 1+(1+\gamma_3(1)) = 1+(1+\gamma_3(0^+)) = 1+(1+\gamma_3(0)^+) = 1+(1+3^+) \\ &= 1+(1+4) = 1+5 = 6.\end{aligned}$$

Theorem 6.24 (Distributive Laws). *For all* $m, n, k \in \omega$, *we have the following;*

(1) $m(n+k) = mn + mk$. (2) $(n+k)m = nm + km$.

Proof. (1) For arbitrary natural numbers m and n, let

$$L_{mn} = \{k \in \omega \mid m(n+k) = mn + mk\}.$$

$\Rightarrow$ (i) $0 \in L_{mn}$ because $m(n+0) = mn = mn + 0 = mn + m0$.

(ii) $k \in L_{mn}$ implies $k^+ \in L_{mn}$.

($\because$) Suppose that $k \in L_{mn}$. i.e., $m(n+k) = mn + mk$.

$$\Rightarrow m(n+k^+) = m(n+k)^+ = m(n+k) + m$$
$$= (mn + mk) + m = mn + (mk + m) = mn + mk^+$$

by Equation 6.2, Equation 6.4 and Theorem 6.19.

$\Rightarrow k^+ \in L_{mn}$.

$\Rightarrow L_{mn} = \omega$ by mathematical induction(Theorem 6.8).

$\Rightarrow m(n+k) = mn + mk$ for all $m, n, k \in \omega$.

(2) The proof of (2) is left as an exercise for the reader. □

Theorem 6.25 (Associative Law for Multiplication).

$(mn)k = m(nk)$ *for all* $m, n, k \in \omega$.

Proof. For arbitrary natural numbers m and n, let

$$L_{mn} = \{k \in \omega \mid (mn)k = m(nk)\}.$$

$\Rightarrow$ (i) $0 \in L_{mn}$ because $(mn)0 = 0 = m0 = m(n0)$ by Equation 6.3.

(ii) $k \in L_{mn}$ implies $k^+ \in L_{mn}$.

($\because$) Suppose $k \in L_{mn}$. i.e., $(mn)k = m(nk)$.

$\Rightarrow (mn)k^+ = (mn)k + mn = m(nk) + mn = m(nk + n) = m(nk^+)$

by Equation 6.4 and Theorem 6.24.

$\Rightarrow k^+ \in L_{mn}$.

$\Rightarrow L_{mn} = \omega$ by mathematical induction(Theorem 6.8).

$\Rightarrow (mn)k = m(nk)$ for all $m, n, k \in \omega$. □

Theorem 6.26 (Commutative Law for Multiplication).

$mn = nm$ *for all* $m, n \in \omega$.

Proof. The proof is left as an exercise for the reader. □

One of the most important aspects of the natural numbers is their ordering. Before proceeding with a formal definition of the order relation in ω, the reader should review the definition of ω. Specifically, it should be noted that for each $n \in \omega$, the natural number n is the set of all the natural numbers preceding n: $n = \{0, 1, \ldots, n-1\}$.

It is clear, now, how we are to define order in ω:

n is to precede m if and only if n is an element of m.

Motivated by this observation, we make the following formal definition.

Definition 6.27 (Order relation in ω). An *order relation* $\leq$ in ω is defined as follows:

$m \leq n$ if and only if $m \in n$ or $m = n$, for all $m, n \in \omega$.

First, it is required to prove that this is indeed an order relation in ω.

Theorem 6.28. *The relation* $\leq$ *is an order relation in* ω.

Proof. (i) $(\omega, \leq)$ is reflexive.

($\because$) For each $m \in \omega$, $m = m$, and hence $m \leq m$.

Thus $\leq$ is reflexive in ω.

(ii) $(\omega, \leq)$ is antisymmetric.

($\because$) Suppose that $m \leq n$ and $n \leq m$.

$\Rightarrow$ ($m \in n$ or $m = n$) and ($n \in m$ or $n = m$).

$\Rightarrow$ ($m = n$) or ($m \in n$ and $n \in m$) by Theorem 1.9.

$\Rightarrow$ $m = n$ or ($m \subseteq n$ and $n \subseteq m$) by Lemma 6.11.

$\Rightarrow$ $m = n$ and hence $\leq$ is antisymmetric in ω.

(iii) $(\omega, \leq)$ is transitive.

($\because$) Suppose that $m \leq n$ and $n \leq p$.

$\Rightarrow$ ($m \in n$ or $m = n$) and ($n \in p$ or $n = p$).

$\Rightarrow$ We have four possible cases.

(1) $m \in n$ and $n \in p$. Then $m \in n$ and $n \subseteq p$, and hence $m \in p$.

(2) $m \in n$ and $n = p$. Then $m \in p$.

(3) $m = n$ and $n \in p$. Then $m \in p$.

(4) $m = n$ and $n = p$. Then $m = p$.

$\Rightarrow$ $m \leq p$ in each case and hence $\leq$ is transitive in ω.

$\Rightarrow$ $\leq$ is an order relation in ω. □

We will show next that ω is well-ordered. This will require the following two lemmas.

Lemma 6.29. $0 \leq m$ *for all natural numbers* $m \in \omega$.

Proof. Let $X = \{m \in \omega \mid 0 \leq m\}$.

$\Rightarrow$ (i) $0 \in X$ because $0 \leq 0$ by the reflexive law.

(ii) $m \in X$ implies $m^+ \in X$.

($\because$) Suppose that $m \in X$. i.e., $0 \leq m$.

$\Rightarrow$ Since $m \in m^+$, $m \leq m^+$.

$\Rightarrow 0 \leq m^+$ by the transitive law.

$\Rightarrow m^+ \in X$.

$\Rightarrow X = \omega$ by mathematical induction(Theorem 6.8).

$\Rightarrow 0 \leq m$ for all natural numbers $m \in \omega$. □

Lemma 6.30. *If $n < m$, then $n^+ \leq m$.*

Proof. For arbitrary natural number $n \in \omega$, let

$L_n = \{m \in \omega \mid n < m \text{ implies } n^+ \leq m\}$.

$\Rightarrow$ (i) $0 \in L_n$.

($\because$) Since $n < 0$ is false by Lemma 6.29,

the statement that $n < 0$ implies $n^+ \leq 0$ is true.

$\Rightarrow 0 \in L_n$.

(ii) $m \in L_n$ implies $m^+ \in L_n$.

($\because$) Let $m \in L_n$. i.e., $n < m$ implies $n^+ \leq m$.

Assume that $n < m^+$. i.e., $n \in m^+$.

$\Rightarrow n \in m$ or $n = m$ by Lemma 6.9.

$\Rightarrow$ If $n = m$, then $n^+ = m^+$ and hence $n^+ \leq m^+$.

If $n \in m$, i.e., $n < m$, then since $m \in L_n$, $n^+ \leq m < m^+$ and hence $n^+ \leq m^+$.

$\Rightarrow m^+ \in L_n$.

$\Rightarrow L_n = \omega$ by mathematical induction(Theorem 6.8).

$\Rightarrow$ For all natural numbers m and n, if $n < m$, then $n^+ \leq m$. □

Theorem 6.31. *ω with the order relation $\leq$ is well-ordered.*

Proof. Suppose that ω is not well-ordered.

$\Rightarrow$ There exists a nonempty subset A of ω without a least element.

Let $X = \{n \in \omega \mid n \leq m \text{ for every } m \in A\}$.

Note that since A has no least element, $X \cap A = \emptyset$.

$\Rightarrow$ (i) $0 \in X$ by Lemma 6.29.

(ii) If $n \in X$, then $n^+ \in X$.

(∵) Let $n \in X$.

$\Rightarrow n \leq m$ for every $m \in A$.

$\Rightarrow$ Since $X \cap A = \emptyset$, $n < m$ for every $m \in A$.

$\Rightarrow n^+ \leq m$ for every $m \in A$ by Lemma 6.30.

$\Rightarrow n^+ \in X$.

$\Rightarrow X = \omega$ by mathematical induction(Theorem 6.8).

$\Rightarrow \emptyset \neq A = \omega \cap A = X \cap A = \emptyset$, which is a contradiction.

$\Rightarrow \omega$ is well-ordered. □

Exercise 6.4

1. Prove Theorem 6.24(2).

2. Prove Theorem 6.26.

3. Prove each of the following.

   (1) If $m = n$, then $m + k = n + k$. (See Exercise 6.2.3.)

   (2) If $m = n$, then $mk = nk$.

4. Prove each of the following.

   (1) If $m < 1$, then $m = 0$.

   (2) There is no natural number k such that $m < k < m^+$.

5. Prove each of the following.

   (1) If $n < k$, then $m + n = m + k$.

   (2) If $m + n = m + k$, then $n = k$.

6. Prove each of the following.

   (1) If $m < n$ and $k \neq 0$, then $mk < nk$.

   (2) If $mk = nk$ and $k \neq 0$, then $m = n$.

7. Prove that if $m \leq n$, then there exists a unique $p \in \omega$ such that $m + p = n$.

8. Prove each of the following.

   (1) If $m + k < n + k$, then $m < n$.

   (2) If $mk < nk$, then $m < n$.

9. Give an inductive definition of exponentiation of natural numbers. That is, define m^n in a manner similar to Equations 6.1 and 6.2 and then justify your definition in terms of the recursion theorem. Then prove each of the following.

   (1) $m^{r+s} = m^r m^s$. (2) $(mn)^k = m^k n^k$. (3) $(m^n)^k = m^{nk}$.

6.5 Concluding remark

(1) Every natural number is a set.

$(\because)$ By Axiom 9 of Infinity, there exists a successor *set* X.

$\Rightarrow \omega \subseteq X$ by Definition 6.6.

$\Rightarrow \omega$ is a set by Axiom 4 of Subset.

$\Rightarrow$ Every natural number is a set by Definition 1.45.

(2) Every finite class is a set.

In the next chapter we will define a class to be *finite* if it is in one-to-one correspondence with a natural number. By Theorem 2.37, *every finite class is a set.* In view of this remark, we will henceforth speak of finite *sets* rather than finite classes.

(3) The classes $\mathbb{N}$, ω, $\mathbb{Z}$, $\mathbb{Q}$, $\mathbb{R}$ and $\mathbb{C}$ of all the positive integers, the natural numbers, the integers, the rational numbers, the real numbers and the complex numbers, respectively, are sets.

   (i) Since there exists a successor set by Axiom 9 of Infinity and ω is the intersection of all the successor sets, ω is a set by Axiom 4 of Subset.

Now since $\mathbb{N} = \omega \setminus \{0\}$, $\mathbb{N}$ is a set by Axiom 4 of Subset and since $\mathbb{Z} = (-\mathbb{N}) \cup \omega$, $\mathbb{Z}$ is a set by Axiom 5 of Union.

(ii) Since ω is a set, $\omega \times \omega$ is a set by Theorem 1.53. Now, if we identify each fraction n/m (where n/m is assumed to be in "lowest terms") with the ordered pair (n, m), then the class of all the positive rational numbers is a subclass of $\omega \times \omega$, hence by Axiom 4 of Subset, it is a set. Analogously, the class of all the negative rational numbers is a set, hence by Axiom 5 of Union, the class $\mathbb{Q}$ of all the rational numbers is a set.

(iii) It is well known that every real number can be regarded as a sequence (called a *Cauchy sequence*) of rational numbers. Hence, roughly speaking, every real number can be regarded as an element of $\mathscr{P}(\mathbb{Q})$. In other words, the class $\mathbb{R}$ of the real numbers is a subclass of $\mathscr{P}(\mathbb{Q})$. Hence, by Axiom 6 of Power Set and Axiom 4 of Subset, the class $\mathbb{R}$ of all the real numbers is a set.

(iv) Since the class $\mathbb{C}$ of the complex numbers is in one-to-one correspondence with $\mathbb{R} \times \mathbb{R}$, $\mathbb{C}$ is a set by Theorem 1.53 and Axiom 7 of Replacement.

(4) The classical construction processes for sets produces a set.

(i) The union and intersection of any set of sets are sets by applying Axiom 5 of Union and by Axiom 4 of Subset, respectively.

(ii) If A is a set, then $\mathscr{P}(A)$ is a set by Axiom 6 of Power Set.

(iii) If $\{A_i\}_{i \in I}$ is a family of sets, where I is a set, then $\prod_{i \in I} A_i$ is a set by Theorem 2.38.

(iv) If A and B are sets, then A^B is a set by Theorem 2.38.

Thus every object we can produce from sets by the classical construction processes is a set.

(5) All the infinite (that is, not finite) classes which occur in traditional mathematics are sets. Thus, in the next chapter, we will confine our attention to finite and infinite *sets*.

Chapter 7

Finite and Infinite Sets

7.1 Introduction

- One of the most fundamental distinctions in mathematics is that between finite and infinite sets. The distinction is so intuitively compelling that, even in the absence of a precise definition, there cannot be any doubt as to whether a given set is finite or infinite. In simple terms, a finite set is one which "has n elements", where n is a natural number, and an infinite set is one which is not finite.

- Although the dichotomy between finite and infinite has always fascinated mathematicians - it has been the source of the most celebrated riddles, paradoxes, and classical errors of mathematicians, a sound theory of infinite sets did not appear until very recent times. It had to await the arrival of the rigorous concepts of set, mapping, and one-to-one correspondence. Once the use of these new tools became familiar to mathematicians, toward the end of the nineteenth century, the modern theory of infinite sets developed rapidly. It was largely the work of Georg Cantor and his successors.

- Using familiar concepts, and arguments which are remarkable for their simplicity, Cantor was able to draw conclusions which surprised mathematicians and laymen alike. Cantor's ideas are well known today. They have been popularized in innumerable expository books and articles, and have entered the lore of modern mathematics. We proceed, in the remainder of this section, to give the bare outlines of Cantor's theory. In this discussion the words "finite" and "infinite" will be used informally. As we remarked earlier, a finite set can be described as one which "has n elements" (n is a natural number), and an infinite set is, simply, one which is not finite.

- Two finite sets A and B have the "same number of elements" if and only if they are in one-to-one correspondence. Even though we cannot speak of two infinite sets as having the "same number of elements", we have the feeling nonetheless, that if A and B are infinite sets and there is a one-to-one correspondence between them, then, in a certain sense, they are of the "same size".

  This intuitive notion is formalized in the following definition.

Definition 7.1. Let A and B be two sets.

(1) A and B are said to be *equipotent*, or to have the *same power*, if there is a one-to-one correspondence from A to B. In this case, we write $A \approx B$.

(2) We say that A is of a *lesser power* than B if there exists a one-to-one correspondence between A and a proper subset of B, but none between A and B. In this case, we write $A \prec B$

(3) We say that the power of A is *less than or equal* to that of B if A is in one-to-one correspondence with a subset of B. In this case, we write $A \preceq B$.

- In Definition 7.1, we have taken our cue from the finite case: for if A and B are finite sets and A has fewer elements than B, then certainly there exists a one-to-one correspondence between A and a part of B, but none between A and all of B.

- It is a curious fact, first proved by Cantor, that many sets which appear to be smaller, or larger, than ω actually have the same power as ω. For example, if E is the set of the even natural numbers, it is easy to see that the function $f(n) = 2n$ is a bijective function from ω to E. Thus, although E is a proper subset of ω (in fact, E appears to have only "half as many" as ω), actually E is equipotent with ω. A more surprising example involves ω and the set $\mathbb{Q}$ of the rational numbers. Our intuition suggests, in the most compelling way, that $\mathbb{Q}$ is a "larger" set than ω; for $\mathbb{Q}$ not only includes ω, but is, in an obvious manner, "infinitely dense" with respect to ω. Yet it can easily be shown that ω and $\mathbb{Q}$ are in one-to-one correspondence. The proof consists in "enumerating" the rational numbers, that is, making a list $r_1, r_2, r_3, \cdots$ of rational numbers which includes all the rational numbers. The correspondence $i \leftrightarrow r_i$ is then a one-to-one correspondence between ω and $\mathbb{Q}$.

Definition 7.2. A set which is in one-to-one correspondence with ω is said to be *denumerable* or *countably infinite*.

Theorem 7.3. *The set $\mathbb{Q}$ of all the rational numbers is denumerable.*

Proof. Let $A_i = \left\{\frac{n}{m} \mid n, m \in (\omega \setminus \{0\}) \text{ and } n + m = i\right\}$ for each $i \in \omega$.
$\Rightarrow \bigcup_{i\in\omega} A_i = \{q \in \mathbb{Q} \mid q > 0\}$.

Order the numbers of $\bigcup_{i\in\omega} A_i$ as follows: For $\frac{n}{m} \in A_i$ and $\frac{n'}{m'} \in A_j$,

$$\frac{n}{m} \leq \frac{n'}{m'} \quad \text{iff} \quad \text{either } (i < j) \text{ or } (i = j \text{ and } n \leq n').$$

Thus the first few fractions in this ordering would be as follows:

$$\frac{1}{1}, \frac{1}{2}, \frac{2}{1}, \frac{1}{3}, \frac{2}{2}, \frac{3}{1}, \frac{1}{4}, \frac{2}{3}, \frac{3}{2}, \frac{4}{1}, \frac{1}{5}, \cdots.$$

Next delete all fractions which are not in "lowest terms". This leaves

$$1, \frac{1}{2}, 2, \frac{1}{3}, 3, \frac{1}{4}, \frac{2}{3}, \frac{3}{2}, 4, \frac{1}{5}, \cdots.$$

$\Rightarrow \{q \in \mathbb{Q} \mid q > 0\}$ is enumerated, say, by $r_1, r_2, r_3, \cdots$.

Similarly we can enumerate $\{q \in \mathbb{Q} \mid q < 0\}$, say, by $t_1, t_2, t_3, \cdots$.

Let $r_0 = 0$.

$\Rightarrow$ The enumeration $r_0, r_1, t_1, r_2, t_2, r_3, t_3, r_4, t_4, \cdots$ is an enumeration of all the rational numbers $\mathbb{Q}$.

$\Rightarrow \mathbb{Q}$ is a denumerable set. $\square$

Faced with unexpected discovery that $\mathbb{Q}$ is denumerable, we are naturally led to wonder whether every infinite set is denumerable. This question was answered by Cantor in the negative. The set of all the real numbers, for example, is not denumerable. To prove this we use the so-called *diagonal method.*

Theorem 7.4. *The set* $\mathbb{R}$ *of all the real numbers is not denumerable.*

Proof. Note that the function $y = \tan\left(\pi x - \frac{\pi}{2}\right)$ is a one-to-one correspondence between the set of all the real numbers and the open interval $(0, 1)$.

Express each real number as a nonterminating decimal.

$\Rightarrow \{x \in \mathbb{R} \mid 0 < x < 1\}$ cannot be enumerated.

($\because$) (Diagonal Method) Suppose that $r_1, r_2, r_3, \cdots$, is an enumeration of all the real numbers in the open interval $(0, 1)$.

$\Rightarrow r_i = 0.r_{i1}r_{i2}r_{i3}\cdots$, where each r_{ij} is a digit $0, 1, \cdots, 9$.

$\Rightarrow r_1 = 0.r_{11}r_{12}r_{13}\cdots,\ r_2 = 0.r_{21}r_{22}r_{23}\cdots,$

$r_3 = 0.r_{31}r_{32}r_{33}\cdots,\ r_4 = 0.r_{41}r_{42}r_{43}\cdots,\ \cdots,$

Define $s = 0.s_1s_2s_3\cdots$, where $s_i = \begin{cases} 1 & \text{if } r_{ii} \neq 1, \\ 5 & \text{if } r_{ii} = 1. \end{cases}$

⇒ Since $0 < s < 1$, s is one of the numbers in the enumeration.

⇒ $s = r_k$ for some k. But since $1 \leq |s_k - r_{kk}| < 9$, $s \neq r_k$.

This is a contradiction.

⇒ There is no way of enumerating the real numbers.

⇒ The open interval $(0, 1)$ is not denumerable.

⇒ $\mathbb{R}$ is not denumerable. □

We have just revealed one of the most significant facts in Cantor's Theory of the infinite.

- While ω and $\mathbb{R}$ are both infinite sets, one of them is strictly larger than the other. i.e., infinite sets, like finite sets, come in different "sizes".

Our next step, naturally, is to find a set which is strictly larger than $\mathbb{R}$. In order to settle this question, we introduce a result of far greater generality, which will provide us at once with a strictly increasing sequence $K_1 \prec K_2 \prec K_3 \prec \cdots$ of infinite sets.

Theorem 7.5. *If A is any set, then there is no surjective function from A to $\mathscr{P}(A)$.*

Proof. (A variant of Diagonal Method) Suppose that there is a surjective function $\phi : A \to \mathscr{P}(A)$. Define a set B by $B = \{x \in A \mid x \notin \phi(x)\}$.

⇒ $B \in \mathscr{P}(A)$.

⇒ Since $\phi : A \to \mathscr{P}(A)$ is surjective, there exists $y \in A$ such that $B = \phi(y)$.

⇒ (i) If $y \in \phi(y)$, then $y \in B$ and hence $y \notin \phi(y)$.

(ii) If $y \notin \phi(y)$, then $y \notin B$ and hence $y \in \phi(y)$.

Both (i) and (ii) are impossible.

⇒ There exists no surjective function from A to $\mathscr{P}(A)$. □

Corollary 7.6. *No subset of a set A can be equipotent with $\mathscr{P}(A)$.*

Proof. Let A be any set and let B be a subset of A.

$\Rightarrow$ There is no surjective function from B to $\mathscr{P}(B)$ by Theorem 7.5.

$\Rightarrow$ Since $\mathscr{P}(B) \subseteq \mathscr{P}(A)$, there is no surjective function from B to $\mathscr{P}(A)$.

$\Rightarrow$ There is no bijective function from B to $\mathscr{P}(A)$.

$\Rightarrow$ B is not equipotent with $\mathscr{P}(A)$. □

Corollary 7.7. *A set A can not be equipotent with any set containing $\mathscr{P}(A)$.*

Proof. Suppose that a set A is equipotent with a set B containing $\mathscr{P}(A)$.

$\Rightarrow$ There exists a bijective function $f : A \to B$.

Consider a function $g : A \to \mathscr{P}(A)$ defined by for all $x \in A$,

$$g(x) = \begin{cases} f(x) & \text{if } f(x) \in \mathscr{P}(A), \\ A & \text{if } f(x) \in B \setminus \mathscr{P}(A). \end{cases}$$

$\Rightarrow$ Since $\mathscr{P}(A) \subseteq B$ and $f : A \to B$ is surjective, $g : A \to \mathscr{P}(A)$ is a surjective function, which is impossible by Theorem 7.5. □

Corollary 7.8. *Let A and B be sets. Then*

(1) *If $B \subseteq A$, then $B \prec \mathscr{P}(A)$.*

(2) *If $\mathscr{P}(A) \subseteq B$, then $A \prec B$.*

Proof. (1) $B \subseteq A$.

$\Rightarrow$ Since $\{x\} \in \mathscr{P}(A)$ for each $x \in B(\subseteq A)$, we can define a function $f : B \to \mathscr{P}(A)$ by setting $f(x) = \{x\}$ for each $x \in B$.

$\Rightarrow$ Clearly, $f : B \to \mathscr{P}(A)$ is an injective function.

$\Rightarrow$ B is of lesser power than $\mathscr{P}(A)$. i.e., $B \preceq \mathscr{P}(A)$.

$\Rightarrow$ Since B is not equipotent with $\mathscr{P}(A)$ by Corollary 7.6, $B \prec \mathscr{P}(A)$.

(2) $\mathscr{P}(A) \subseteq B$.

$\Rightarrow$ Since $\{x\} \in \mathscr{P}(A)(\subseteq B)$ for each $x \in A$, we can define a function $g : A \to B$ by setting $g(x) = \{x\}$ for each $x \in A$.

$\Rightarrow$ Clearly, $g : A \to B$ is an injective function.

$\Rightarrow$ A is of lesser power than B. i.e., $A \preceq B$.

$\Rightarrow$ Since A is not equipotent with B by Corollary 7.7, $A \prec B$. □

Recall the following theorem.

Theorem 2.35 If A is a set, then $\mathscr{P}(A)$ and 2^A are in one-to-one correspondence.

Remark 7.9. (1) By Theorem 2.35, all of the above statements(Theorem 7.5, Corollaries 7.6, 7.7, 7.8) hold true when we replace $\mathscr{P}(A)$ by 2^A.

(2) There exists a strictly increasing sequence of infinite sets. Speaking informally, there are many more "sizes" of infinite sets.

($\because$) Let $K_1 = \omega$, $K_2 = \mathscr{P}(K_1)$, $K_3 = \mathscr{P}(K_2)$ and so forth.

$\Rightarrow K_1 \prec K_2 \prec K_3 \prec \cdots$ is a strictly increasing sequence of infinite sets. Let $L_1 = \bigcup_{i\in\omega} K_i$.

$\Rightarrow$ For each $i \in \omega$, since $K_{i+1} \subseteq L_1$, $\mathscr{P}(K_i) = K_{i+1} \subseteq L_1$.

$\Rightarrow K_i \prec L_1$ for all $i \in \omega$ by Corollary 7.8(2).

Let $L_2 = \mathscr{P}(L_1)$, $L_3 = \mathscr{P}(L_2)$ and so forth.

$\Rightarrow K_1 \prec K_2 \prec K_3 \prec \cdots \prec L_1 \prec L_2 \prec L_3 \prec \cdots$ is a strictly increasing sequence of infinite sets.

(3) The set $\mathbb{R}$ of the real numbers is equipotent with 2^ω.

($\because$) Note that $\mathbb{R}$ is equipotent with the open interval $(0,1) \subseteq \mathbb{R}$.

Express each element $r \in (0,1)$ by binary notation

$r = 0.r_1r_2r_3\cdots$, where each r_i is either 0 or 1 and $r_0 = 0$.

$\Rightarrow$ The expression for $r \in (0,1)$ can be identified with the function $\phi_r : \omega \to \{0,1\}$ given by $\phi_r(i) = r_i$, for all $i \in \omega$.

Define $\phi : (0,1) \to 2^\omega$ by $\phi(r) = \phi_r$, for all $r \in (0,1)$.

$\Rightarrow$ Clearly the function $\phi : (0,1) \to 2^\omega$ is bijective.

(4) All of classical mathematics deal with only two sizes of infinite sets, namely, sets equipotent with ω and sets equipotent with 2^ω.

(i) The set $\mathbb{Z}$ of the integers is equipotent with ω.

(ii) The set $\mathbb{Q}$ of the rational numbers is equipotent with ω.

(iii) The set $\mathbb{R}$ of the real numbers is equipotent with 2^ω.

(iv) The set $\mathbb{C}$ of the complex numbers is equipotent with $\mathbb{R}$.

Definition 7.10. The power of 2^ω is often called the *power of the continuum.*

Question 7.11. Is there a power between that of ω and that of 2^ω.

In other words, does there exist a set A such that $\omega \prec A \prec 2^\omega$?

Since no such set of Question 7.11 occurs anywhere in classical mathematics and there appears to be no way of constructing one, it was conjectured by Cantor and his contemporaries that the answer to Question 7.11 must be "no". This conjecture is known as the *continuum hypothesis.* A closely related conjecture is the *generalized continuum hypothesis*, which proposes that for every infinite set B, there is no set A such that $B \prec A \prec 2^B$. These hypotheses have never been either proved or disproved. We shall have more to say about them in Chapter 9.

The aim of this Chapter is to exploit the various ideas which have been motivated in our introduction. We will define rigorously the concepts of finite and infinite set, "power", and cardinality, and give the classical results of Cantor's theory.

Conjecture 7.12 (Continuum Hypothesis)**.** There exists no set A such that $\omega \prec A \prec 2^\omega$.

Conjecture 7.13 (Generalized Continuum Hypothesis)**.** For every infinite set B, there exists no set A such that $B \prec A \prec 2^B$.

7.2 Equipotence of sets

In Definition 7.1, we have defined the symbols $\approx, \prec$ and $\preceq$ as follows:

$A \approx B$ iff A is in one-to-one correspondence with B.

$A \preceq B$ iff A is in one-to-one correspondence with a subset of B.

$A \prec B$ iff A is in one-to-one correspondence with a subset of B and

A is not in one-to-one correspondence with B.

From the second of these statements, we have the following

Proposition 7.14. *$A \preceq B$ iff there exists an injective function $f : A \to B$.*

Proof. It is an immediate consequence of Definition 7.1. □

Lemma 7.15. *There exists an injective function $f : A \to B$ if and only if there exists a surjective function $g : B \to A$.*

Proof. **(only if)** $f : A \to B$ is an injective function.

$\Rightarrow$ There exists a function $g : B \to A$ such that $g \circ f = 1_A$ by Theorem 2.24.

$\Rightarrow$ $g : B \to A$ is surjective by Theorem 5.9.

(if) $g : B \to A$ is a surjective function.

$\Rightarrow$ There exists a function $f : A \to B$ such that $f \circ g = 1_B$ by Theorem 5.9.

$\Rightarrow$ $f : A \to B$ is injective by Theorem 2.24. □

Corollary 7.16. *$A \preceq B$ iff there exists a surjective function $g : B \to A$.*

Proof. It follows directly from Proposition 7.14 and Lemma 7.15. □

Theorem 7.17. *Let A, B, C and D be sets where $A \cap C = \emptyset$ and $B \cap D = \emptyset$. If $f : A \to B$ and $g : C \to D$ are bijective functions, then $f \cup g : A \cup C \to B \cup D$ is a bijective function.*

Proof. Let A, B, C and D be sets. Assume that $A \cap C = \emptyset$ and $B \cap D = \emptyset$ and let $f : A \to B$ and $g : C \to D$ be bijective functions.

$\Rightarrow$ (i) $f \cup g : A \cup C \to B \cup D$ is a function.

($\because$) Since $f : A \to B \cup D$ and $g : C \to B \cup D$ are functions, $f \cup g : A \cup C \to B \cup D$ is a function by Theorem 2.15(1).

(ii) $f^{-1} : B \to A$ and $g^{-1} : D \to C$ are bijective functions by Theorem 2.19.

(iii) $f^{-1} \cup g^{-1} : B \cup D \to A \cup C$ is a function by a similar argument.

$\Rightarrow$ $(f \cup g)^{-1} : B \cup D \to A \cup C$ is a function and $f^{-1} \cup g^{-1} = (f \cup g)^{-1}$.

$\Rightarrow$ $f \cup g : A \cup C \to B \cup D$ is an invertible function.

$\Rightarrow$ $f \cup g : A \cup C \to B \cup D$ is a bijective function by Theorem 2.19(2). $\square$

Corollary 7.18. *Assume that $A \cap C = \emptyset$ and $B \cap D = \emptyset$. If $A \approx B$ and $C \approx D$, then $A \cup C \approx B \cup D$.*

Proof. It follows directly from Theorem 7.17. $\square$

Theorem 7.19. *If $A \approx B$ and $C \approx D$, then $A \times C \approx B \times D$.*

Proof. Assume that $A \approx B$ and $C \approx D$.

$\Rightarrow$ There exists bijective functions $f : A \to B$ and $g : C \to D$.

Define $h : A \times C \to B \times D$ by $h(x, y) = (f(x), g(y))$ for all $(x, y) \in A \times C$.

$\Rightarrow$ It can be shown routinely that $h : A \times C \to B \times D$ is a bijective function and hence $A \times C \approx B \times D$. $\square$

Theorem 7.20. *If $A \approx B$ and $C \approx D$, then $C^A \approx D^B$.*

Proof. Assume that $A \approx B$ and $C \approx D$.

$\Rightarrow$ There exist bijective functions $f : B \to A$ and $g : C \to D$.

Define $h : C^A \to D^B$ by $h(\alpha) = g \circ \alpha \circ f$ for all $\alpha \in C^A$.

i.e., the following diagram is commutative;

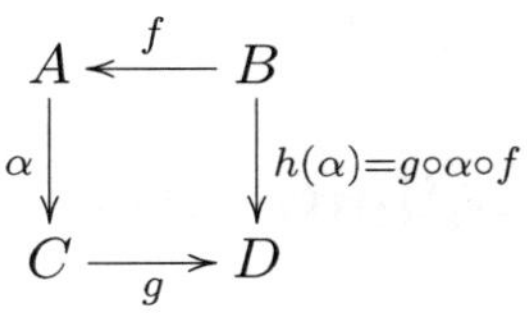

⇒ The following can be proved easily and left to the readers.

(i) $h : C^A \to D^B$ is a well-defined function.

(ii) $h : C^A \to D^B$ is bijective.

⇒ $C^A \approx D^B$. □

Corollary 7.21. *If $A \approx B$, then $\mathscr{P}(A) \approx \mathscr{P}(B)$.*

Proof. Since $\mathscr{P}(X) \approx 2^X$ for any set X by Theorem 2.35, it follows directly from Theorem 7.20. □

Remark 7.22. Corollary 7.21 can be proved directly by showing that if $f : A \to B$ is a bijective function, then the function $\overline{f} : \mathscr{P}(A) \to \mathscr{P}(B)$ defined by $\overline{f}(C) = f(C)$ for all $C \in \mathscr{P}(A)$, is a bijective function.

Exercise 7.2

1. Complete the proof of Theorem 7.19.

2. Complete the proof of Theorem 7.20.

3. Prove that if $(A \setminus B) \approx (B \setminus A)$, then $A \approx B$.

4. Suppose that $A \approx B, a \in A$ and $b \in B$. Prove that $(A \setminus \{a\}) \approx (B \setminus \{b\})$.

5. Suppose that $A \approx B, C \approx D, C \subseteq A$ and $D \subseteq B$. Prove that $(A \setminus C) \approx (B \setminus D)$.

6. Let $\{B_i\}_{i \in I}$ and $\{C_i\}_{i \in I}$ be families of mutually disjoint sets. If $B_i \approx C_i$ for each $i \in I$, prove that $\bigcup_{i \in I} B_i \approx \bigcup_{i \in I} C_i$.

7. Let $\{B_i\}_{i\in I}$ and $\{C_i\}_{i\in I}$ be families of mutually disjoint sets. If $B_i \approx C_i$ for each $i \in I$, prove that $\prod_{i\in I} B_i \approx \prod_{i\in I} C_i$.

7.3 Properties of infinite sets

Definition 7.23. Let A be a set.

(1) A is said to be *finite* if A is in one-to-one correspondence with a natural number n, i.e., there exists $n \in \omega$ such that $A \approx n$.

(2) A is said to be *infinite* if it is not finite. i.e., A is not in one-to-one correspondence with any natural number.

Several other definitions of "finite" and "infinite" are to be found in the mathematical literature. Foremost among them are the following:

(i) A set A is infinite if and only if A has a denumerable subset. A is finite if it is not an infinite set.

(ii) A set A is infinite if and only if A is equipotent with a proper subset of itself. A is finite if it is not an infinite set.

It will be shown next that each of (i) and (ii) is equivalent to our definition of "infinite" above. (See Theorem 7.26 and Theorem 7.30.)

Lemma 7.24. *If A is a denumerable set and $x \in A$, then $A \setminus \{x\}$ is a denumerable set.*

Proof. Let A be a denumerable set and let $x \in A$.

$\Rightarrow$ There exists a bijective function $f : \omega \to A$ and then there exists $n \in \omega$ such that $f(n) = x$.

$\Rightarrow$ Define $g : \omega \to A \setminus \{x\}$ by for all $m \in \omega$,

$$g(m) = \begin{cases} f(m) & \text{if } m < n, \\ f(m+1) & \text{if } m \geq n. \end{cases}$$

$\Rightarrow$ It is easy to see that $g : \omega \to A \setminus \{x\}$ is a bijective function.

$\Rightarrow$ $A \setminus \{x\}$ is a denumerable set. □

Lemma 7.25. *Each natural number $n \in \omega$ cannot have a denumerable subset.*

Proof. Let $L = \{n \in \omega \mid n \text{ has no denumerable subset}\}$.

$\Rightarrow$ (i) $0 \in L$ because $\emptyset$ is not a denumerable set.

(ii) $n \in L$ implies $n^+ \in L$.

($\because$) Assume that $n \in L$ and note that $n = \{0, 1, \cdots, n-1\}$.

Suppose that n^+ has a denumerable subset S. i.e., $n^+ \notin L$.

$\Rightarrow$ If $n \notin S$, then S is a denumerable subset of n.

This is impossible by the hypothesis $n \in L$.

If $n \in S$, then $S \setminus \{n\} (\subseteq n)$ is denumerable by Lemma 7.24.

Again, this is impossible because $n \in L$.

$\Rightarrow$ n^+ cannot have a denumerable subset. i.e., $n^+ \in L$.

$\Rightarrow$ $L = \omega$ by mathematical induction(Theorem 6.8).

$\Rightarrow$ Each natural number $n \in \omega$ cannot have a denumerable subset. □

Theorem 7.26. *A set A is an infinite set if and only if A has a denumerable subset.*

Proof. **(only if)** Let A be an infinite set.

Suppose that A does not have a denumerable subset.

$\Rightarrow$ By the well ordering theorem(Theorem 5.24), A can be well-ordered.

Well order A.

$\Rightarrow$ By Theorem 4.66, exactly one of the following cases holds:

(α) ω is isomorphic with A.

(β) ω is isomorphic with an initial segment of A.

(γ) A is isomorphic with an initial segment of ω.

$\Rightarrow$ Since A does not have a denumerable subset, (α) and (β) cannot hold, and hence

A is isomorphic with an initial segment S_n of ω.

$\Rightarrow$ Since $S_n = \{0, 1, 2, \cdots, n-1\} = n$ by Definition 6.2[1], A is finite, which contradicts to our assumption.

$\Rightarrow$ A has a denumerable subset.

(if) Assume that A has a denumerable subset B and suppose that A is finite.

$\Rightarrow$ There exists $n \in \omega$ such that $A \approx n$, $B \subseteq A$ and $B \approx \omega$.

$\Rightarrow$ There exists an injective function $i : \omega \to n$ as a composition of injective functions $\omega \to B \to A \to n$.

$\Rightarrow$ n has the denumerable subset $i(\omega)$, which is impossible by Lemma 7.25.

$\Rightarrow$ A is an infinite set. □

Corollary 7.27. *Every set which has an infinite subset is infinite.*

Proof. A set A has an infinite subset B.

$\Rightarrow$ B has a denumerable subset C by Theorem 7.26.

$\Rightarrow$ A has the denumerable subset C.

$\Rightarrow$ A is infinite by Theorem 7.26. □

Corollary 7.28. *Every subset of a finite set is finite.*

Proof. B is a subset of a finite set A.

$\Rightarrow$ A cannot have a denumerable subset by Theorem 7.26.

$\Rightarrow$ B cannot have a denumerable subset.

$\Rightarrow$ B is finite by Theorem 7.26. □

Corollary 7.29. *If A is an infinite set and B is a nonempty set, then $A \times B$ and $B \times A$ are infinite sets.*

---

[1]By definition 6.2, the natural number n is the set $\{0, 1, 2, \cdots, n-1\}$. i.e., n is exactly the initial segment S_n of ω.

Proof. A is an infinite set and $B \neq \emptyset$.

$\Rightarrow$ A has a denumerable subset by Theorem 7.26 and there exists $y \in B$.

$\Rightarrow$ There exists an injective function $i : \omega \to A$ by Proposition 7.14.

Define $f : \omega \to A \times B$ by $f(n) = (i(n), y)$ for all $n \in \omega$.

$\Rightarrow$ $f : \omega \to A \times B$ is an injective function.

($\because$) Assume that $f(n) = f(m)$ for $m, n \in \omega$.

$\Rightarrow$ $(i(n), y) = (i(m), y)$.

$\Rightarrow$ $i(n) = i(m)$ by Theorem 1.25.

$\Rightarrow$ Since $i : \omega \to A$ is injective, $n = m$.

$\Rightarrow$ $f : \omega \to A \times B$ is injective.

$\Rightarrow$ $A \times B$ has a denumerable subset $f(\omega)$.

$\Rightarrow$ $A \times B$ is infinite by Theorem 7.26.

Similarly, we can show that $B \times A$ is infinite. □

Theorem 7.30. *A is an infinite set if and only if A is equipotent with a proper subset of itself.*

Proof. **(only if)** A is infinite.

$\Rightarrow$ A has a denumerable subset $B = \{a_0, a_1, a_2, \cdots\}$ by Theorem 7.26.

Define a function $f : A \to A$ by for all $x \in A$,

$$f(x) = \begin{cases} x & \text{if } x \in A \setminus B, \\ a_{m+1} & \text{if } x = a_m \in B. \end{cases}$$

$\Rightarrow$ Clearly, f is a one-to-one correspondence between A and $A \setminus \{a_0\}$.

(if) A is equipotent with a proper subset B of A.

$\Rightarrow$ There exists a bijective function $f : A \to B$.

$\Rightarrow$ $f : A \to A$ is a function by Corollary 2.9 and there exists $c \in A \setminus B$.

$\Rightarrow$ By the recursion theorem(Theorem 6.14), there exists a function $\gamma : \omega \to A$ which satisfies the conditions

I. $\gamma(0) = c$ and II. $\gamma(n^+) = f(\gamma(n))$,

where $c \notin \text{ran}(f)$ because $\text{ran}(f) = B$ and $c \in A \setminus B$.

$\Rightarrow$ $\gamma : \omega \to A$ is injective by Corollary 6.15.

$\Rightarrow$ A has a denumerable subset $\gamma(\omega)$.

$\Rightarrow$ A is infinite by Theorem 7.26. □

Exercise 7.3

1. Let A and B be a pair of disjoint finite sets. Use induction to prove that if $A \approx m$ and $B \approx n$, then $A \cup B \approx m + n$.

2. Using the result of Exercise 7.3.1above, prove that if A is an infinite set and B is a finite subset of A, then $A \setminus B$ is infinite.

3. Prove that a natural number is not equipotent with a proper subset of itself. Conclude that if $A \approx m$ and $n > m$, then A is not equipotent with n.

4. Prove that A is an infinite set if and only if for each $n \in \omega$, A has a subset B such that $B \approx n$.

5. Let A and B be finite sets. Use induction to prove that if $A \approx m$ and $B \approx n$, then $A \times B \approx mn$. Conclude that the Cartesian product of two finite sets is finite.

6. Assuming that if A is an infinite set and B is denumerable, prove that $A \approx (A \cup B)$.

7. Suppose that $x \in A$. Prove that A is an infinite set if and only if $A \approx (A \backslash \{x\})$.

8. Use induction to prove that $A \approx n$, then $\mathscr{P}(A) \approx 2^n$. Conclude that if A is a finite set, then $\mathscr{P}(A)$ is a finite set.

7.4 Properties of denumerable sets

Recall that a set is called *denumerable* if it is in one-to-one correspondence with ω (Definition 7.2). The fundamental properties of denumerable sets are presented in this section.

Lemma 7.31. *Every subset of ω is finite or denumerable.*

Proof. $E \subseteq \omega$.

$\Rightarrow$ E and ω are well-ordered sets by Theorem 6.31.

$\Rightarrow$ Either $E \cong \omega$ or $E \cong S_n = n$ for some $n \in \omega$ by Corollary 4.67.

$\Rightarrow$ Either E is denumerable or finite. □

Theorem 7.32. *Every subset of a denumerable set is finite or denumerable.*

Proof. A is a denumerable set and $B \subseteq A$.

$\Rightarrow$ There exists a bijective function $f : A \to \omega$.

$\Rightarrow$ Since $f(B) \subseteq \omega$, $f(B)$ is finite or denumerable by Lemma 7.31.

$\Rightarrow$ Since $f : A \to \omega$ is bijective, $B \approx f(B)$ and hence B is finite or denumerable. □

Theorem 7.33. $\omega \times \omega \approx \omega$.

Proof. We will use the recursion theorem(Theorem 6.8) to establish the existence of a bijective function from ω to $\omega \times \omega$.

Let $A = \omega \times \omega$.

(Step 1) Define a function $f : A \to A$ as follows:

$$\text{for all } (k, m) \in A,\ \ f(k, m) = \begin{cases} (0, k+1) & \text{if } \ m = 0, \\ (k+1, m-1) & \text{if } \ m \neq 0. \end{cases}$$

First we will show that $f : A \to A$ is injective.

Suppose that $f(k, m) = f(n, p) = (r, s)$.

(Case 1) Assume that $r = 0$.

$\Rightarrow$ $m = 0$ and $p = 0$ by the way f is defined.

$\Rightarrow$ $(0, s) = f(k, m) = (0, k+1)$ and $(0, s) = f(n, p) = (0, n+1)$.

$\Rightarrow$ $n + 1 = s = k + 1$ by Theorem 1.25 and $m = p = 0$.

$\Rightarrow$ $k = n$ and $m = p$ by Theorem 6.12.

$\Rightarrow$ $(k, m) = (n, p)$.

(Case 2) Assume that $r \neq 0$.

$\Rightarrow$ $m \neq 0$ and $p \neq 0$ by the way f is defined.

$\Rightarrow (r,s) = f(k,m) = (k+1, m-1)$ and

$(r,s) = f(n,p) = (n+1, p-1)$.

$\Rightarrow k+1 = r = n+1$ and $m-1 = s = p-1$ by Theorem 1.25.

$\Rightarrow k = n$ by Theorem 6.12 and $m = (m-1)^+ = (p-1)^+ = p$.

$\Rightarrow (k,m) = (n,p)$.

$\Rightarrow (k,m) = (n,p)$ in both cases and hence $f : A \to A$ is injective.

(Step 2) Define a function $\gamma : \omega \to A$ by the two conditions

$\gamma(0) = (0,0)$, and $\gamma(n^+) = f(\gamma(n))$.

$\Rightarrow \gamma : \omega \to A$ is a well-defined function by Recursion Theorem(Theorem 6.14). We will show that γ is bijective.

$\Rightarrow$ (1) Since $(0,0) \notin \mathrm{ran}(f)$, $\gamma : \omega \to A$ is injective by Corollary 6.15.

(2) To show that $\gamma : \omega \to A$ is surjective, let $(k,m) \in A$.

We will show that there exists $n \in \omega$ such that $(k,m) = \gamma(n)$.

The proof is by induction on $k+m$.

(i) If $k+m = 0$, then $(k,m) = (0,0) = \gamma(0)$.

(ii) Assume that $(r,s) \in \mathrm{ran}(\gamma)$ for $r+s = n$ and let $k+m = n^+$.

(Case 1) Assume that $k = 0$.

$\Rightarrow m = n+1$ and $(k,m) = (0,m) = f(m-1, 0)$.

$\Rightarrow$ Since $(m-1, 0) = \gamma(q)$ for some $q \in \omega$ by the hypothesis of induction, $(k,m) = f(m-1,0) = f(\gamma(q)) = \gamma(q^+) \in \mathrm{ran}(\gamma)$.

(Case 2) Assume that $k \neq 0$.

$\Rightarrow k+m-1 = n$ and $(k,m) = f(k-1, m+1)$.

$\Rightarrow (k+m-1, 0) = \gamma(p)$ for some $p \in \omega$ by the hypothesis of induction.

$$\Rightarrow \gamma(p+1) = \gamma(p^+) = f(\gamma(p)) = f(k+m-1, 0) = (0, k+m),$$

$$\begin{aligned}\gamma(p+2) &= \gamma((p+1)^+) = f(\gamma(p+1)) \\ &= f(0, k+m) = (1, k+m-1),\end{aligned}$$

$$\vdots$$

$$\gamma(p+k+1) = f(\gamma(p+k)) = \cdots = (k,m).$$

$\Rightarrow (k,m) = \gamma(p+k+1) \in \mathrm{ran}(\gamma)$.

Thus $\gamma : \omega \to A$ is surjective by induction(Theorem 6.8).

$\Rightarrow \gamma : \omega \to \omega \times \omega$ is bijective by (1) and (2) above. □

Corollary 7.34. *If A and B are denumerable sets, then $A \times B$ is a denumerable set.*

Proof. A and B are denumerable sets.

$\Rightarrow A \approx \omega$ and $B \approx \omega$.

$\Rightarrow A \times B \approx \omega \times \omega$ by Theorem 7.19.

$\Rightarrow$ Since $\omega \times \omega \approx \omega$, $A \times B \approx \omega$.

$\Rightarrow A \times B$ is denumerable. □

Theorem 7.35. *Let $\{A_n\}_{n\in\omega}$ be a denumerable family of denumerable sets, and let $A = \bigcup_{n\in\omega} A_n$. Then A is denumerable. In other words, a denumerable union of denumerable sets is denumerable.*

Proof. A_n is a denumerable set for each $n \in \omega$ and $A = \bigcup_{n\in\omega} A_n$.

$\Rightarrow$ For each $n \in \omega$, there exists a bijective function $f_n : \omega \to A_n$.

Define $\sigma : \omega \times \omega \to A$ by $\sigma(k, m) = f_k(m)$ for all $(k, m) \in \omega \times \omega$.

$\Rightarrow \sigma : \omega \times \omega \to A$ is surjective.

($\because$) $x \in A$.

$\Rightarrow$ Since $A = \bigcup_{n\in\omega} A_n$, $x \in A_n$ for some $n \in \omega$.

$\Rightarrow$ Since $f_n : \omega \to A_n$ is surjective, $x = f_n(m) = \sigma(n, m)$ for some $m \in \omega$.

$\Rightarrow \sigma : \omega \times \omega \to A$ is surjective.

$\Rightarrow$ Since $\omega \approx \omega \times \omega$ by Theorem 7.33, there exists a surjective function from ω to A.

$\Rightarrow$ There exists an injective function from A to ω by Lemma 7.15.

$\Rightarrow A \approx E$ for some subset $E \subseteq \omega$ by Corollary 7.16.

$\Rightarrow$ Since E is either finite or denumerable by Theorem 7.32, A is either finite or denumerable.

$\Rightarrow$ Since $A = \bigcup_{n\in\omega} A_n$ has a denumerable subset, A is not finite by

Corollary 7.27 and hence A is denumerable. □

Corollary 7.36. *The union of two denumerable sets is denumerable.*

Proof. It is a special case of Theorem 7.35, where $A_2 = A_i$ for each $i \geq 3$. □

Exercise 7.4

1. Prove that the union of two denumerable sets is denumerable. (Corollary 7.36.)

2. Let A be a denumerable set. Prove that A has a denumerable subset B such that $A \setminus B$ is denumerable.

3. Prove that $\omega^n \approx \omega$. [Hint: Use the definition $\omega^1 = \omega$ and $\omega^{n^+} = \omega^n \times \omega$. Use Theorems 7.19, 7.33 and induction.]

   Conclude that if A is a denumerable set, then A^n is a denumerable set.

4. Prove that $\omega \cup \omega^2 \cup \omega^3 \cup \cdots$ is a denumerable set.

5. Prove that the set of all finite subsets of ω is denumerable. Then prove that the set of all finite subsets of a denumerable set is denumerable.

6. Let A be an infinite set. Prove that A is denumerable if and only if $A \approx B$ for every infinite subset B of A.

7. Prove that if A is a nonempty finite set and B is denumerable, then $A \times B$ is denumerable.

8. Let $\mathscr{L}$ be the set of all polynomials $a_0 + a_1x + \cdots + a_nx^n$ with integer coefficients. Prove that $\mathscr{L}$ is denumerable. [Hint: This may be proved by using an argument of the kind used in the introduction to prove that $\mathbb{Q}$ is denumerable.]

9. An *algebraic number* is any real root of an equation $a_0 + a_1x + \cdots + a_nx^n = 0$ where the coefficients a_i are integers. Prove that the set of all algebraic numbers is denumerable.

10. A real number is called *transcendental* if it is not algebraic. Prove that the set of all transcendental numbers is nondenumerable.

11. Use the results of Exercises 7.4.1 and 7.4.5 above to prove that the set of all infinite subsets of ω is equipotent with 2^ω.

Chapter 8

Arithmetic of Cardinal Numbers

8.1 Introduction

- In the preceding chapter we defined what is meant by a finite set. It follows from our definition that *every finite set is equipotent with exactly one natural number* n. This fact has an important consequence, namely, that the natural numbers may be used as a set of standards -a scale, as it were- to measure the size of finite sets. If A is any finite set, then A may be "measured" by comparison with the natural numbers, and will be found to correspond -that is, to be equipotent- with exactly one of them. When the natural numbers serve in this capacity -as standards to measure the size of sets- they are commonly called *cardinal numbers*.

- A natural and fascinating question arises now: Can we find a way of extending our system of cardinal numbers so as to create a set of standards for measuring the size of all sets? To put it another way: Can we define "infinite cardinal numbers", and can we construct a sufficient supply of them so that every set has a cardinal number (if A has n elements, we say that A *has cardinal number* n)? The answer is "yes". We can generalize the concept of cardinal number with such remarkable ease that almost all of the properties of the finite cardinals -their ordering, their arithmetic, and so forth- apply as naturally to the infinite cardinals as they did to the finite ones. To take one example: What is meant by the sum of two cardinal numbers? The idea, clearly, is that if A has m elements and B has n elements -and if A and B are disjoint- then $m+n$ is the cardinal number of $A \cup B$. What could be more natural than to extend this notion of cardinal sum to all sets (or rather, to all "set sizes")?

- Before giving a general definition of cardinal numbers, let us take one more look at the natural numbers and see why they can be used as standards to measure the size of finite sets. As we stated earlier, every finite set is equipotent with exactly one natural number; that is, the natural numbers are well defined sets, and there is a unique natural number for each and every finite "set size". It is clear, now, what we expect of our definition of cardinal numbers: The cardinal numbers are to be *well-defined sets*, and *every set is to be equipotent with exactly one cardinal number*. It is immaterial what sets the cardinal numbers are; the only requirement is that there be exactly one cardinal number of each "size".

- A simple way of constructing the cardinal numbers would be the following. We observe that the relation "A is equipotent with B" $(A \approx B)$ is an equivalence relation among sets. Thus we might partition the class of all sets into equipotence classes, and select one representative of each class: the representatives would be our cardinal numbers. This process seems quite natural, and will, indeed, serve as the intuitive basis of our definition. However, it cannot be applied literally. Note, for example, that if A is a set, the equipotence class $\{B \mid B \approx A\}$ may be a proper class. Hence it is not legitimate to speak of the "class of all the equipotence classes". Furthermore, even if we could speak of the "class of all the equipotence classes", it would not be legitimate to use the Axiom of Choice to pick a representative of each class. Indeed, the Axiom of Choice (see the statement Ch 1 of Chapter 5) allows us to pick representatives from a *set of sets*, not from an arbitrary class of classes.

- Since we cannot literally "select" our cardinals by using the Axiom of Choice, how are we to proceed? A simple way, and one which is sanctioned by mathematical tradition, is to *posit* their existence (that is, to posit the existence of a representative set from each "equipotence class") by means of a new axiom.

Axiom 10 (Axiom of Cardinality). There is a class CD of sets, called *cardinal numbers*, with the following properties:

K1 If A is any set, there exists a cardinal number a such that $A \approx a$.

K2 If A is a set and a, b are cardinal numbers, then $A \approx a$ and $A \approx b$ imply $a = b$.

Remark 8.1. (1) We will add the Axiom of Cardinality to our list of axioms for set theory, but only on a provisional basis, for in the next chapter we will describe a method for constructing sets with properties K1 and K2, that is, we will produce actual sets (in much the same way as we produced the natural numbers) which will serve as cardinal numbers.

(2) We will use lower-case Roman letters, such as a, b, c, d, etc., to denote cardinal numbers.

(3) It is worth noting, incidentally, that the class CD of all the cardinal numbers is a proper class.

(∵) Assume that CD is a set and let $V = \bigcup_{a \in \mathrm{CD}} a$.

⇒ Since each cardinal number is a set, $V = \bigcup_{a \in \mathrm{CD}} a$ is a set by Axiom 5 of Union.

⇒ $\mathscr{P}(V)$ is a set by Axiom 6 of Power Set.

⇒ By condition K1 of Axiom 10 of Cardinality, there is a cardinal number e such that $e \approx \mathscr{P}(V)$.

⇒ Since $e \in \mathrm{CD}$, $e \subseteq V$, which is impossible by Corollary 7.6.

⇒ CD is not a set, but a proper class.

8.2 Operations on cardinal numbers

Definition 8.2. If A is a set, a is a cardinal number and $A \approx a$, then we say that a is the *cardinal number* of A. We denote this by writing $a = \sharp A$.

Now conditions K1 and K2 can be conveniently restated as follows:

K1 If A is any set, then there exists a cardinal number a such that $a = \sharp A$.

K2 If A is a set and a, b are cardinal numbers, then $a = \sharp A$ and $b = \sharp A$ imply $a = b$.

Lemma 8.3. *If a and b are cardinal numbers and $a \approx b$, then $a = b$.*

Proof. The proof is an immediate consequence of K2. □

Lemma 8.4. *Let A and B be sets. If $A \approx B$, then $\sharp A = \sharp B$.*

Proof. Let A and B be sets and assume that $A \approx B$.

⇒ There are cardinals a, b such that $a = \sharp A$ and $b = \sharp B$ by K1.

⇒ $a \approx A$ and $b \approx B$.

⇒ Since $A \approx B$, $a \approx A$ and $b \approx A$.

⇒ By the condition K2 of Axiom 10 of Cardinality, $a = b$.

⇒ $\sharp A = \sharp B$. □

We now proceed to define the addition, multiplication and order relation of cardinal numbers. Our definitions require no comment; they correspond in the most natural way to our intuitive understanding of the process of adding and multiplying whole numbers.

Definition 8.5 (Addition and multiplication of cardinal numbers). Let a and b be two cardinals and let A and B be disjoint sets such that $a = \sharp A$ and $b = \sharp B$. Define $a + b$ and ab as follows.

(1) $a + b$ is the cardinal number defined by

$$a + b = \sharp(A \cup B).$$

(2) ab is the cardinal number defined by

$$ab = \sharp(A \times B).$$

Remark 8.6. (1) In Definition 8.5, it has been assumed that we can always find disjoint sets A and B such that $a = \sharp A$ and $b = \sharp B$. This is obviously true. For example, take $A = a \times \{0\}$ and $B = b \times \{1\}$; then A consists of pairs $(x, 0)$ for $x \in a$, whereas B consists of pairs $(y, 1)$ for $y \in b$.

(2) Since a and b are sets, we can write $ab = \sharp(a \times b)$.

The usual algebraic laws for addition and multiplication follow from the elementary properties of sets.

Theorem 8.7. *If a, b, c are cardinal numbers, the following laws hold.*

(1) $a + b = b + a$. (2) $ab = ba$. (3) $a + (b + c) = (a + b) + c$.

(4) $a(bc) = (ab)c$. (5) $a(b + c) = ab + ac$.

Proof. Properties (1), (3), and (5) are immediate consequences of Theorem 1.20(6), (8), and Theorem 1.27 respectively.

(2) $ab = ba$.

($\because$) Let A and B be any sets such that $a = \sharp A$ and $b = \sharp B$.

$\Rightarrow$ The function $\phi : A \times B \to B \times A$ defined by $\phi(x, y) = (y, x)$ for all $(x, y) \in A \times B$, is obviously bijective.

$\Rightarrow \sharp(A \times B) = \sharp(B \times A)$ by Lemma 8.4.

$\Rightarrow ab = \sharp(A \times B) = \sharp(B \times A) = ba$.

(4) $a(bc) = (ab)c$.

($\because$) Let A, B and C be any sets such that $a = \sharp A$, $b = \sharp B$ and $c = \sharp C$.

$\Rightarrow$ The function $\phi : A \times (B \times C) \to (A \times B) \times C$ defined by $\phi(x, (y, z)) = ((x, y), z)$ for all $(x, (y, z)) \in A \times (B \times C)$, is obviously bijective.

$\Rightarrow \sharp(A \times (B \times C)) = \sharp((A \times B) \times C)$ by Lemma 8.4.

$\Rightarrow a(bc) = \sharp(A \times (B \times C)) = \sharp((A \times B) \times C) = (ab)c$. □

Let A and B be finite sets. In Chapter 2 we defined A^B to be the set of all functions from B to A. Now suppose that A has m elements and B has n elements, and consider the process of constructing an arbitrary function from B to A. Since B has n elements, and each element can be assigned an image in one of m possible ways, this means there are exactly m^n distinct functions from B to A. This simple observation suggests the following definition of cardinal exponentiation.

Definition 8.8. Let a and b be two cardinals and let A and B be sets such that $a = \sharp A$ and $b = \sharp B$. Then the cardinal number a^b is defined by

$$a^b = \sharp(A^B).$$

For notational convenience, we agree that $0^a = 0$ and $a^0 = 1$ for any cardinal number a (See Exercise 8.2.1).

Theorem 8.9 (Exponential Laws). *For any cardinals a, b and c, the following rules hold.*

(1) $a^{b+c} = a^b a^c$. (2) $(ab)^c = a^c b^c$. (3) $(a^b)^c = a^{bc}$.

Proof. Let A, B and C be sets such that $a = \sharp A, b = \sharp B$ and $c = \sharp C$.

(1) $a^{b+c} = a^b a^c$. For this we assume that $B \cap C = \emptyset$.

Define $\sigma : A^{B\cup C} \to A^B \times A^C$ by $\sigma(f) = (f|_B, f|_C)$ for all $f \in A^{B\cup C}$, where $f|_B$ and $f|_C$ are the restrictions of f to B and C, respectively.

$\Rightarrow$ (i) $\sigma : A^{B\cup C} \to A^B \times A^C$ is a well-defined function.

(ii) $\sigma : A^{B\cup C} \to A^B \times A^C$ is injective.

($\because$) Suppose that $\sigma(f) = \sigma(g)$ for $f, g \in A^{B\cup C}$.

$\Rightarrow (f|_B, f|_C) = (g|_B, g|_C)$.

$\Rightarrow f|_B = g|_B$ and $f|_C = g|_C$ by Theorem 1.25.

$\Rightarrow f = f|_B \cup f|_C = g|_B \cup g|_C = g$ by Theorem 2.14 and hence σ is injective.

(iii) $\sigma : A^{B\cup C} \to A^B \times A^C$ is surjective.

($\because$) Suppose that $(f_1, f_2) \in A^B \times A^C$.

$\Rightarrow$ Since $B \cap C = \emptyset$, $f = f_1 \cup f_2 \in A^{B\cup C}$, $f_1 = f|_B$ and $f_2 = f|_C$ by Theorem 2.15.

$\Rightarrow (f_1, f_2) = (f|_B, f|_C) = \sigma(f)$ and hence σ is surjective.

$\Rightarrow \sigma : A^{B\cup C} \to A^B \times A^C$ is a bijective function.

$\Rightarrow a^{b+c} = \sharp(A^{B\cup C}) = \sharp(A^B \times A^C) = a^b a^c$.

(2) $(ab)^c = a^c b^c$.

Define $(\sigma(f_1, f_2))(c) = (f_1(c), f_2(c))$ for all $c \in C$ and for all $(f_1, f_2) \in A^C \times B^C$ so that $\sigma(f_1, f_2) \in (A \times B)^C$.

$\Rightarrow$ (i) $\sigma : A^C \times B^C \to (A \times B)^C$ is a well-defined function.

(ii) $\sigma : A^C \times B^C \to (A \times B)^C$ is injective.

($\because$) Suppose that $f = \sigma(f_1, f_2) = \sigma(f_1', f_2') = f'$.

$\Rightarrow (f_1(c), f_2(c)) = f(c) = f'(c) = (f_1'(c), f_2'(c))$ for all $c \in C$.

$\Rightarrow f_1(c) = f_1'(c)$ and $f_2(c) = f_2'(c)$ for all $c \in C$ by Theorem 1.25.

$\Rightarrow f_1 = f_1'$ and $f_2 = f_2'$ by Theorem 2.10.

$\Rightarrow (f_1, f_2) = (f_1', f_2')$ and hence σ is injective.

(iii) $\sigma : A^C \times B^C \to (A \times B)^C$ is surjective.

($\because$) Let $f \in (A \times B)^C$.

$\Rightarrow$ Since $f(c) \in A \times B$ for each $c \in C$, we can write

$f(c) = (f_1(c), f_2(c))$ where $f_1(c) \in A$ and $f_2(c) \in B$.

$\Rightarrow f_1 \in A^C$, $f_2 \in B^C$ so that $(f_1, f_2) \in A^C \times B^C$ and

$f(c) = (f_1(c), f_2(c)) = (\sigma(f_1, f_2))(c)$ for all $c \in C$.

$\Rightarrow f = \sigma(f_1, f_2)$ by Theorem 2.10 and hence σ is surjective.

$\Rightarrow \sigma : A^C \times B^C \to (A \times B)^C$ is a bijective function.

$\Rightarrow (ab)^c = \sharp((A \times B)^C) = \sharp(A^C \times B^C) = a^c b^c$.

(3) $\left(a^b\right)^c = a^{bc}$.

We will define a bijective function $\sigma : \left(A^B\right)^C \to A^{B \times C}$. Note that for $f \in \left(A^B\right)^C$, $c \in C$ and $b \in B$, $f(c) \in A^B$ and $(f(c))(b) \in A$. For $f \in \left(A^B\right)^C$, define $\sigma(f)$ to be the function $\widehat{f} : B \times C \to A$, given by $\widehat{f}(b, c) = (f(c))(b)$ for all $(b, c) \in B \times C$.

$\Rightarrow \sigma(f) = \widehat{f} \in A^{B \times C}$.

$\Rightarrow$ (i) $\sigma : \left(A^B\right)^C \to A^{B \times C}$ is a well-defined function.

(ii) $\sigma : \left(A^B\right)^C \to A^{B \times C}$ is injective.

($\because$) Suppose that $\widehat{f} = \sigma(f) = \sigma(f') = \widehat{f'}$.

$\Rightarrow (f(c))(b) = \widehat{f}(b, c) = (f'(c))(b)$ for all $(b, c) \in B \times C$.

$\Rightarrow f(c) = f'(c)$ for all $c \in C$ by Theorem 2.10.

$\Rightarrow f = f'$ by Theorem 2.10 again and hence σ is injective.

(iii) $\sigma : \left(A^B\right)^C \to A^{B \times C}$ is surjective.

($\because$) Let $g \in A^{B \times C}$. First, for each $c \in C$, define

$f_c : B \to A$ by $f_c(b) = g(b, c)$ for all $b \in B$ and then define

$f : C \to A^B$ by $f(c) = f_c$ for all $c \in C$.

$\Rightarrow g(b, c) = f_c(b) = (f(c))(b) = \widehat{f}(b, c) = \sigma(f)(b, c)$ for all

$(b,c) \in B \times C$, and $f \in \left(A^B\right)^C$.

$\Rightarrow g = \sigma(f)$ by Theorem 2.10 and hence σ is surjective.

$\Rightarrow \sigma : \left(A^B\right)^C \to A^{B\times C}$ is a bijective function.

$\Rightarrow \left(a^b\right)^c = \sharp\left(\left(A^B\right)^C\right) = \sharp\left(A^{B\times C}\right) = a^{bc}$. □

Notation 8.10. (1) The finite cardinal numbers are designated, as usual, by the symbols 0, 1, 2, and so forth. It is to be especially noted that 0 is the cardinal number of the empty set, and 1 is the cardinal number of any singleton.

(2) The cardinal number of ω - that is, the cardinal number of any denumerable set - is customarily designated by the symbol $\aleph_0$ ("aleph-null").

Remark 8.11. It is useful to distinguish between *finite cardinal numbers* which are cardinal numbers of finite sets, and *infinite, or transfinite cardinal numbers*, which are the cardinal numbers of infinite sets. We will see, shortly, that infinite cardinals have several properties which do not hold for finite cardinals.

Exercise 8.2

1. Prove each of the following, where a is any cardinal number.

   (1) $a + 0 = a$. (2) $a0 = 0$. (3) $0^a = 0$.

2. Prove each of the following, where a is any cardinal number.

   (1) $1a = a$. (2) $a^1 = a$. (3) $1^a = 1$.

3. If a, b are arbitrary cardinal numbers, prove that $ab = 0$ if and only if $a = 0$ or $b = 0$.

4. If a, b are arbitrary cardinal numbers, prove that $ab = 1$ if and only if $a = 1$ and $b = 1$.

5. Give a counterexample to the rule: $a+b=a+c$ implies $b=c$ for cardinal numbers a, b and c.

6. Give a counterexample to the rule: $ab=ac$ implies $b=c$ for cardinal numbers a, b and c.

7. If n is a finite cardinal number, use induction to prove that $na=a+a+\cdots+a$, where the right hand side of the equality has n terms.

8. If n is a finite cardinal number, use induction to prove that $a^n=aa\cdots a$, where the right hand side of the equality has n factors.

9. Let a, b be cardinals and let A, B be sets such that $a=\sharp A$ and $b=\sharp B$. Prove that $a+b=\sharp(A\cup B)+\sharp(A\cap B)$.

10. Prove that if a is an infinite cardinal number and n is a finite cardinal number, then $a+n=a$.

11. Prove that if $a+1=a$, then a is an infinite cardinal number.

12. If b is an infinite cardinal number, prove that $\aleph_0+b=b$.

8.3 Ordering of the cardinal numbers

Since cardinal numbers measure the size of sets, we naturally expect the cardinal number of a smaller set to be "less than" the cardinal number of a larger set. This suggests a natural ordering of the cardinal numbers.

Definition 8.12. Let a and b be cardinals, and let A and B be sets such that $a=\sharp A$ and $b=\sharp B$. The relation $\leqslant$ on CD is defined by

$$a \leqslant b \text{ if and only if } A \preceq B.$$

Remark 8.13. Recall that $A \preceq B$ means that A is one-to-one correspondence with a subset of B(Definition 7.1(3)). Let $a, b \in$ CD. Then $a \leqslant b$ if and only if $a \preceq b$ if and only if there exists an injective function $f: a \to b$ if and only if there exists a surjective function $g: b \to a$.

Our goal in this section is to show that the relation $\leqslant$ defined above is an order relation among the cardinal numbers, and, in particular, that the class of all the cardinal numbers is well-ordered with respect to this relation.

Theorem 8.14 (Schröder-Bernstein). *Let a and b be cardinal numbers. If $a \leqslant b$ and $b \leqslant a$, then $a = b$.*

Proof. Suppose that $a \leqslant b$ and $b \leqslant a$ and let A and B be sets such that

$a = \sharp A$ and $b = \sharp B$.

$\Rightarrow$ $A \preceq B$ and $B \preceq A$.

$\Rightarrow$ There exist injective functions $f : A \to B$ and $g : B \to A$.

For $C \subseteq A$, define $\Delta(C) = A \setminus g(B \setminus f(C))$.

$\Rightarrow$ For subsets C and D of A, $C \subseteq D$ implies $\Delta(C) \subseteq \Delta(D)$.

($\because$) Assume that $C \subseteq D$.

$\Rightarrow$ $f(C) \subseteq f(D)$ (this is half of Theorem 2.30).

$\Rightarrow$ $B \setminus f(D) \subseteq B \setminus f(C)$ by elementary class algebra.

$\Rightarrow$ $g(B \setminus f(D)) \subseteq g(B \setminus f(C))$.

$\Rightarrow$ $A \setminus g(B \setminus f(C)) \subseteq A \setminus g(B \setminus f(D))$.

$\Rightarrow$ $\Delta(C) \subseteq \Delta(D)$.

Now, let $S = \{C \mid C \subseteq A \text{ and } C \subseteq \Delta(C)\}$ and let $A_1 = \bigcup_{C \in S} C$.

$\Rightarrow$ $A_1 \subseteq \Delta(A_1)$.

($\because$) Let $x \in A_1$.

$\Rightarrow$ $x \in C$ for some $C \in S$.

$\Rightarrow$ Since $C \subseteq A_1$, $\Delta(C) \subseteq \Delta(A_1)$.

$\Rightarrow$ Since $C \in S$, $x \in C \subseteq \Delta(C) \subseteq \Delta(A_1)$.

$\Rightarrow$ $A_1 \subseteq \Delta(A_1)$.

$\Rightarrow$ $\Delta(A_1) \subseteq \Delta(\Delta(A_1))$.

$\Rightarrow$ Since $\Delta(A_1) = A \setminus g(B \setminus f(A_1)) \subseteq A$, $\Delta(A_1) \in S$.

$\Rightarrow$ Since $A_1 = \bigcup_{C \in S} C$, $\Delta(A_1) \subseteq A_1$.

$\Rightarrow$ Since $A_1 \subseteq \Delta(A_1)$, $A_1 = \Delta(A_1) = A \setminus g(B \setminus f(A_1))$.

$\Rightarrow$ $A \setminus A_1 = g(B \setminus f(A_1))$.

$\Rightarrow$ Since f is injective, $A_1 \approx f(A_1)$ and since g is injective,

$B \setminus f(A_1) \approx g(B \setminus f(A_1)) = A \setminus A_1$.

$\Rightarrow A = A_1 \cup (A \setminus A_1) \approx f(A_1) \cup (B \setminus f(A_1) = B$ by Corollary 7.18.

$\Rightarrow a = b$. □

Remark 8.15. It is immediate that the relation $\leqslant$ between cardinal numbers is reflexive and transitive and by Theorem 8.14, it is antisymmetric, hence it is an order relation.

Theorem 8.16. *Every nonempty class of cardinal numbers has a least element.*

Proof. Let $\mathscr{A}$ be an arbitrary nonempty class of cardinal numbers, and let $a \in \mathscr{A}$.

$\Rightarrow$ If a is the least element of $\mathscr{A}$, we are done.

Assume that a is not the least element of $\mathscr{A}$ and let

$\mathscr{B} = \{b \in \mathscr{A} \mid b < a\}$.

$\Rightarrow$ Since a is not the least element of $\mathscr{A}$, $a \neq \emptyset$ and hence $\mathscr{B} \neq \emptyset$.

Now using the well-ordering theorem(Theorem 5.24), well-order a.

$\Rightarrow$ For each $b \in \mathscr{B}$, $\{x \in a \mid b \approx S_x\} \neq \emptyset$.

($\because$) Let $b \in \mathscr{B}$ and well-order b.

$\Rightarrow$ Since $b < a$, b is not isomorphic with a.

$\Rightarrow$ b is isomorphic with an initial segment of a by Theorem 4.66.

$\Rightarrow \{x \in a \mid b \approx S_x\} \neq \emptyset$.

$\Rightarrow$ Since a is well-ordered, $\{x \in a \mid b \approx S_x\}$ has the least element $\phi(b)(\in a)$ for each $b \in \mathscr{B}$.

$\Rightarrow$ Since $\{\phi(b) \mid b \in \mathscr{B}\}$ is a nonempty subset of a,

$\{\phi(b) \mid b \in \mathscr{B}\}$ has a least element $\phi(d)$ for some $d \in \mathscr{B}$.

$\Rightarrow$ d is the least element of $\mathscr{B}$.

($\because$) Let b be an arbitrary element of $\mathscr{B}$.

$\Rightarrow \phi(d) \leqslant \phi(b)$.

$\Rightarrow S_{\phi(d)} \subseteq S_{\phi(b)}$.

$\Rightarrow$ There exists injective functions $d \to S_{\phi(d)} \xrightarrow{\lambda} S_{\phi(b)} \to b$,

where λ is the inclusion function.

$\Rightarrow d \leqslant b$.

$\Rightarrow d$ is the least element of $\mathscr{A}$ by the construction of $\mathscr{B}$. □

We are able to conclude the following corollary.

Corollary 8.17. *The class* CD *of all the cardinal numbers, ordered by* $\leqslant$, *is well-ordered.*

Proof. It follows directly from Theorem 8.16. □

Corollary 8.18. *If A and B are arbitrary sets, then $\sharp A \leqslant \sharp B$ or $\sharp B \leqslant \sharp A$. Thus the class* CD *of all the cardinal numbers, ordered by* $\leqslant$, *is fully ordered.*

Proof. It follows directly from Proposition 4.51(1) and Corollary 8.17.

It can also be proved by using Theorem 5.24 as follows:

(another proof). Let A and B be sets.

$\Rightarrow A$ and B can be well-ordered by Theorem 5.24 and Theorem 4.66.

$\Rightarrow \sharp A \leqslant \sharp B$ or $\sharp B \leqslant \sharp A$ by Theorem 4.66. □

Corollary 8.19 (Cantor-Bernstein)**.** *If $\sharp A \leqslant \sharp B$ and $\sharp B \leqslant \sharp A$, then $\sharp A = \sharp B$.*

Proof. It follows directly from Theorem 8.14. □

Remark 8.20. It is important to note that Corollary 8.18 and Corollary 8.19 give us valuable new information on the relation $A \preceq B$ among sets.

(1) If A and B are arbitrary sets, then $A \preceq B$ or $B \preceq A$.

(2) If $A \preceq B$ and $B \preceq A$, then $A \approx B$.

The item (2) is especially useful when we need to prove that two sets are in one-to-one correspondence, for it is now sufficient to show that there is an injective function from A to B and an injective function from B to A

(alternatively, a surjective function from A to B and a surjective function from B to A).

The familiar "rules of inequality" apply to the cardinal numbers, as we shall see next.

Theorem 8.21. *Let a and b be cardinal numbers. Then $a \leqslant b$ if and only if there exists a cardinal number c such that $b = a + c$.*

Proof. **(if)** Let a and b be cardinal numbers. Assume that $b = a + c$ for some cardinal number c and let A, B and C be sets such that $A \cap C = \emptyset$, $a = \sharp A, b = \sharp B$, and $c = \sharp C$.

$\Rightarrow$ Since $b = a + c$ and $A \cap C = \emptyset$, $\sharp(A \cup C) = \sharp B$.

$\Rightarrow$ There exists a bijective function $f : A \cup C \to B$.

$\Rightarrow$ $f|_A : A \to B$ is an injective function.

$\Rightarrow$ $A \preceq B$.

$\Rightarrow$ $a \leqslant b$.

(only if) Assume that $a \leqslant b$ and let A, B be disjoint sets such that $a = \sharp A$ and $b = \sharp B$.

$\Rightarrow$ There exists an injective function $f : A \to B$.

$\Rightarrow$ $A \approx f(A)$, and so $a = \sharp f(A)$. Let $C = B \setminus f(A)$ and $c = \sharp C$.

$\Rightarrow$ Since $C \cap f(A) = \emptyset$, $\sharp(f(A) \cup C) = \sharp f(A) + \sharp C$.

$\Rightarrow$ $b = \sharp B = \sharp(f(A) \cup C) = \sharp f(A) + \sharp C = a + c$. □

Theorem 8.22. *Let a, b, c, d be cardinal numbers. If $a \leqslant c$ and $b \leqslant d$, then we have the following:*

(1) $a + b \leqslant c + d$. (2) $ab \leqslant cd$. (3) $a^b \leqslant c^d$.

Proof. Let a, b, c, d be cardinal numbers. Assume that $a \leqslant c$ and $b \leqslant d$.

$\Rightarrow$ By Theorem 8.21, there exist r and s such that $c = a + r$ and $d = b + s$.

$\Rightarrow$ (1) $a + b \leqslant c + d$.

($\because$) $c + d = (a + r) + (b + s) = (a + b) + (r + s)$ by Theorem 8.7.

$\Rightarrow a + b \leqslant c + d$ by Theorem 8.21.

(2) $ab \leqslant cd$.

($\because$) $cd = (a + r)(b + s) = ab + as + rb + rs = ab + (as + rb + rs)$

by applying Theorem 8.7.

$\Rightarrow ab \leqslant cd$ by Theorem 8.21.

(3) $a^b \leqslant c^d$.

($\because$) Let A, B and R be sets such that $A \cap R = \emptyset$, $a = \sharp A$,

$b = \sharp B$ and $r = \sharp R$. Define $\sigma : A^B \to (A \cup R)^B$ by

$\sigma(f) = f$ as a function from B to $A \cup R$ for all $f \in A^B$.

$\Rightarrow$ (i) $\sigma : A^B \to (A \cup R)^B$ is well-defined because every function $f : B \to A$ can be regarded as a function $f : B \to A \cup R$ by Corollary 2.9.

(ii) Clearly $\sigma : A^B \to (A \cup R)^B$ is injective.

$\Rightarrow A^B \preceq (A \cup R)^B$.

$\Rightarrow a^b = \sharp\left(A^B\right) \leqslant \sharp\left((A \cup R)^B\right) = (a + r)^b$.

$\Rightarrow$ Since $a + r = c$, $a^b \leqslant c^b$.

$\Rightarrow$ Since $1 \leqslant c^s$ and $b + s = d$, $a^b = a^b 1 \leqslant c^b c^s = c^{b+s} = c^d$

by applying Exercise 8.2.2, Theorem 8.9 and (2) above. □

Exercise 8.3

1. Prove that if A and B are arbitrary sets, then $A \preceq B$ or $B \preceq A$.

2. Prove that if $A \preceq B$ and $B \preceq A$ are arbitrary sets, then $A \approx B$.

3. Prove the following, where a, b, c, d are cardinal numbers.

   (1) If $a^c < b^c$, then $a < b$.

   (2) If $a^c < a^d$, then $c < d$.

4. Prove that if $a + b = c$, then $(r + s)^c \geqslant r^a s^b$.

5. Prove that there exists a strictly increasing sequence $a_1 < a_2 \cdots$ of cardinal numbers, each with the property $a_i^{\aleph_0} = a_i$. [Hint: Take $a_1 = \aleph_0^{\aleph_0}$. Then take $a_2 = 2^{a_1 \aleph_0}, a_3 = 2^{a_2 \aleph_0}, \cdots$.]

6. Let A be a denumerable set. Prove each of the following.

   (1) $A^A \subseteq \mathscr{P}(A \times A)$. Conclude that $A^A \preceq \mathscr{P}(A \times A)$ and hence $A^A \preceq \mathscr{P}(A)$.

   (2) Verify that the function $\phi : A^A \to \mathscr{P}(A)$ given by $\phi(f) = \text{ran}(f)$ for all $f \in A^A$, is a surjective function. Conclude that $\mathscr{P}(A) \preceq A^A$.

   (3) $A^A \approx \mathscr{P}(A)$. That is $A^A \approx 2^A$. Conclude that $\aleph_0^{\aleph_0} = 2^{\aleph_0}$.

7. Use the argument outlined in the preceding problem to prove that the set of all injective functions $A \to A$ is equivalent with 2^A.

8.4 Special properties of infinite cardinal numbers

A few remarkable arithmetic rules hold exclusively for infinite cardinals. As a result of these rules, the arithmetic of infinite cardinal numbers is a very simple matter.

Theorem 8.23. *If a is an infinite cardinal number, then $aa = a$.*

Proof. Let A be an infinite set such that $a = \sharp A$.

$\Rightarrow$ A has a denumerable subset D by Theorem 7.26.

$\Rightarrow$ $D \approx D \times D$ by Corollary 7.34.

$\Rightarrow$ There exists a bijective function $\phi : D \to D \times D$.

Now let $\mathscr{A}$ be the set of all pairs (B, f) which satisfy the following three conditions:

(i) $D \subseteq B \subseteq A$.

(ii) $f : B \to B \times B$ is a bijective function.

(iii) $\phi \subseteq f$.

$\Rightarrow$ $\mathscr{A}$ is nonempty because $(D, \phi) \in \mathscr{A}$.

We order $\mathscr{A}$ by the relation $\leqslant$ defined by

$(B_1, f_1) \leqslant (B_2, f_2)$ if and only if $B_1 \subseteq B_2$ and $f_1 \subseteq f_2$.

$\Rightarrow$ Clearly $(\mathscr{A}, \leqslant)$ is an order relation and $\mathscr{A}$ satisfies the hypotheses of Zorn's Lemma(Theorem 5.20).

($\because$) Let $\mathscr{C} = \{(B_i, f_i) \mid i \in I\} \subseteq \mathscr{A}$ be a chain of $\mathscr{A}$ and let

$B = \bigcup_{i \in I} B_i$ and $f = \bigcup_{i \in I} f_i$.

$\Rightarrow (B, f) \in \mathscr{A}$ and (B, f) is an upper bound of $\mathscr{C}$ and hence $\mathscr{A}$ is inductive.

(The details are left as an exercise for the reader).

$\Rightarrow \mathscr{A}$ has a maximal element (C, g) by Zorn's Lemma(Theorem 5.20).

To show that $\sharp C = a$, suppose that $\sharp C < a$ and let $b = \sharp C (< a)$.

$\Rightarrow$ Since $C \times C \approx C$, we have $b = bb$, C is not finite and hence $2 < b$.

$\Rightarrow b = 0 + b \leqslant b + b = 1b + 1b = (1 + 1)b = 2b \leqslant bb = b$ by applying Exercise 8.2.1, Theorem 8.22, Exercise 8.2.2 and Theorem 8.7.

$\Rightarrow b = b + b$. Now let $d = \sharp(A \setminus C)$.

Since $C \cap (A \setminus C) = \emptyset$, $a = \sharp A = \sharp(A \setminus C) + \sharp C = d + b$.

$\Rightarrow b < d$.

($\because$) Suppose that $d \leqslant b$.

$\Rightarrow a = d + b \leqslant b + b = b$ which contradicts to our assumption that $b < a$.

$\Rightarrow A \setminus C$ has a subset E such that $\sharp E = b$.

$\Rightarrow (C \cup E) \times (C \cup E) = (C \times C) \cup (C \times E) \cup (E \times C) \cup (E \times E)$, where $C \times C, C \times E, E \times C, E \times E$ are mutually disjoint sets, each of which has the cardinal $bb = b$.

$\Rightarrow \sharp((C \times E) \cup (E \times C) \cup (E \times E)) = b + b + b = b + b = b$.

$\Rightarrow$ There exists a bijective function $h : E \to (C \times E) \cup (E \times C) \cup (E \times E)$.

$\Rightarrow$ By Theorem 7.17, $g \cup h$ is a bijective function from $C \cup E$ to $(C \times C) \cup ((C \times E) \cup (E \times C) \cup (E \times E)) = (C \cup E) \times (C \cup E)$.

$\Rightarrow$ Since $D \subseteq C \subseteq C \cup E \subseteq A$ and $\phi \subseteq g \subseteq g \cup h$, $(C \cup E, g \cup h) \in \mathscr{A}$ and $(C \cup E, g \cup h) > (C, g)$ which contradicts to the maximality of (C, g) and hence $a = b = \sharp C$.

$\Rightarrow a = \sharp C = \sharp(C \times C) = aa.$ □

Corollary 8.24. *Let a and b be cardinal numbers, where a is infinite. If $1 \leqslant b \leqslant a$, then $ab = a$.*

Proof. Assume that $1 \leqslant b \leqslant a$ and that a is infinite.

$\Rightarrow a = a1 \leqslant ab \leqslant aa = a$ by Theorem 8.22 and Theorem 8.23.

$\Rightarrow ab = a.$ □

Corollary 8.25. *If a is an infinite cardinal number, $a + a = a$.*

Proof. Since $1 < 2 < a$, $a = 1a \leqslant 2a \leqslant aa = a$ and $2a = (1+1)a = a + a$ by applying Theorem 8.7, Theorem 8.21 and Theorem 8.23.

$\Rightarrow a + a = a.$ □

Corollary 8.26. *Let a and b be cardinal numbers, where a is infinite. If $b \leqslant a$, then $a + b = a$.*

Proof. Since $a = a + 0 \leqslant a + b \leqslant a + a = a$ by applying Theorem 8.21, Theorem 8.23 and Corollary 8.25, $a + b = a$. □

Corollary 8.27. *Let a and b be infinite cardinal numbers. Then $a + b = ab = \max\{a, b\}$.*

Proof. It follows from Corollaries 8.24 and 8.26. □

Theorem 8.28. *Let a be a cardinal number and let b be an infinite cardinal number. If $1 < a \leqslant b$, then $a^b = 2^b$.*

Proof. Assume that $1 < a \leqslant b$ and that b is infinite.

$\Rightarrow$ (i) Since $a \leqslant 2^a$ by Corollary 7.8 and $ab = b$ by Corollary 8.24, $a^b \leqslant (2^a)^b = 2^{ab} = 2^b$ by Theorem 8.22.

(ii) Since $2 \leqslant a$, $2^b \leqslant a^b$ by Theorem 8.22.

$\Rightarrow a^b = 2^b.$ □

Remark 8.29. Theorem 8.23 and its corollaries can be interpreted very profitably in terms of sets and the relation $A \preceq B$ among sets. For example, Theorem 8.23 tells us that if A is an infinite set, then $A \approx A \times A$. This has an interesting consequence: $A \times A$ has a partition $\{B_x\}_{x \in A}$ where $B_x = \{(x, y) \mid y \in A\}$. Hence the bijective function from $A \times A$ to A induces a corresponding partition $\{C_x\}_{x \in A}$ of A, where A is the index set and each member of the partition is equipotent with A.

Exercise 8.4

1. If a is an infinite cardinal number and $a \leqslant bc$, prove that $a \leqslant b$ or $a \leqslant c$.
2. Let a be a cardinal number, and let b be an infinite cardinal number. Prove that if $1 < a$ and $a = a^b$, then $b < a$.
3. An infinite cardinal number a is said to be *dominant* if it satisfies the following condition:

   if b and c are cardinal numbers such that $b < a$ and $c < a$, then $b^c < a$.

   Prove that a is a dominant cardinal number if and only if $d < a$ implies $2^d < a$.
4. If a, c, and d are arbitrary cardinal numbers and b is an infinite cardinal number, prove that $a + b \leqslant c + d$ implies $a \leqslant c$ or $b \leqslant d$.
5. Let a, b, c, d be cardinal numbers. Prove that if $a < b$ and $c < d$, then $ac < bd$ and $a+c < b+d$. [Hint: For the case where b and d are both finite, this result has been proved in Exercises 6.4.5 and 6.4.6. Ignore this case and assume that b is infinite or d is infinite. (This assumption includes three cases.)]
6. Let a and b be cardinal numbers. Prove that if $a + a = a + b$, then $a \geqslant b$.
7. Let a, b and c be cardinal numbers. Prove that if $a + b < a + c$, then $b < c$.
8. Let a, b and c be cardinal numbers. Prove that if $ab < ac$, then $b < c$.

8.5 Infinite sums and products of cardinal numbers

Early in this book we spoke of the union of two classes and later we extended this notion by defining the union of an arbitrary family of classes. Similarly, we introduced the Cartesian product of two classes and later generalized this to the product of a family of classes. In both cases, extending our original definition seemed like a perfectly natural thing to do, for the intuitive concepts of union and product can be applied as easily to a family of classes as to a pair of classes. The same holds true for the process of adding and multiplying cardinal numbers. They lend themselves to the following obvious generalization.

Definition 8.30. Let $\{a_i\}_{i\in I}$ be a family of cardinal numbers and let $\{A_i\}_{i\in I}$ be a family of sets such that $a_i = \sharp A_i$ for each $i \in I$.

(1) Assume that the family $\{A_i\}_{i\in I}$ are pairwise disjoint.

$\sum\limits_{i\in I} a_i$ is, by definition, the cardinal number defined by $\sum\limits_{i\in I} a_i = \sharp\big(\bigcup\limits_{i\in I} A_i\big)$.

(2) $\mathop{\mathsf{X}}\limits_{i\in I} a_i$ is, by definition, the cardinal number defined by $\mathop{\mathsf{X}}\limits_{i\in I} a_i = \sharp\big(\prod\limits_{i\in I} A_i\big)$.

In elementary arithmetic we learn that ab is the result of "adding a to itself b times", and that a^b is the result of "multiplying a by itself b times". It is useful to know that this holds true for all cardinal numbers a and b.

Theorem 8.31. *Let a and b be cardinal numbers, and let I be a set such that $b = \sharp I$. If $a = a_i$ for all $i \in I$, then*

(1) $ab = \sum\limits_{i\in I} a_i$ *and* (2) $a^b = \mathop{\mathsf{X}}\limits_{i\in I} a_i$.

Proof. Let a and b be cardinal numbers, and let I be a set such that $b = \sharp I$. Assume that $a_i = a$ for all $i \in I$.

(1) Let $\{A_i\}_{i\in I}$ be a family of disjoint sets such that $a_i = \sharp A_i$, for all

$i \in I$ and let A be a set such that $a = \sharp A$.

$\Rightarrow$ Since $a = a_i$, $A_i \approx A$ for each $i \in I$.

$\Rightarrow$ There exists a bijective function $f_i : A \to A_i$ for each $i \in I$.

Define $f : A \times I \to \bigcup_{i \in I} A_i$ by $f(x, i) = f_i(x)$ for all $(x, i) \in A \times I$.

$\Rightarrow$ It is elementary to verify that $f : A \times I \to \bigcup_{i \in I} A_i$ is a bijective function, which is left for the reader.

$\Rightarrow$ $A \times I \approx \bigcup_{i \in I} A_i$.

$\Rightarrow$ $ab = \sharp(A \times I) = \sharp(\bigcup_{i \in I} A_i) = \sum_{i \in I} a_i$.

(2) Let A be a set and let $A_i = A$ for each $i \in I$. Write $a = \sharp A$ and $a_i = \sharp A_i$ for each $i \in I$.

$\Rightarrow$ $\bigcup_{i \in I} A_i = A$ and $a = a_i$ for all $i \in I$.

$$\Rightarrow \text{Since } \prod_{i \in I} A_i = \left\{ f \mid f : I \to \bigcup_{i \in I} A_i \text{ is a function, } f(i) \in A_i \text{ for all } i \in I \right\}$$
$$= \{ f \mid f : I \to A \text{ is a function} \} = A^I,$$

$a^b = \sharp(A^I) = \sharp(\prod_{i \in I} A_i) = \mathsf{X}_{i \in I}\, a_i$. □

Theorem 8.22 has the following analogue for infinite sums and products.

Theorem 8.32. *Let $\{a_i\}_{i \in I}$ and $\{b_i\}_{i \in I}$ be families of cardinal numbers. If $a_i \leqslant b_i$ for each $i \in I$, then*

(1) $\sum_{i \in I} a_i \leqslant \sum_{i \in I} b_i$ *and* (2) $\mathsf{X}_{i \in I}\, a_i \leqslant \mathsf{X}_{i \in I}\, b_i$.

Proof. Assume that $a_i \leqslant b_i$ for each $i \in I$ and let $\{A_i\}_{i \in I}$ and $\{B_i\}_{i \in I}$ be families of pairwise disjoint sets with $a_i = \sharp A_i$ and $b_i = \sharp B_i$.

$\Rightarrow$ There exists an injective function $f_i : A_i \to B_i$ for each $i \in I$.

(1) It is easy to verify that $f = \bigcup_{i \in I} f_i$ is an injective function from $\bigcup_{i \in I} A_i$ to $\bigcup_{i \in I} B_i$. (The details are left for the reader.)

$\Rightarrow$ $\sum_{i \in I} a_i = \sharp(\bigcup_{i \in I} A_i) \leqslant \sharp(\bigcup_{i \in I} B_i) = \sum_{i \in I} b_i$.

(2) Define a function $f : \prod_{i \in I} A_i \to \prod_{i \in I} B_i$ as follows:

for $x \in \prod_{i \in I} A_i$ and $y \in \prod_{i \in I} B_i$, $f(x) = y$ iff $f_i(x_i) = y_i$ for all $i \in I$.

$\Rightarrow f : \prod_{i\in I} A_i \to \prod_{i\in I} B_i$ is injective.

(∵) Assume that $f(u) = f(v)$ for $u, v \in \prod_{i\in I} A_i$.

$\Rightarrow f_i(u_i) = f_i(v_i)$ for all $i \in I$.

$\Rightarrow$ Since each f_i is injective, $u_i = v_i$ for all $i \in I$.

$\Rightarrow u = v$ and hence f is injective.

$\Rightarrow \bigtimes_{i\in I} a_i = \sharp(\prod_{i\in I} A_i) \leqslant \sharp(\prod_{i\in I} B_i) = \bigtimes_{i\in I} b_i$. □

Theorems 8.31 and 8.32 have the following useful corollary.

Corollary 8.33. *Let $\{a_i \mid i \in I\}$ be a set of cardinal numbers, and let b and c be cardinal numbers. If $a_i \leqslant b$ for each $i \in I$ and if $\sharp I = c$, then*

(1) $\sum_{i\in I} a_i \leqslant bc$ *and* (2) $\bigtimes_{i\in I} a_i \leqslant b^c$.

Proof. The proof follows immediately from Theorems 8.31 and 8.32 and is left for the reader. □

Exercise 8.5

1. Prove that $\bigtimes_{i\in I} a_i = 0$ if and only if $a_i = 0$ for some $i \in I$.
2. Suppose $a_i \leqslant a$ for all $i \in I$ and $\sharp I \leqslant a$, where a is some fixed infinite cardinal number. Prove that $\sum_{i\in I} a_i \leqslant a$. [Hint: Use Theorems 8.31 and 8.32.]
3. Suppose $a_i \leqslant a$ for all $i \in I$ and $\sharp I \leqslant a$, where a is some fixed infinite cardinal number. Prove that $\bigtimes_{i\in I} a_i \leqslant 2^a$.
4. Let $\{a_i\}_{i\in I}$ be a set of cardinal numbers and suppose that there is no greatest element in $\{a_i\}_{i\in I}$. Prove that $a_j < \sum_{i\in I} a_i$ for all $j \in I$.
5. Let a and $b_i (i \in I)$ be cardinal numbers. Prove that $a\sum_{i\in I} b_i = \sum_{i\in I} ab_i$.
6. Use Theorem 8.32 to justify each of the following. (Each sum is understood to have $\aleph_0$ terms.)

   (1) $1+2+3+\cdots = \aleph_0$. (2) $n+n+n+\cdots = \aleph_0$. (3) $\aleph_0+\aleph_0+\aleph_0+\cdots = \aleph_0$.

7. Let $f : A \to B$ be a surjective function, where B is an infinite set. If $f^{-1}(y)$ is finite or denumerable for each $y \in B$, prove that $A \approx B$.

8. Let A be an infinite set and let $F(A)$ designate the family of all finite subsets of A. Prove that $F(A) \approx A$. [Hint: For each $n \in \omega$, let F_n designate the family of n-element subsets of A. There exists an obvious surjective function from A^n to F_n. There are $\aleph_0$ sets F_n.]

9. If $\{C_i\}_{i \in I}$ is a family of sets, prove that $\sharp\big(\bigcup_{i \in I} C_i\big) \leqslant \sum_{i \in I} \sharp C_i$.

10. Let $\{a_i\}_{i \in I}$ and $\{a_i\}_{i \in J}$ be families of cardinal numbers. Prove that $\sum_{i \in I} a_i \leqslant \sum_{i \in I \cup J} a_i$.

Chapter 9

Arithmetic of Ordinal Numbers

9.1 Introduction

- In elementary school we learn that there are cardinal numbers and ordinal numbers. The cardinal numbers, we are told, are the "counting" numbers: 1, 2, 3, and so on. The ordinal numbers are the "ranking" numbers: first, second, third, etc. The distinction may appear to be somewhat pedantic, for the natural numbers serve in both capacities, as ordinals and as cardinals, and there is no need in elementary arithmetic to differentiate between the two. However, one of the unexpected discoveries of modern set theory is that, just as the infinite cardinals behave differently from the finite ones, so the infinite ordinals exhibit a strikingly different behavior from the cardinals. It is the purpose of this chapter to introduce the ordinal numbers and explore their properties, especially those of the infinite ordinals.

- The dichotomy between cardinal and ordinal, from the scholastic point of view, arises from two different ways of using the natural numbers.

In their role as cardinals, the natural numbers measure the "size", or power, of sets; in their role as ordinals, they serve to designate the rank, or position, of an object in a linearly ordered array. We will use this insight, although it is somewhat outdated, as the starting point of our discussion.

- When we speak of ranking elements in some order, what kind of order do we have in mind? There must be a first element, a second element, and so on. In other words, the order is that of the natural numbers, which is a well-ordering. Now the reader should note that the general notion of well-ordering is an extension of the order of the natural numbers. Every infinite well-ordered set has a first element, a second element, and for each natural number n, an nth element. But it may also have elements which are "beyond the reach" of the finite ordinals. Thus the set

$$\{x_1, x_2, x_3, \cdots, y_1, y_2, y_3, \cdots\}$$

has a first element x_1, a second element x_2, and so forth. But y_1, for example, though it has a perfectly well-defined "position" in the set, cannot be described as the "nth element" for any finite n.

- Situations of this kind arise frequently in almost every branch of mathematics. For example, in Remark 7.9(1), we defined a sequence of sets by these conditions: $\omega = K_1$, $\mathscr{P}(K_1) = K_2$, and so on; $\bigcup_{i\in\omega} K_i = L_1$, $\mathscr{P}(L_1) = L_2$, etc. Continuing in this manner, we get the well-ordered family of sets:

$$K_1 \prec K_2 \prec K_3 \prec \cdots \prec L_1 \prec L_2 \prec L_3 \prec \cdots.$$

Now K_1 is the first element of this family, and, in general, K_n is the nth element; but what of (say) L_1? Its position in the family is unambiguous: L_1 immediately follows all of the sets K_i. Yet classical mathematics has not provided us with any ordinal number o which would enable us to describe L_1 as the "oth element of the set".

Thus, as in our study of cardinal numbers, we are led to ask an intriguing question:

Question 9.1. Can we find a way of extending our system of ordinal numbers so as to create a set of standards for designating the position of any element in any well-ordered set?

The answer, once again, is "yes". We can generalize the concept of ordinal number with such remarkable ease that no barrier, either logical or intuitive, seems to separate the finite ordinals from the infinite ones.

- We will approach the ordinals in much the same way that we approached the cardinals. We will begin by defining a relation of "having the same ordinal number", and later define the ordinals, essentially, to be representative of the distinct classes induced by this relation.

  If A and B are well-ordered sets, and if $x \in A$ and $y \in B$, then to say that "x has the *same rank* as y" (for example, x is the third element of A and y is the third element of B) is the same as saying that the initial segment S_x is isomorphic with the initial segment S_y. To say that "x has a *lower rank* than y" is the same as saying that S_x is isomorphic with an initial segment of S_y. The reader should stop here until he has thoroughly understood this fact, for it is the point of departure for achieving an understanding of the modern approach to the ordinal numbers. Isomorphism plays the same role in the study of ordinal numbers that one-to-one correspondence plays in the study of cardinal numbers.

- An important warning needs to be given here. The alert reader may question the necessity of introducing the concept of isomorphism. After all, he may ask, why not say that x and y have the same rank if and only if S_x is equipotent with S_y? Surely if ten elements precede x and ten elements precede y, then x and y are both eleventh in their class. This is true when we are dealing with finite rank, but untrue in the case of infinite rank. A simple example will convince the reader. In the set

$$\{x_1, x_2, x_3, \cdots, y_1, y_2, y_3, \cdots\},$$

both y_1 and y_2 are preceded by a denumerable number of elements - that is, S_{y_1} is equipotent with S_{y_2}. Yet y_2 clearly follows y_1.

- When we say that x has the same rank as y, or x has a lower rank than y, we are only apparently speaking of x and y. Actually, we are comparing the initial segments S_x and S_y. Hence we lose nothing if we confine our attention to the study of initial segments of well-ordered sets. But we can go a step further. An initial segment (of a well-ordered set) is a well-ordered set and conversely, every well-ordered set A is an initial segment. (adjoin a last element x to A, then A is S_x). Hence the study of ordinality is, essentially, the study of well-ordered sets.

Motivated by the foregoing remarks, we introduce the following definitions:

Definition 9.2. Let A and B be well-ordered sets.

(1) We say that A and B are *similar*(or have the *same ordinality*) if A is isomorphic with B as partially ordered sets. (See Definition 4.12(3).) In this case, we also write $A \cong B$.

(2) We say that B is a *continuation* of A, or A has a *lower ordinality* than B if A is isomorphic with an initial segment of B. In this case, we write $A \lhd B$.

It follows from Theorem 4.66 that if A and B are any two well-ordered sets, then $A \cong B$, or $A \lhd B$, or $B \lhd A$. In conclusion,

(1) $S_x \cong S_y$ if and only if x has the same rank as y and

(2) $S_x \lhd S_y$ if and only if x has a lower rank than y.

Thus we have completely captured, and formalized, the intuitive concept of "rank", and have extended it beyond the unnatural confines of finite ordinality. As for the ordinal numbers, we simply imitate the procedure we followed for the cardinals by introducing the following axiom.

Axiom 11 (Axiom of Ordinality)**.** There is a class OR of well-ordered sets, called *ordinal numbers*, with the following properties:

O1 If A is any well-ordered set, then there exists an ordinal number α such that $A \cong \alpha$.

O2 If A is a well-ordered set and α, β are ordinal numbers, then $A \cong \alpha$ and $A \cong \beta$ imply $\alpha = \beta$.

Remark 9.3. We will add the Axiom of Ordinality to our list of axioms for set theory, but only on a provisional basis, for in the last section of this chapter we will describe a method for constructing sets with properties O1 and O2. Those sets will then serve the purpose of ordinal numbers.

Proposition 9.4. *The class of all the ordinal numbers is a proper class.*

Proof. Let OR be the class of all ordinal numbers, and suppose that OR is a set. Let $B = \bigcup_{\alpha \in \mathrm{OR}} \alpha$ and let $A = \mathscr{P}(B)$.

$\Rightarrow$ Since each ordinal number α is a set and OR is a set under our assumption, B is a set by Axiom 5 of Union.

$\Rightarrow$ By Axiom 6 of Power Set, A is a set. Let us well-order A.

$\Rightarrow$ By O1, there is an ordinal number $\alpha \in \mathrm{OR}$ such that $\alpha \cong A$.

But, since $\alpha \subseteq B$ and $A = \mathscr{P}(B)$, α cannot be equipotent with A by Corollary 7.6.

$\Rightarrow$ This contradiction proves that OR is a proper class. □

9.2 Operations on ordinal numbers

Following our "naive" introduction to ordinal numbers in the preceding section, we now proceed to study the ordinals from a formal point of view. We will henceforth consider the ordinals to be the objects defined by Conditions O1 and O2. The reader should adjust his thinking accordingly. He should cease thinking of ordinals as "symbols for designating rank," and begin to think of them as certain well-ordered sets.

Definition 9.5. We say that α is the *ordinal number* of A if A is a well-ordered set, α is an ordinal number, and $A \cong \alpha$. We denote this by writing

$$\alpha = \oslash A.$$

Now Conditions O1 and O2 can be conveniently restated as follows:

O1 If A is a well-ordered set, then there exists an ordinal number α such that $\alpha = \oslash A$.

O2 If A is a well-ordered set and α, β are ordinal numbers, then $\alpha = \oslash A$ and $\beta = \oslash A$ imply $\alpha = \beta$.

Lemma 9.6. *If α and β are ordinal numbers and $\alpha \cong \beta$, then $\alpha = \beta$.*

Proof. The proof is an immediate consequence of O2. □

Lemma 9.7. *If $A \cong B$, then $\oslash A = \oslash B$.*

Proof. The proof is analogous to that of Lemma 8.4. □

Before defining the addition and multiplication of ordinal numbers, we need to introduce two new operations on well-ordered sets.

Definition 9.8. Let A and B be disjoint, well-ordered sets. The *ordinal sum* of A and B, denoted by $A \oplus B$, is the set $A \cup B$ ordered as follows.

For $x, y \in A \cup B$, $x \leqslant y$ if and only if

(i) $x \in A$ and $y \in A$ and $x \leqslant y$ in A, or

(ii) $x \in B$ and $y \in B$ and $x \leqslant y$ in B, or (iii) $x \in A$ and $y \in B$.

Having defined the ordinal sum of two well-ordered sets, it is natural to define the ordinal sum of an arbitrary family of well-ordered sets.

Definition 9.9. Let I be a well-ordered set and let $\{A_i\}_{i \in I}$ be a family of disjoint well-ordered sets. The *ordinal sum* of the family $\{A_i\}_{i \in I}$, denoted by $\bigoplus_{i \in I} A_i$, is defined to be the set $\bigcup_{i \in I} A_i$ ordered in the following way:

For $x, y \in \bigcup_{i \in I} A_i$, $x \leqslant y$ if and only if

(i) for some $i \in I$, $x \in A_i$ and $y \in A_i$ and $x \leqslant y$ in A_i, or

(ii) $x \in A_i$ and $y \in A_j$ and $i < j$.

Remark 9.10. (1) In $A \oplus B$, the elements of A are ordered as before, the elements of B are ordered as before, and every element of B is greater than every element of A.

(2) In $\bigoplus_{i \in I} A_i$, each set A_i is ordered as before, and, for $i < j$, every element of A_i is less than every element of A_j.

An easy step leads us, now, to the notion of ordinal product. To put it simply, the ordinal product $A \odot B$ is the result of "adding A to itself B times". More precisely, if $\{A_i\}_{i \in B}$ is a family of disjoint, well-ordered sets, indexed by B, where each A_i is similar to A, then $A \odot B$ is the well-ordered set $\bigoplus_{i \in B} A_i$.

The only remaining difficulty is to produce the *disjoint* family $\{A_i\}_{i \in B}$. To do so is easy enough: for each $i \in B$, we define A_i to be the set $\{(x, i) \mid x \in A\}$, that is, $A_i = A \times \{i\}$. But a happy thought strikes us now, as we realize that the well-ordered set $\bigoplus_{i \in B} A_i$ is none other than the Cartesian product $A \times B$ ordered by the antilexicographic ordering(Definition 4.5). We exploit this fortunate coincidence to give the following elegant definition of ordinal product:

Definition 9.11. Let A and B be well-ordered sets. The *ordinal product* of A and B, denoted by $A \odot B$, is defined to be the set $A \times B$ ordered by the antilexicographic ordering.

Example 9.12. Let $A = \{a, b, c\}$ be well-ordered by $a < b < c$ and let $B = \{1, 2, 3\}$ be well-ordered by $1 < 2 < 3$. Then $A \odot B$ is the set $A \times B$ well-ordered as follows. (see Figure 9.1):

$$(a, 1) < (b, 1) < (c, 1) < (a, 2) < (b, 2) < (c, 2) < (a, 3) < (b, 3) < (c, 3).$$

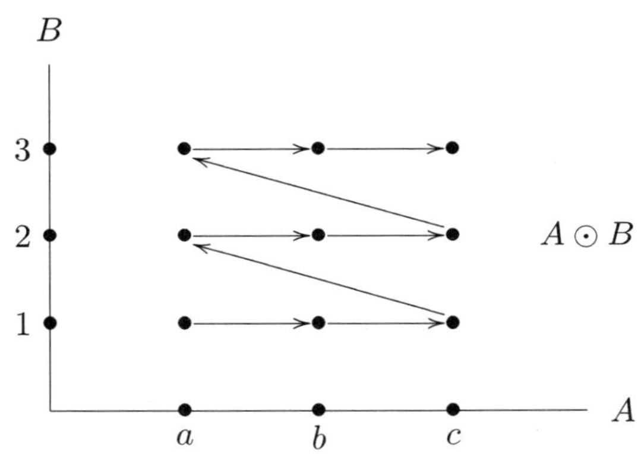

Figure 9.1: Ordering of $A \odot B$

Now, back to the ordinal numbers.

Definition 9.13 (Addition and Multiplication of Ordinals). Let α and β be ordinal numbers, and let A and B be well-ordered sets such that $\alpha = \oslash A$ and $\beta = \oslash B$. The operations, addition and multiplication of ordinal numbers, are defined as follows.

(1) Assume that A and B are disjoint. The *sum* $\alpha+\beta$ is defined to be the ordinal number given by

$$\alpha+\beta=\oslash(A\oplus B).$$

Note that as a set, $A\oplus B$ is not the product $A\times B$ but the disjoint union $A\cup B$.

(2) The *product* $\alpha\beta$ is defined to be the ordinal number given by

$$\alpha\beta=\oslash(A\odot B).$$

The elementary properties of ordinal addition and multiplication are given in following theorem.

Theorem 9.14. *Let α,β and γ be ordinal numbers. Then we have the following.*

(1) *Associative Law for Addition:* $\alpha+(\beta+\gamma)=(\alpha+\beta)+\gamma$.

(2) *Associative Law for Multiplication:* $\alpha(\beta\gamma)=(\alpha\beta)\gamma$.

(3) *Left Distributive Law:* $\alpha(\beta+\gamma)=\alpha\beta+\alpha\gamma$.

Proof. Let A, B and C be disjoint, well-ordered sets such that $\alpha=\oslash A$, $\beta=\oslash B$, and $\gamma=\oslash C$.

$\Rightarrow$ (1) $A\oplus(B\oplus C)\cong(A\oplus B)\oplus C$.

($\because$) By Definition 9.9 of ordinal sums, both $A\oplus(B\oplus C)$ and $(A\oplus B)\oplus C$ designate the set $A\cup B\cup C$ with the following order:

(i) If x and y are both in A, or both in B, or both in C, then they are ordered according to their order in A, or B or C, respectively.

(ii) Every element of C is greater than every element of B.

(iii) Every element of B is greater than every element of A.

$\Rightarrow \alpha+(\beta+\gamma) = \oslash(A \oplus (B \oplus C)) = \oslash((A \oplus B) \oplus C) = (\alpha+\beta)+\gamma.$

(2) $A \odot (B \odot C) \cong (A \odot B) \odot C.$

($\because$) The function $f : A \times (B \times C) \to (A \times B) \times C$ defined by $f(x,(y,z)) = ((x,y),z)$ is a one-to-one correspondence.

$\Rightarrow$ Since $(x,(y,z)) \leqslant (x',(y',z'))$ iff $((x,y),z)) \leqslant ((x',y'),z')$, f is an isomorphism. (The details, which are immediate from Definition 9.13, are left for the reader.)

$\Rightarrow \alpha(\beta\gamma) = \oslash(A \odot (B \odot C)) = \oslash((A \odot B) \odot C) = (\alpha\beta)\gamma.$

(3) $A \odot (B \oplus C) \cong (A \odot B) \oplus (A \odot C).$

($\because$) Both $A \odot (B \oplus C)$ and $(A \odot B) \oplus (A \odot C)$ designate the same set, $A \times (B \cup C) = (A \times B) \cup (A \times C)$ with certain orderings. It is easy to show that the two orderings are the same. (The details are left for the reader.)

$\Rightarrow \alpha(\beta+\gamma) = \oslash(A \odot (B \oplus C)) = \oslash((A \odot B) \oplus (A \odot C)) = \alpha\beta+\alpha\gamma.$ □

Remark 9.15. (1) The addition of ordinal numbers is not commutative.

($\because$) $1+\omega$ is similar to ω but $\omega+1$ is not similar to ω because ω has no last element but $\omega+1$ has a last element!

$\Rightarrow \omega+1 \neq 1+\omega.$

(2) The multiplication of ordinal numbers is not commutative.

($\because$) Since $2\omega = \{(0,0),(1,0),(0,1),(1,1),(0,2)(1,2),\cdots\}$ and $\omega 2 = \{(0,0),(1,0),(2,0),\cdots,(0,1),(1,1),(2,1),\cdots\}$ are obviously not isomorphic, $\omega 2 \neq 2\omega$.

(3) The "right distributive law" $(\alpha+\beta)\gamma = \alpha\gamma+\beta\gamma$ does not hold.

($\because$) Since $(1+1)\omega = 2\omega \neq \omega 2 = \omega(1+1) = \omega+\omega = 1\omega+1\omega$, $(1+1)\omega \neq 1\omega+1\omega$.

As usual, 0 is the ordinal number of the empty set, and 1 is the ordinal number of any singleton and 2 is the ordinal number of any doubleton. And so on.

Definition 9.16. Let μ be an ordinal number.

(1) μ is said to be *finite* if μ is similar to a natural number n.

(2) μ is said to be *infinite* or *transfinite* if μ is not a finite ordinal.

Exercise 9.2

1. Let A_1, A_2, B_1, B_2 be well-ordered sets. Prove that if $A_1 \cong A_2$ and $B_1 \cong B_2$, then

   (1) $A_1 \oplus B_1 \cong A_2 \oplus B_2$, and (2) $A_1 \odot B_1 \cong A_2 \odot B_2$.

2. If A and B are well-ordered sets, prove that $A \oplus B$ and $A \odot B$ are well-ordered sets.

3. If α is an ordinal number, prove that $1 + \alpha = \alpha$ iff α is an infinite ordinal number.

4. Let α and β be two nonzero ordinal numbers.

   (1) Prove that if $\alpha + \beta = \omega$, then α is a finite ordinal number (that is, similar to a natural number) and $\beta = \omega$.

   (2) Now assume that $\beta \neq 1$. Prove that if $\alpha\beta = \omega$, then α is a finite ordinal number and $\beta = \omega$.

   [Hint: Consider the well-ordered set $A \oplus B$ and $A \odot B$, where $\alpha = \oslash A$ and $\beta = \oslash B$.]

5. Prove each of the following, where μ designates a finite ordinal number.

   (1) $\mu + \omega = \omega$. (2) $\mu\omega = \omega$.

   (3) If α is an infinite ordinal number, then $\mu + \alpha = \alpha$.

6. Prove that $(\omega + \omega)\omega = \omega\omega$.

7. Give a counterexample to the (false) rule: $\alpha + \gamma = \beta + \gamma \Rightarrow \alpha = \beta$.

8. Prove the following, for every ordinal number α.

   (1) $0 + \alpha = \alpha + 0 = \alpha$. (2) $\alpha 0 = 0\alpha = 0$. (3) $\alpha 1 = 1\alpha = \alpha$.

9. Prove that $\alpha\beta = 0$ if and only if $\alpha = 0$ or $\beta = 0$.

10. Prove each of the following, where μ, ν, π are finite ordinal numbers.

    (1) $m \cong \mu$ iff $m \approx \mu$, where $m \in \omega$. (2) $\mu + \nu = \nu + \mu$.

    (3) $\mu\nu = \nu\mu$. (4) $(\mu + \nu)\pi = \mu\pi + \nu\pi$.

11. Give a complete proof of the isomorphism of Theorem 9.14(2)

$$A \odot (B \odot C) \cong (A \odot B) \odot C.$$

12. Give a complete proof of the isomorphism of Theorem 9.14(3)

$$A \odot (B \oplus C) \cong (A \odot B) \oplus (A \odot C).$$

9.3 Ordering of the ordinal numbers

In the introduction to this chapter, we spoke of comparing the ordinality of well-ordered sets.

Definition 9.17. Let A and B be well-ordered sets.

(1) We say that A has a *lower ordinality* than B, denoted by $A \lhd B$, if A is isomorphic with an initial segment of B.

(2) The ordinality of A is *less than or equal* to the ordinality of B if and only if A is isomorphic with B or an initial segment of B. In this case, we write $A \unlhd B$.

Note. Note that A is less than or equal to the ordinality of B if and only if there exists an injective, order-preserving function from A to B, whose range is a section of B. (See Theorems 4.49 and 4.56.)

Lemma 9.18. *Let A and B be well-ordered sets. Then $A \unlhd B$ if and only if there exists an injective, order preserving function $f : A \to B$.*

Proof. **(only if)** Assume that $A \trianglelefteq B$.

$\Rightarrow$ Clearly there exists an injective, order preserving function from A to B by definition.

(if) Assume that there exists an injective, order preserving function $f : A \to B$. Let $C = f(A)$.

$\Rightarrow$ $f : A \to C$, as a function from A to C, is an isomorphism.

$\Rightarrow$ Since $C \subseteq B$, there exists an isomorphism $g : C \to D$, where D is B or an initial segment of B, by Corollary 4.67.

$\Rightarrow$ $g \circ f : A \to D$ is an isomorphism and hence $A \trianglelefteq B$. □

Corollary 9.19. *If $A \trianglelefteq B$ and $B \trianglelefteq C$ then $A \trianglelefteq C$.*

Proof. Clearly the composite of two injective, order preserving functions is injective and order preserving. □

If A and B are well-ordered sets and A has a lower ordinality than B, we quite naturally expect the ordinal number of A to be "less than" the ordinal number of B. Accordingly, we define the "natural" ordering of the ordinal numbers as follows.

Definition 9.20 (Ordering of Ordinal Numbers). Let α and β be ordinal numbers and let A and B be well-ordered sets such that $\alpha = \oslash A$ and $\beta = \oslash B$. The relation $\leqslant$ is defined by

$$\alpha \leqslant \beta \text{ if and only if } A \trianglelefteq B.$$

Remark 9.21. (1) We note that $\alpha \leqslant \beta$ iff $\alpha \trianglelefteq \beta$ for all ordinal numbers α and β.

(2) The relation $\leqslant$, which we have just defined, is obviously reflexive and it is antisymmetric by Lemma 4.64. It is transitive by Corollary 9.19. Hence it is an order relation among the ordinal numbers.

We can say more:

Theorem 9.22. *Every nonempty class of ordinal numbers has a least element.*

Proof. Let $\mathscr{O}$ be a nonempty class of ordinal numbers, and let α be an arbitrary element of $\mathscr{O}$.

$\Rightarrow$ If α is the least element of $\mathscr{O}$, we are done.

Otherwise, let $\mathscr{B} = \{\beta \in \mathscr{O} \mid \beta < \alpha\}$.

$\Rightarrow$ By definition of the relation $<$, every $\beta \in \mathscr{B}$ is similar to an initial segment of α.

For each $\beta \in \mathscr{B}$, let $\phi(\beta)$ be the least element of $\{x \in \alpha \mid \beta \cong S_x\}$.

$\Rightarrow$ Since $\{\phi(\beta) \mid \beta \in \mathscr{B}\} \subseteq \alpha$, $\{\phi(\beta) \mid \beta \in \mathscr{B}\}$ has a least element $\phi(\delta)$.

$\Rightarrow$ δ is the least element of $\mathscr{B}$.

($\because$) Let $\beta \in \mathscr{B}$.

$\Rightarrow$ By the choice of $\phi(\beta)$ and $\phi(\delta)$, $\beta \cong S_{\phi(\beta)}$, $\delta \cong S_{\phi(\delta)}$ and $\phi(\delta) \leqslant \phi(\beta)$.

$\Rightarrow$ $S_{\phi(\delta)} \subseteq S_{\phi(\beta)}$ and hence $S_{\phi(\delta)} \trianglelefteq S_{\phi(\beta)}$ by Definition 9.20.

$\Rightarrow$ $\delta \cong S_{\phi(\delta)} \trianglelefteq S_{\phi(\beta)} \cong \beta$ and hence $\delta \leqslant \beta$.

$\Rightarrow$ δ is the least element of $\mathscr{B}$. $\square$

Thus, **the class of all the ordinal numbers is well-ordered.**

Theorem 9.23. *Let α and β be ordinal numbers.*

(1) *If $\beta > 0$, then $\alpha < \alpha + \beta$.* (2) $\beta \leqslant \alpha + \beta$.

Proof. Let A and B be will-ordered sets such that $\alpha = \oslash A$ and $\beta = \oslash B$.

(1) Assume that $\beta > 0$ and let b be the least element of B.

$\Rightarrow$ A is the initial segment S_b of $A \oplus B$.

$\Rightarrow$ $A \triangleleft (A \oplus B)$. i.e., $\alpha < \alpha + \beta$.

(2) Since $B \subseteq A \oplus B$, B is isomorphic with $A \oplus B$ or an initial segment of $A \oplus B$ by Corollary 4.67.

$\Rightarrow$ $B \trianglelefteq A \oplus B$ and hence $\beta \leqslant \alpha + \beta$. $\square$

Theorem 9.24. *Let α and β be ordinal numbers such that $\alpha < \beta$. Then there exists a unique ordinal number $\gamma > 0$ such that $\alpha + \gamma = \beta$.*

Proof. Let α and β be ordinal numbers such that $\alpha < \beta$ and let A and B be well-ordered sets such that $\alpha = \oslash A$ and $\beta = \oslash B$.

(Existence) Since $\alpha < \beta$, $A \lhd B$.

$\Rightarrow A \cong S_x$ for some $x \in B$. Let $C = B \setminus S_x$ and let $\gamma = \oslash C$.

$\Rightarrow C$ is well-ordered, $C \neq \emptyset$ and $\gamma > 0$.

$\Rightarrow$ Since $B = S_x \oplus C$ and $\alpha = \oslash A = \oslash S_x$, $\beta = \alpha + \gamma$.

(Uniqueness) Suppose that $\beta = \alpha + \gamma = \alpha + \gamma'$, where say, $\gamma < \gamma'$.

$\Rightarrow$ By the above argument, $\gamma' = \gamma + \delta$ for some ordinal number $\delta > 0$.

$\Rightarrow \beta = \alpha + \gamma' = \alpha + (\gamma + \delta) = (\alpha + \gamma) + \delta = \beta + \delta (\delta > 0)$,

which is in contradiction with the result of Theorem 9.23(1).

$\Rightarrow \gamma = \gamma'$. □

Theorem 9.25. *For any ordinal numbers α, β, γ, the following rules hold.*

(1) *If $\alpha < \beta$, then $\gamma + \alpha < \gamma + \beta$.* (2) *If $\gamma + \alpha < \gamma + \beta$, then $\alpha < \beta$.*

(3) *If $\alpha \leqslant \beta$, then $\alpha + \gamma \leqslant \beta + \gamma$.* (4) *If $\alpha + \gamma < \beta + \gamma$, then $\alpha < \beta$.*

(5) *If $\alpha < \beta$ and $\gamma > 0$, then $\gamma\alpha < \gamma\beta$.* (6) *If $\gamma\alpha < \gamma\beta$, then $\alpha < \beta$.*

(7) *If $\alpha \leqslant \beta$, then $\alpha\gamma \leqslant \beta\gamma$.* (8) *If $\alpha\gamma < \beta\gamma$, then $\alpha < \beta$.*

Proof. In (1), (3), (5), and (7), we assume that $\alpha < \beta$, hence we can assume that there exists $\delta > 0$ such that $\beta = \alpha + \delta$ by Theorem 9.24.

[Note: In (3) and (7), the case $\alpha = \beta$ is easily disposed of. Indeed, if $\alpha = \beta$, then $\alpha + \gamma = \beta + \gamma$ and $\alpha\gamma = \beta\gamma$ by Exercise 9.2.1.]

(1) Assume that $\alpha < \beta$.

$\Rightarrow$ Since $\beta = \alpha + \delta$, $\gamma + \beta = \gamma + (\alpha + \delta) = (\gamma + \alpha) + \delta > \gamma + \alpha$.

(The last relation is a consequence of Theorem 9.23(1).)

(2) Assume that $\gamma + \alpha < \gamma + \beta$.

$\Rightarrow$ If $\alpha = \beta$, then $\gamma + \alpha = \gamma + \beta$ by Exercise 9.2.1.

If $\beta < \alpha$, then $\gamma + \beta < \gamma + \alpha$ by (1) above.

$\Rightarrow$ Each of the cases contradicts to the assumption and hence $\alpha < \beta$.

(3) Assume that $\alpha < \beta$ and suppose that $\beta + \gamma < \alpha + \gamma$.

$\Rightarrow$ Since $\beta = \alpha + \delta$, $\alpha + (\delta + \gamma) = (\alpha + \delta) + \gamma = \beta + \gamma < \alpha + \gamma$.

$\Rightarrow$ $\delta + \gamma < \gamma$ by (2) above. This is impossible by Theorem 9.23(2).

Thus $\alpha + \gamma \leqslant \beta + \gamma$.

(4) Assume that $\alpha + \gamma < \beta + \gamma$.

$\Rightarrow$ If $\alpha = \beta$, then $\alpha + \gamma = \beta + \gamma$.

If $\beta < \alpha$, then $\beta + \gamma \leqslant \alpha + \gamma$ by (3) above.

$\Rightarrow$ Each of the cases contradicts to the assumption and hence $\alpha < \beta$.

(5) Assume that $\alpha < \beta$ and $\gamma > 0$.

$\Rightarrow$ Since $\beta = \alpha + \delta$ with $\delta > 0$, $\gamma\beta = \gamma(\alpha + \delta) = \gamma\alpha + \gamma\delta > \gamma\alpha$.

(The last relation holds by Theorem 9.23(1).)

(6) Assume that $\gamma\alpha < \gamma\beta$.

$\Rightarrow$ If $\alpha = \beta$, then $\gamma\alpha = \gamma\beta$.

If $\beta < \alpha$, then $\gamma\beta < \gamma\alpha$ by (5) above.

$\Rightarrow$ Each of the cases contradicts to the assumption and hence $\alpha < \beta$.

(7) Assume that $\alpha \leqslant \beta$.

$\Rightarrow$ There exists an ordinal number δ such that $\beta = \alpha + \delta$

by Theorem 9.24. Let $\alpha = \oslash A$, $\gamma = \oslash C$, $\delta = \oslash D$.

$\Rightarrow$ Since $A \subseteq A \oplus D$, $A \odot C \subseteq (A \oplus D) \odot C$.

$\Rightarrow$ $A \odot C \trianglelefteq (A \oplus D) \odot C$ by Corollary 4.67.

$\Rightarrow$ $\alpha\gamma \leqslant (\alpha + \delta)\gamma = \beta\gamma$.

(8) Assume that $\alpha\gamma < \beta\gamma$.

$\Rightarrow$ If $\alpha = \beta$, then $\alpha\gamma = \beta\gamma$.

If $\beta < \alpha$, then $\beta\gamma \leqslant \alpha\gamma$ by (7) above.

$\Rightarrow$ Each of the cases contradicts to the assumption and hence $\alpha < \beta$. □

Theorem 9.26. *Let α, β, γ be ordinal numbers.*

(1) *If $\gamma + \alpha = \gamma + \beta$, then $\alpha = \beta$*

(2) *If* $\gamma > 0$ *and* $\gamma\alpha = \gamma\beta$, *then* $\alpha = \beta$.

Proof. The proof, an immediate consequence of Theorem 9.25(1) and (5), is left as an exercise for the reader. □

Lemma 9.27. *Let* α, β, γ *be ordinal numbers. If* $\gamma < \beta\alpha$, *then there exist ordinal numbers* δ *and* ε *such that* $\gamma = \beta\delta + \varepsilon, \delta < \alpha$ *and* $\varepsilon < \beta$.

Proof. Assume that $\gamma < \beta\alpha$ and let A, B and C be well-ordered sets such that $\alpha = \oslash A, \beta = \oslash B$ and $\gamma = \oslash C$.

$\Rightarrow \beta\alpha = \oslash(B \odot A)$ and $C \lhd B \odot A$.

$\Rightarrow C \cong S_{(b,a)}$ for some $(b, a) \in B \odot A$. Let $E = \{(x, a) \mid x < b\}$.

$\Rightarrow$ (i) $E = S_b \odot \{a\} \cong S_b$, clearly.

(ii) $S_{(b,a)} = (B \times S_a) \cup E$ as sets,

($\because$) For $x \in B$ and $y \in A$, we have the following:

$(x, y) \in S_{(b,a)} \Leftrightarrow (x, y) < (b, a) \Leftrightarrow [y < a]$ or $[y = a$ and $x < b]$

$\Leftrightarrow [y \in S_a]$ or $[y = a$ and $x < b] \Leftrightarrow (x, y) \in B \times S_a$ or $(x, y) \in E$

$\Leftrightarrow (x, y) \in (B \times S_a) \cup E$.

(iii) The ordering of $(B \odot S_a) \oplus E$ is the same as the ordering of $S_{(b,a)}$. (The details are left for the reader).

$\Rightarrow S_{(b,a)} = (B \odot S_a) \oplus E$ as well-ordered sets. (See Figure 9.2.)

Let $\varepsilon = \oslash E = \oslash S_b$ and $\delta = \oslash S_a$.

$\Rightarrow$ By the definition of the relation $<$, we have $\varepsilon < \beta$, $\delta < \alpha$ and $\gamma = \oslash C = \oslash S_{(b,a)} = \oslash((B \odot S_a) \oplus E) = \beta\delta + \varepsilon$. □

It is very useful to note that the "division algorithm" for the natural numbers can be generalized to all the ordinal numbers.

Theorem 9.28. *If* α *and* β *are ordinal numbers with* $\beta > 0$, *then there exist unique ordinal numbers* ζ *and* ρ *such that* $\alpha = \beta\zeta + \rho$ *and* $\rho < \beta$.

Proof. Assume that α and β are ordinal numbers with $\beta > 0$.

(Existence) Since $1 \leqslant \beta$, we have $\alpha \leqslant \beta\alpha$ by Theorem 9.25(5).

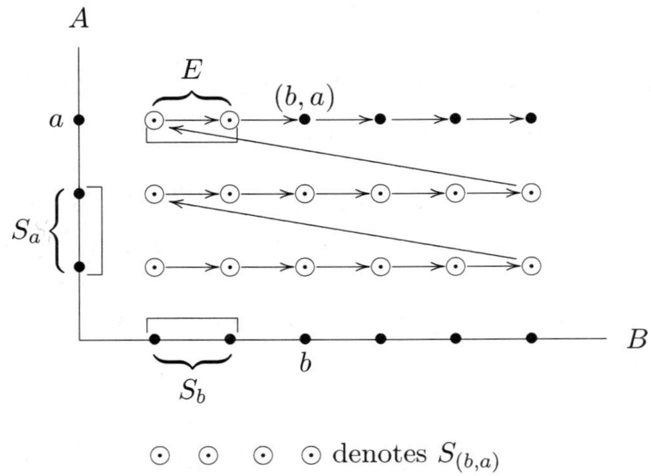

Figure 9.2: $S_{(b,a)} = (B \odot S_a) \oplus E$

$\Rightarrow$ If $\beta\alpha = \alpha$, we are done.

Otherwise, $\alpha < \beta\alpha$. Then by Lemma 9.27, there exist ordinal numbers $\delta < \alpha$ and $\varepsilon < \beta$ such that $\alpha = \beta\delta + \varepsilon$.

Hence again we are done.

(Uniqueness) Assume that $\alpha = \beta\zeta + \rho = \beta\zeta' + \rho'$, where (say) $\zeta' < \zeta$.

$\Rightarrow$ There exists $\mu > 0$ such that $\zeta = \zeta' + \mu$ by Theorem 9.24.

$\Rightarrow \beta\zeta' + \rho' = \beta\zeta + \rho = \beta(\zeta' + \mu) + \rho = \beta\zeta' + \beta\mu + \rho.$

$\Rightarrow \rho' = \beta\mu + \rho$ by Theorem 9.26(1).

$\Rightarrow \beta \leqslant \beta\mu \leqslant \beta\mu + \rho = \rho'$, which contradicts our assumption that $\rho' < \beta$.

$\Rightarrow$ We cannot have $\zeta' < \zeta$. By symmetry, we cannot have $\zeta < \zeta'$.

$\Rightarrow \zeta = \zeta'$ and hence $\rho = \rho'$ by Theorem 9.26(1). □

Definition 9.29. Let α be a nonzero ordinal number.

(1) α is called a *limit ordinal* if α has no immediate predecessor.

(2) α is called a *nonlimit ordinal* if α has an immediate predecessor.

Remark 9.30. (1) α is a limit ordinal iff $\alpha \neq \beta + 1$ for any ordinal β.

(2) α is a nonlimit ordinal iff $\alpha = \beta + 1$ for some ordinal β.

Theorem 9.31. *Let α be an ordinal number.*

(1) *If α is a limit ordinal, there exists a unique ordinal ζ such that $\alpha = \omega\zeta$.*

(2) *If α is a nonlimit ordinal, there exist a unique ordinal ζ and a unique finite ordinal $n \neq 0$ such that $\alpha = \omega\zeta + n$.*

Proof. We will prove the existence assertion. The uniqueness assertion is left as an exercise for the reader.

(1) Assume that α is a limit ordinal.

$\Rightarrow$ Since $\omega > 0$ is an ordinal number, there exist unique ordinal numbers ζ and ρ such that $\alpha = \omega\zeta + \rho$ and $\rho < \omega$ by Theorem 9.28.

$\Rightarrow$ ρ is finite.

$\Rightarrow$ $\rho = 0$.

($\because$) Suppose $\rho \neq 0$.

$\Rightarrow$ $\rho = m + 1$ for some finite m.

$\Rightarrow$ $\alpha = \omega\zeta + m + 1$ would not be a limit ordinal which contradicts to our assumption that α is a limit ordinal and hence $\rho = 0$.

(2) Assume that α is a nonlimit ordinal.

$\Rightarrow$ $\alpha = \omega\zeta + \rho$, where ρ is finite, as above.

$\Rightarrow$ $\rho \neq 0$.

($\because$) Suppose that $\rho = 0$.

$\Rightarrow$ $\alpha = \omega\zeta$, which is impossible because $\omega\zeta$ is a limit ordinal by Exercise 9.3.6 and hence $\rho \neq 0$. □

Exercise 9.3

1. Prove that $1 + \alpha = \alpha$ if and only if $\alpha \geqslant \omega$.

2. An ordinal number $\rho > 0$ is called *irreducible* if there exists no pair of ordinal numbers α, β such that $\alpha < \rho, \beta < \rho$ and $\alpha + \beta = \rho$. Prove the following.
   (1) An ordinal number ρ is irreducible if and only if $\pi + \rho = \rho$ for every ordinal $\pi < \rho$.
   (2) Assume that $\rho > 1$ and $\epsilon > 0$. If $\epsilon\rho$ is irreducible, then ρ is irreducible.
   (3) If ρ is irreducible and $0 < \mu < \rho$, then there exists an irreducible ordinal number ξ such that $\rho = \mu\xi$.
   (4) Assume that $\alpha > 0$. Then the set of all irreducible ordinal numbers $\leqslant \alpha$ has a greatest element. [Hint: Consider the set of all β such that $\alpha = \rho + \beta$ for some irreducible ρ.]
3. Show that an ordinal number α is a limit ordinal if and only if $\beta < \alpha$ implies $(\beta + 1) < \alpha$.
4. Let γ be a nonlimit ordinal. Prove the following.
   (1) If $\alpha < \beta$, then $\alpha\gamma < \beta\gamma$. (2) If $\alpha\gamma = \beta\gamma$, then $\alpha = \beta$.
5. Let $\beta \neq 0$. Prove that $\alpha + \beta$ is a limit ordinal if and only if β is a limit ordinal.
6. Let α and β be two nonzero ordinal numbers. Prove that $\alpha\beta$ is a limit ordinal if and only if α is a limit ordinal or β is a limit ordinal. [Hint: Use Exercise 9.3.3 above and Lemma 9.27.]
7. Prove that $n\omega = \omega$ for all $n \in \omega$. [Hint: Use Theorem 9.28 to "divide" ω by n.]
8. Let $\alpha \neq 0$. Prove that α is a limit ordinal if and only if $n\alpha = \alpha$, for every finite n. [Hint: Use Exercise 9.3.7 above and Theorem 9.31(1). For the converse, use Exercise 9.3.4(2) above.]
9. (1) Use induction to prove that if γ is an infinite ordinal, then $(\gamma+1)n = \gamma n+1$ for all finite n. [Use Exercise 9.3.1 above.]
   (2) Prove that $(\gamma + 1)\omega = \gamma\omega$ for all $\gamma > 0$. [Hint: If γ is infinite, assume that $\gamma\omega < (\gamma + 1)\omega$ and use Lemma 9.27 to arrive at a contradiction. If γ is finite, use Exercise 9.3.7 above.]
   (3) Conclude that if β is any limit ordinal, then $(\gamma + 1)\beta = \gamma\beta$ for all $\gamma > 0$.

10. Let α be a nonlimit ordinal. Prove that $(\gamma+1)\alpha > \gamma\alpha$ for all $\gamma > 0$. [Hint: Use Exercise 9.3.4 above.]

11. If α is an infinite ordinal and $\beta \neq 0$ is a nonlimit ordinal, prove that $(\alpha+1)\beta = \alpha\beta+1$. [Hint: Use Theorem 9.31(2) and assume $\alpha\beta+1 < (\alpha+1)\beta$ to arrive at a contradiction. Use Exercise 9.3.9 above.]

12. (1) If α is a limit ordinal, prove that $\alpha = \sup\{\beta \mid \beta < \alpha\}$.

    [Hint: Use Exercise 9.3.3 above.]

    (2) If α is a limit ordinal and β is any ordinal number, prove that $\beta + \alpha = \sup\{\beta + \mu \mid \mu < \alpha\}$. [Hint: If γ is an upper bound of $\{\beta + \mu \mid \mu < \alpha\}$, then $\gamma > \beta$, that is, $\gamma = \beta + \delta$. δ proves to be an upper bound of $\{\mu \mid \mu < \alpha\}$.]

13. If α is a limit ordinal and β is any ordinal number, prove that $\beta\alpha = \sup\{\beta\mu \mid \mu < \alpha\}$. [Hint: If γ is an upper bound of $\{\beta\mu \mid \mu < \alpha\}$, then $\gamma = \alpha\delta + \rho (\rho < \alpha)$. Note that $\mu < \alpha$ implies $\gamma \geqslant \alpha(\mu+1)$ and conclude that δ is an upper bound of $\{\mu \mid \mu < \alpha\}$.]

14. Prove Theorem 9.26.

15. Prove the uniqueness assertion of Theorem 9.31.

9.4 The alephs and the continuum hypothesis

In Chapter 8 it was proven that the relation $\leqslant$ among cardinals is a well-ordering. Hence there is a smallest infinite cardinal, a next greater infinite cardinal, and so on; every infinite cardinal has a uniquely determined immediate successor. It follows that the infinite cardinal numbers can be ranked in "first, second, third, $\cdots$" order. This opportunity of making the cardinals and using the ordinals to designate the ranks, has valuable applications in mathematics. We will proceed to show that there is an isomorphism between the class of all the infinite cardinals and the class of all the ordinals.

Lemma 9.32. *Let* IC *be the class of all the infinite cardinal numbers and let* OR *be the class of all the ordinal numbers.*

(1) *Every initial segment of* IC *is a set.*

(2) *Every initial segment of* OR *is a set.*

Proof. (1) Let IC be the class of all the infinite cardinal numbers.

Let a be an infinite cardinal number and let

$B = \{b \mid b \in \text{IC and } b \leqslant a\} \cup \{\text{all finite cardinal numbers}\}$.

$\Rightarrow$ Well-order a, and then define a function $\phi : a \cup \{a\} \to B$ by

$\phi(x) = \sharp S_x$ for all $x \in a$ and $\phi(a) = a$.

$\Rightarrow$ (i) $\phi : a \cup \{a\} \to B$ is well-defined.

($\because$) Let $x \in a$.

$\Rightarrow S_x = \{y \in a \mid y < x\} \subseteq a$.

$\Rightarrow$ Since a is an infinite cardinal number, $\sharp S_x$ is finite or otherwise $\sharp S_x \leqslant \sharp a = a$.

$\Rightarrow \sharp S_x \in B$.

$\Rightarrow$ Since $\phi(a) = a \in B$, $\phi : a \cup \{a\} \to B$ is well-defined.

(ii) $\phi : a \cup \{a\} \to B$ is surjective.

($\because$) Let $c \in B$.

$\Rightarrow c$ is an infinite cardinal number with $c \leqslant a$.

$\Rightarrow c$ is isomorphic with a or an initial segment S_x of a as well-ordered sets by Corollary 4.67.

$\Rightarrow c$ is equipotent with a or S_x.

$\Rightarrow c = a$ by Lemma 8.3 or c is equipotent with S_x.

$\Rightarrow c = a = \phi(a)$ or $c = \sharp S_x = \phi(x)$ and hence $\phi : a \cup \{a\} \to B$ is surjective.

$\Rightarrow$ Since $a \cup \{a\}$ is a set, B is a set by Axiom 7 of Replacement.

$\Rightarrow$ Since the initial segment $S_a = \{b \mid b \in \text{IC and } b < a\}$ is a subset of B, S_a is a set by Axiom 4 of Subset.

(2) can be proved using an argument analogous to the above and is left for the reader. □

Theorem 9.33. *Let* IC *be the class of all the infinite cardinals and let* OR *be the class of all the ordinals. There exists an isomorphism* $\Omega : \mathrm{IC} \to \mathrm{OR}$.

Proof. Since both IC and OR are well-ordered classes, by Theorem 4.66 exactly one of the following three cases must hold.

(a) $\mathrm{IC} \cong \mathrm{OR}$, or

(b) IC is similar to an initial segment of OR, or

(c) OR is similar to an initial segment of IC.

Suppose for a moment that the case (c) holds.

$\Rightarrow$ Since every initial segment of IC is a set by Lemma 9.32(2), OR is a set by Theorem 2.37. But OR is a proper class by Proposition 9.4.

$\Rightarrow$ The case (c) cannot hold.

Analogously, if the case (b) holds, then we arrive at contradiction.

$\Rightarrow$ The case (a) $\mathrm{IC} \cong \mathrm{OR}$ must hold. □

Definition 9.34. We have an isomorphism $\Omega : \mathrm{IC} \to \mathrm{OR}$ by Theorem 9.33.

(1) For an infinite cardinal number a, the ordinal number $\Omega(a)$ is called the *ordinal rank* of a.

(2) The infinite cardinal numbers are often called *alephs*.

(3) If a is an infinite cardinal number and $\alpha = \Omega(a)$, we frequently write $a = \aleph_\alpha$.

Remark 9.35. If a is the least infinite cardinal number, then $\Omega(a) = \aleph_0$. If a is the next greater infinite cardinal, then $\Omega(a) = \aleph_1$, and so on. Thus the first few infinite cardinal numbers are

$$\aleph_0, \aleph_1, \aleph_2, \aleph_3, \cdots .$$

Theorem 9.33 has the following simple consequences.

Theorem 9.36. *Let a and b be infinite cardinal numbers and let $\alpha = \Omega(a)$ and $\beta = \Omega(b)$ so that $a = \aleph_\alpha$ and $b = \aleph_\beta$.*

(1) *If* $\aleph_\alpha = \aleph_\beta$, *then* $\alpha = \beta$. (2) *If* $\alpha = \beta$, *then* $\aleph_\alpha = \aleph_\beta$.

(3) *If* $\aleph_\alpha < \aleph_\beta$, *then* $\alpha < \beta$. (4) *If* $\alpha < \beta$, *then* $\aleph_\alpha < \aleph_\beta$.

Proof. It follows directly from Theorem 9.33. □

We have seen that every infinite cardinal number a has an immediate successor. But what exactly is the immediate successor of a?

We know that 2^a is greater than a. But is there any cardinal number between a and 2^a?

Let us ask a more specific question. We have seen that $\aleph_0$ is the cardinal number of denumerable sets (for the cardinal number of ω is the least infinite cardinal), and that $2^{\aleph_0}$ is the cardinal number of the real numbers. Is there a cardinal number between these two?

The early set theorists proposed the hypothesis that there is no cardinal between $\aleph_0$ and $2^{\aleph_0}$, and named it the *continuum hypothesis* for it is equivalent to saying that every set of real numbers which is not denumerable has the power of the real numbers, called the "*power of the continuum*". An obvious extension of this conjecture is the statement: For every infinite cardinal number a, there is no cardinal between a and 2^a ; this is called the *generalized continuum hypothesis.* Recall the following Conjectures 7.12 and 7.13.

Conjecture 7.12(Continuum Hypothesis) There does not exist any cardinal c such that $\aleph_0 < c < 2^{\aleph_0}$.

Conjecture 7.13(Generalized Continuum Hypothesis) If a is any infinite cardinal, there does not exist any cardinal c such that $a < c < 2^a$.

It was proved that the continuum hypothesis and the generalized continuum hypothesis cannot be proven from the other axioms of set theory, and do not contradict these. Hence their status is analogous to that of Euclid's "Fifth Postulate" in geometry. We may postulate them or deny them, in each case getting a consistent theory of cardinal numbers.

Exercise 9.4

1. Prove that the generalized continuum hypothesis is equivalent to $2^{\aleph_\alpha} = \aleph_{\alpha+1}$.
2. Assuming the generalized continuum hypothesis, prove the following.

   If $\alpha \leqslant \beta$, then $\aleph_\alpha^{\aleph_\beta} = \aleph_{\beta+1}$.
3. Assuming the generalized continuum hypothesis, prove the following.

   For arbitrary cardinals a, b, $a < b \Rightarrow 2^a < 2^b$.
4. Prove Lemma 9.32(2).

9.5 Construction of the ordinals and cardinals*

We said, in the introduction to this chapter, that it is possible to construct sets which satisfy Conditions O1 and O2 of Axiom 11 of Ordinality. The chief purpose of this construction is to prove that we can dispense with Axiom 11 of Ordinality by actually producing the sets whose existence the axiom asserts. Our process of construction is based upon the same idea – outlined in the introduction of Chapter 6 – that we used to construct the natural numbers. Recall that the *successor* of a set A is defined to be the set $A^+ = A \cup \{A\}$(Definition 6.3).

We begin by defining

$$0 = \emptyset,$$
$$1 = 0^+ = 0 \cup \{0\} = \emptyset \cup \{\emptyset\} = \{\emptyset\},$$
$$2 = 1^+ = 1 \cup \{1\} = \{0, 1\} = \{\emptyset, \{\emptyset\}\},$$
$$3 = 2^+ = 2 \cup \{2\} = \{0, 1, 2\} = \{\emptyset, \{\emptyset\}, \{\emptyset, \{\emptyset\}\}\},$$
$$4 = 3^+ = 3 \cup \{3\} = \{0, 1, 2, 3\} = \{\emptyset, \{\emptyset\}, \{\emptyset, \{\emptyset\}\}, \{\emptyset, \{\emptyset\}, \{\emptyset, \{\emptyset\}\}\}\},$$
$$\vdots$$
$$\omega = \{0, 1, 2, 3, 4, \cdots\}.$$

This time, however, we will go further than we did in Chapter 7.

[0] *This section may be omitted without loss of continuity.

Note that $n \notin n$ for any $n \in \omega$ and $\omega \notin \omega$.

Starting with ω, we define

$$\begin{aligned} &\omega + 1 = \omega^+, \\ &\omega + 2 = (\omega + 1)^+, \\ &\vdots \\ &\omega + \omega. \end{aligned}$$

Then, starting with $\omega + \omega = \omega 2$, we get

$$\begin{aligned} &\omega 2 + 1 = (\omega 2)^+, \\ &\omega 2 + 2 = (\omega 2 + 1)^+, \\ &\vdots \\ &\omega 3, \\ &\omega 3 + 1 = (\omega 3)^+, \\ &\vdots \\ &\omega 4, \\ &\vdots \\ &\omega\omega, \\ &\vdots \end{aligned}$$

and so on.

This is the basic idea of our construction process, but we will not proceed exactly in this fashion. Instead of starting with 0 and constructing successive sets one by one, we will define all the sets simultaneously. This can be accomplished in the following way. The "elementhood" relation $\in$ is not, generally speaking, an order relation. For example, if $x \in A$ and $A \in B$, it does not necessarily follow that $x \in B$. However, there are special cases where $\in$ does behave as if it were an order relation. One of these cases concerns us here. It will be convenient to consider, rather than $\in$, the relation $\underline{\in}$ defined as follows:

Definition 9.37. (1) The relation $\underline{\in}$ on the class of sets is defined by

$x \underline{\in} y$ if and only if $x = y$ or $x \in y$.

(2) A class A is said to be $\in$-*ordered* if it is ordered by the relation $\underline{\in}$ and if $x \notin x$ for all $x \in A$.

(3) A class A is said to be $\in$-*well-ordered* if it is well-ordered by the relation $\underline{\in}$ and if $x \notin x$ for all $x \in A$.

(4) Recall that a set A is *transitive* if $x \in A$ implies $x \subseteq A$ (Definition 6.10).

(5) A transitive $\in$-well-ordered set will (temporarily) be called an *ord.*

Remark 9.38. In an $\in$-ordered set A,

(1) "x is less than or equal to y" means that $x \in y$ or $x = y$.

(2) "x is strictly less than y" means that $x \in y$.

(3) It is clear that any subclass of an $\in$-well-ordered class is $\in$-well-ordered.

Examples of ords are the natural numbers. They are transitive sets by Lemma 6.11. It follows immediately from Definition 6.27 and Theorem 6.31 that every natural number is $\in$-well-ordered (note simply that every natural number is a subset of ω). But ω, too, is an ord, and so are the sets $\omega + 1, \cdots, \omega 2, \cdots, \omega 3, \cdots$, and so on.

The name "ord" is a provisional one. The purpose of this section is to prove that the ords satisfy O1 and O2. Once this has been shown, they will be renamed "ordinal numbers".

Lemma 9.39. *Let α be an ord and let $x \in \alpha$. Then we have the following.*

(1) *x is the initial segment S_x of α.*

(2) *x is an ord.*

Proof. Let α be an ord and let $x \in \alpha$.

(1) (i) $x \subseteq S_x$.

(∵) Let $u \in x$.

⇒ Since α is $\in$-well-ordered, $u \in \alpha$.

Since $\underline{\in}$ is the order relation in α, $u \in x$ means that u is "strictly less than" x.

⇒ $u \in S_x$ and hence $x \subseteq S_x$.

(ii) $S_x \subseteq x$.

(∵) Let $u \in S_x$.

⇒ u is "strictly less than" x.

⇒ $u \in x$ and hence $S_x \subseteq x$.

⇒ $x = S_x$ by (i) and (ii) above.

(2) Since $x \in \alpha$, $x = S_x$ by (1) above.

⇒ (i) Since $S_x \subseteq \alpha$ and α is $\in$-well-ordered and $x = S_x$, x is $\in$-well-ordered.

(ii) x is a transitive set.

(∵) Suppose that $v \in x$. To show that $v \subseteq x$, let $u \in v$.

⇒ Since $v \in x$ and $x \in \alpha$ and α is a transitive set, $v \in \alpha$.

Analogously, $u \in \alpha$.

⇒ u, v and x are elements of α with $u \in v$ and $v \in x$.

⇒ Since α is ordered by $\underline{\in}$ and $u \neq x$, $u \in x$ and hence $v \subseteq x$.

⇒ x is a transitive set.

⇒ x is an ord by (i) and (ii) above. □

Lemma 9.40. *Let α and β be ords. Let $f : \alpha \to \beta$ be an isomorphism and let $A = \{x \in \alpha \mid f(x) = x\}$. Then $A = \alpha$.*

Proof. Suppose that $A \neq \alpha$ and let c be the least element of $\alpha \setminus A$.

Note that $f(c) \neq c$ because $c \notin A$.

⇒ $f(x) = x$ for all $x \in c$.

(∵) Let $x \in c$.

⇒ Since x and c are ords by Lemma 9.39, $x \underline{\in} c$ and $c \in \alpha$, $x \in \alpha$.

$\Rightarrow$ Since c is the least element of $\alpha \setminus A$ and $x \in c$, $x \notin \alpha \setminus A$.

$\Rightarrow x \in A$. i.e., $f(x) = x$.

$\Rightarrow f(c) = c$ which contradicts to our choice of c ($c \in \alpha \setminus A$).

($\because$) Let $x \in c$.

$\Rightarrow x = f(x) \in f(c)$ and hence $f(c) \subseteq x$.

Conversely, let $x \in f(c)$.

$\Rightarrow x \in \beta$, because $f(c) \in \beta$ and β is a transitive set.

$\Rightarrow x = f(y)$ for some $y \in \alpha$.

$\Rightarrow f(y) = x \in f(c)$.

$\Rightarrow y \in c$ because f is an isomorphism.

$\Rightarrow y \notin \alpha \setminus A$ because c is the least element of $\alpha \setminus A$.

$\Rightarrow y \in A$ and hence $f(y) = y$.

$\Rightarrow x = f(y) = y \in c$ and hence $x \subseteq f(c)$.

The above contradiction yields that $A = \alpha$. □

Theorem 9.41. *Let α and β be ords. If $\alpha \cong \beta$ then $\alpha = \beta$.*

Proof. Let α and β be ords. Assume that $\alpha \cong \beta$.

$\Rightarrow$ There exists an isomorphism $f : \alpha \to \beta$.

Let $A = \{x \in \alpha \mid f(x) = x\}$.

$\Rightarrow A = \alpha$ by Lemma 9.40.

$\Rightarrow \alpha = \beta$.

($\because$) Let $x \in \alpha$.

$\Rightarrow$ Since $A = \alpha$, $x = f(x) \in \beta$ and hence $\alpha \subseteq \beta$.

Conversely, let $y \in \beta$.

$\Rightarrow$ There exists $x \in \alpha$ such that $f(x) = y$.

$\Rightarrow y = f(x) = x \in \alpha$ and hence $\beta \subseteq \alpha$. □

Let $\langle \mathrm{ORD}, \in \rangle$, or briefly ORD, designate the class of all the ords with the relation $\in$.

Lemma 9.42. *For each $\alpha \in \mathrm{ORD}$, α is the initial segment S_α of* ORD.

Proof. Let $\alpha \in$ ORD and let $S_\alpha = \{x \in \text{ORD} \mid x \in \alpha\}$.

$\Rightarrow$ Since $x \in \alpha$ iff $x \in S_\alpha$, $\alpha = S_\alpha$. □

Theorem 9.43. ORD *is an* $\in$*-well-ordered class.*

Proof. We must prove that $\underline{\in}$ is a well-ordering of the class ORD.

(i) $<$ ORD, $\underline{\in}$ $>$ is reflexive.

($\because$) For every ord α, $\alpha = \alpha$ and hence $\alpha \underline{\in} \alpha$.

(ii) $<$ ORD, $\underline{\in}$ $>$ is antisymmetric.

($\because$) Suppose that $\alpha \underline{\in} \beta$ and $\beta \underline{\in} \alpha$.

$\Rightarrow$ If $\alpha = \beta$, we are done. Now suppose that $\alpha \neq \beta$.

$\Rightarrow \alpha \in \beta$ and $\beta \in \alpha$.

$\Rightarrow \alpha \subseteq \beta$ and $\beta \subseteq \alpha$.

$\Rightarrow \alpha = \beta$. This contradicts the assumption that $\alpha \neq \beta$.

$\Rightarrow \alpha = \beta$.

(iii) $<$ ORD, $\underline{\in}$ $>$ is transitive.

($\because$) Suppose that $\alpha \underline{\in} \beta$ and $\beta \underline{\in} \gamma$.

$\Rightarrow$ Since γ is a transitive set, $\alpha \underline{\in} \gamma$.

(iv) For all $\alpha \in$ ORD, $\alpha \notin \alpha$.

($\because$) Let α be an ord and suppose that $\alpha \in \alpha$.

$\Rightarrow \alpha$ is an element of the $\in$-ordered set α.

$\Rightarrow$ By Definition 9.37(2), $\alpha \notin \alpha$. A contradiction!

Thus $\alpha \notin \alpha$ for every $\alpha \in$ ORD.

(v) Any two ords α and β are comparable.

($\because$) Let α, β be two ords.

$\Rightarrow$ By Theorem 4.66, $\alpha \cong \beta$, or

α is similar to an initial segment of β or

β is similar to an initial segment of α.

$\Rightarrow \alpha \trianglelefteq \beta$ or $\beta \trianglelefteq \alpha$ by Definition 9.17.

$\Rightarrow \alpha \underline{\in} \beta$ or $\beta \underline{\in} \alpha$.

($\because$) We will show that $\alpha \trianglelefteq \beta$ implies $\alpha \underline{\in} \beta$.

For this, assume that $\alpha \lhd \beta$.

$\Rightarrow$ There exists an initial segment S_x of β such that $\alpha \cong S_x$.

$\Rightarrow$ $x = S_x$ by Lemma 9.42.

$\Rightarrow$ Since x is an ord by Lemma 9.39(2) and $\alpha \cong x$, $\alpha = x$ by Theorem 9.41.

$\Rightarrow$ Since $x \in \beta$, $\alpha \in \beta$ by axiom 1 of Extent.

Similarly $\alpha \lhd \beta$ implies $\alpha \in \beta$.

$\Rightarrow$ α and β are comparable.

(vi) Every nonempty subclass of ORD has a least element.

($\because$) Let A be a nonempty subclass of ORD, and let $\alpha \in A$.

$\Rightarrow$ If α is the least element of A, we are done.

Assume that α is not the least element of A.

$\Rightarrow$ There exists an element $\delta \in A$ such that δ is "less than" α.

$\Rightarrow$ $\delta \in \alpha$.

$\Rightarrow$ Since $A \cap \alpha \neq \emptyset$ and $A \cap \alpha \subseteq \alpha$, $A \cap \alpha$ has a least element γ.

$\Rightarrow$ γ is the least element of A.

($\because$) Suppose that γ is not the least element A.

$\Rightarrow$ There exists $\beta \in A$ such that β is "less than" γ.

$\Rightarrow$ $\beta \in \gamma$. Note that by (v) above, γ is comparable with every elements of A.

$\Rightarrow$ Since $\gamma \in \alpha$ and α is a transitive set, $\beta \in \alpha$.

$\Rightarrow$ $\beta \in A \cap \alpha$ and $\beta \in \gamma$. This is impossible because γ was chosen to be the least element of $A \cap \alpha$.

$\Rightarrow$ γ is the least element of A.

$\Rightarrow$ ORD is an $\in$-well-ordered class. $\square$

Theorem 9.44. ORD *is a proper class.*

Proof. Suppose that ORD is a set.

$\Rightarrow$ By Theorem 9.43 and Lemma 9.42, ORD is an ord.

$\Rightarrow$ ORD $\in$ ORD.

$\Rightarrow$ Since ORD is $\in$-ordered, ORD $\notin$ ORD. This is a contradiction.

$\Rightarrow$ ORD is a proper class. □

Theorem 9.45. *Let A be an arbitrary well-ordered set. Then there exists an ord α such that $A \cong \alpha$.*

Proof. Let A be a well-ordered set.

$\Rightarrow$ By Theorem 4.66,
(i) $A \cong$ ORD, or (ii) ORD is similar to an initial segment of A, or
(iii) A is similar to an initial segment of ORD.

$\Rightarrow$ Since Cases (i) and (ii) are impossible by Theorem 9.44,
Case (iii) must hold.

$\Rightarrow$ A is similar to an initial segment S_α of ORD for some $\alpha \in$ ORD.

$\Rightarrow$ Since $x \in S_\alpha$ iff $x \in \alpha$, $S_\alpha = \alpha \in$ ORD and hence A is similar to an element $\alpha \in$ ORD. □

Remark 9.46. (1) Axiom 11 of Ordinality follows from Axiom 9 of Infinity. Thus Axiom 11 can now be eliminated from our list of axioms.

(2) Axiom 10 of Cardinality follows from Axiom 9 of Infinity. Thus Axiom 10 can also be eliminated from our list of axioms.

Proof. (1) Since there exists ω by Axiom 9 of Infinity and $\omega \in$ ORD, ORD is a nonempty class. Moreover the class ORD satisfies the condition O1 by Theorem 9.41 and satisfies the condition O2 by Theorem 9.45.

(2) The class of all the initial ordinals is a nonempty class and satisfies the conditions K1 and K2 of the Axiom 10 of Cardinality by Exercise 9.5.5. □

Since the class ORD satisfies the conditions O1 and O2, we are justified in renaming the ords ordinal numbers and writing ORD = OR.

Since OR = ORD, every ordinary number is an "ord" as defined in Definition 9.37. In particular, since every natural number is an "ord" and

ω is an ordinal number, every natural number is an ordinal number and ω itself is an "ord". It is easy to prove that if X is an "ord" then $X^+ = X \cup \{X\}$ is an "ord", that is, if α is an ordinal number then α^+ is an ordinal number. (Note that α^+ is obtained by adjoining a single element to α, hence $\alpha^+ = \alpha + 1$.) Thus, the class OR begins with the following elements: $0, 1, 2, \cdots, \omega, \omega+1, \omega+2, \cdots \omega+\omega, \cdots$.

Definition 9.47. Let A be a class.

(1) Consider the class of all the ordinals equipotent with A. This class has a unique least element, which we call the *initial ordinal* equipotent with A.

(2) Since the class of all the initial ordinals satisfies the conditions K1 and K2 of the Axiom 10 of Cardinality(See Exercise 9.5.5), we are justified in making the following definition. By a *cardinal number* we mean an initial ordinal.

Remark 9.48. (1) The class CD of the cardinal numbers is the class of all the initial ordinals. We have thus fulfilled our promise of actually producing sets to serve as the cardinal numbers and the ordinal numbers. We noted earlier that every natural number is a transitive, $\in$-well-ordered set, that is, an ordinal number. It is immediate, too, that every natural number is an initial ordinal number. Thus, in our construction, the natural numbers coincide with the finite ordinals, as well as with the finite cardinals.

(2) The reader should note that everything we have proved in Chapters 8 and 9 about the class CD of the cardinal numbers has depended solely upon Conditions K1 and K2 of Axiom 10 of Cardinality. Thus everything we have already said about the cardinals holds, without any alteration, for the class CD defined by Definition 9.47.

In particular, Theorem 9.36 still holds, that is, the correspondence $\alpha \leftrightarrow \aleph_\alpha$ is an isomorphism between OR and the class of all the infinite cardinals.

(3) It is important to note that by Definition 9.47, $\aleph_\alpha$ is both a cardinal and an ordinal (specifically, an initial ordinal). In order to avoid any confusion, it is common practice in mathematics to write

$$\aleph_\alpha = \omega_\alpha$$

for every infinite cardinal $\aleph_\alpha$, and to treat $\aleph_\alpha$ as a cardinal number and ω_α as an ordinal number. In other words, the number in question is denoted by $\aleph_\alpha$ when it is used as a cardinal, and by ω_α when it is used as an ordinal. For example, $\aleph_\alpha + \aleph_\beta$ designates the cardinal sum of the two numbers, whereas $\omega_\alpha + \omega_\beta$ designates their ordinal sum.

Exercise 9.5

1. For each ordinal number α, prove that $\alpha \leqslant \omega_\alpha$. Conclude that $\sharp\alpha \leqslant \aleph_\alpha$. [Hint: Use Lemma 4.61.]
2. Prove that if β is a limit ordinal, then $\aleph_\beta = \sup\{\aleph_\gamma \mid \gamma < \beta\}$. [Hint: Use Theorem 9.36.]
3. Prove that $\sum_{\gamma \leqslant \mu} \aleph_\gamma = \aleph_\mu$.
4. If μ is a limit ordinal, prove that $\sum_{\gamma < \mu} \aleph_\gamma = \aleph_\mu$.
5. Show that the class of all the initial ordinals satisfies the conditions K1 and K2 of the Axiom 10 of Cardinality.

Chapter 10

Transfinite Recursion and Selected Topics in the Theory of Ordinals and Cardinals*

10.1 Transfinite recursion

In Chapter 6 we discussed the notions of finite induction and finite recursion; finite induction is a method of **proof by induction** in ω, and finite recursion is a method of *definition by induction* in ω. Let us briefly review these two concepts. Proof by induction in ω works as follows: A theorem is first shown to be true for 0. Then it is shown that if the theorem holds for n, it must hold for $n+1$. We are able to conclude that the theorem holds for all the natural numbers. In order to define a function by induction we proceed in much the same manner. First, we define the value of the function at 0. Then we use its value at n to define its value at $n+1$. In this way its

[0] *This chapter may be omitted without loss of continuity.

value is defined at every natural number.

We have already noted that OR, the class of the ordinal numbers, is an extension of ω. The finite ordinal numbers are, respectively, $0, 1, 2, 3, \cdots$; after these comes ω, then $\omega + 1, \omega + 2, \cdots, \omega 2, \omega 2 + 1, \cdots$ and so on.

Some ordinal numbers (other than 0) have no immediate predecessor; we call them *limit ordinals*. Because of them, finite induction cannot be used to prove theorems for all the elements of OR. Indeed, suppose we are able to show that a theorem holds for 0, and that whenever it holds for α it holds also for $\alpha + 1$; we may then conclude that the theorem holds for all the finite ordinals $0, 1, 2, \cdots$, but not necessarily for ω (because finite recursion permits us to make an inference from any number α to its immediate successor $\alpha + 1$, but ω is not the immediate successor of any number). Furthermore, we may not conclude that the theorem necessarily holds for any ordinal greater than ω. Proof by induction in OR must use the principle of *transfinite induction*: We show that if a theorem holds for all the elements less than any given β, then it must hold also for β; we may then conclude that it holds for all the ordinals.

Theorem 4.57(Principle of Transfinite Mathematical Induction).
Let A be a well-ordered class and let $P(x)$ be a statement which is either true or false for each element $x \in A$. Suppose that the following condition(induction hypothesis) holds:

Ind. If $P(y)$ is true for every $y < x$, then $P(x)$ is true.

Then $P(x)$ is true for every element $x \in A$.

Note that Theorem 4.57 can also be applied for the well-ordered class OR of all ordinal numbers. The induction hypothesis Ind of Theorem 4.57 can be restated as the following three conditions:

(1) $P(x_0)$ is true for the least element x_0 of A.

(2) For all $y \in A$, if $P(y)$ is true, then $P(y+1)$ is true, where $y+1$ is the

immediate successor of y.

(3) For all limit ordinal $x \neq x_0$, if $P(y)$ is true for all $y \in A$ with $y < x$, then $P(x)$ is true.

The situation is analogous when we wish to use transfinite induction to define a function f on OR. We may begin by defining the value of f at 0. Next, we may use its value at α to define its value at $\alpha+1$; this will define the values of f at $0, 1, 2, \cdots$. But in order to "reach beyond" the finite ordinals we need an additional rule of definition such as the following:
If β is any limit ordinal, then

$$f(\beta) = \sup\{f(\alpha) \mid \alpha < \beta\}.$$

With this additional rule it would appear to us that all the values of f (at all the ordinal numbers) are determined. This is indeed true, and the statement of this fact is called the *transfinite recursion theorem*; it will be proven next.

In order to clearly understand the meaning of the transfinite recursion theorem we advise the reader to compare it with Theorem 6.14 - the finite recursion theorem - and reread the comments which precede Theorem 6.14.

Theorem 10.1 (Transfinite Recursion Theorem). *Let A be a well-ordered class, let $f : A \to A$ be a function, and let $a \in A$. Then there exists a unique function $u : \mathrm{OR} \to A$ such that*

(1) $u(0) = a$,

(2) $u(\alpha^+) = f(u(\alpha))$ *for all* $\alpha \in \mathrm{OR}$,

(3) $u(\alpha) = \sup\{u(\gamma) \mid \gamma < \alpha\}$ *if α is a limit ordinal.*

Proof. Let A be a well-ordered class, $f : A \to A$ a function, and let $a \in A$. The proof is an adaptation of the argument we used to prove Theorem 6.14. If $G \subseteq \mathrm{OR} \times A$, we will call G *perfect* if it satisfies the following conditions:

(1′) $(0, a) \in G$.

(2′) For all $\alpha \in \mathrm{OR}$, if $(\alpha, x) \in G$, then $(\alpha^+, f(x)) \in G$.

(3′) For every limit ordinal α, if $(\gamma, x_\gamma) \in G$ for all $\gamma < \alpha$ and if $x = \sup\{x_\gamma \mid \gamma < \alpha\}$, then $(\alpha, x) \in G$.

It is clear that (1′), (2′), (3′) are not different from (1), (2), (3).

There are perfect subclasses of $\mathrm{OR} \times A$, for example $\mathrm{OR} \times A$ itself.

Now, let u be the intersection of all the perfect subclasses of $\mathrm{OR} \times A$; that is, $u = \{(\alpha, x) \mid \text{ if } G \text{ is perfect, then } (\alpha, x) \in G\}$,

that is, $u = \{(\alpha, x) \mid (\alpha, x) \in G \text{ for all perfect subclass } G \text{ of } \mathrm{OR} \times A\}$.

It is trivial to show that u is perfect. To show that u is a function, let $S = \{\gamma \in \mathrm{OR} \mid (\gamma, x) \in u \text{ for exactly one } x \in A\}$.

Using induction, we will prove that $S = \mathrm{OR}$, and it will follow that u is a function from OR to A.

Now suppose that for all $\gamma < \beta$, $\gamma \in S$.

That is, for all $\gamma < \beta$, $(\gamma, x_\gamma) \in u$ for exactly one element $x_\gamma \in A$.

We will consider two cases:

(a) β is a nonlimit ordinal.

$\Rightarrow$ $\beta = \delta^+$ for some $\delta \in \mathrm{OR}$.

$\Rightarrow$ Since $\delta < \beta$, $(\delta, x_\delta) \in u$ for exactly one $x_\delta \in A$ by induction hypothesis.

$\Rightarrow$ $(\beta, f(x_\delta)) = (\delta^+, f(x_\delta)) \in u$ and $(\delta^+, f(x_\delta)) \in u$ by (2′).

To show that $f(x_\delta)$ is a unique element of A such that $(\beta, f(x_\delta)) \in u$, suppose, on the contrary, that $(\beta, y) \in u$ where $y \neq f(x_\delta)$ and let $u' = u \setminus \{(\beta, y)\}$.

$\Rightarrow$ • u' satisfies (1′).

($\because$) Since $u' = u \setminus \{(\beta, y)\}$ and $\beta \neq 0$, $(\beta, y) \neq (0, a)$.

$\Rightarrow$ Since u is perfect, $(0, a) \in u$ and hence $(0, a) \in u'$.

• u' satisfies (2′).

($\because$) Suppose that $(\alpha, x) \in u'$ for $\alpha \in \mathrm{OR}$.

$\Rightarrow (\alpha, x) \in u$, and so $(\alpha^+, f(x)) \in u$ by (2′).

$\Rightarrow$ If $\alpha^+ \neq \beta$, then $(\alpha^+, f(x)) \neq (\beta, y)$ and hence

$(\alpha^+, f(x)) \in u'$ for $\alpha^+ \neq \beta$.

If $\alpha^+ = \beta = \delta^+$, then $\alpha = \delta$ and $(\alpha, x) = (\delta, x_\delta)$, and hence

$(\alpha^+, f(x)) = (\delta^+, f(x_\delta)) \in u$ for $\alpha^+ \neq \beta$.

$\Rightarrow (\alpha^+, f(x)) \in u$ in any case.

$\Rightarrow$ Since $y \neq f(x_\delta)$ by assumption, $(\beta, y) \neq (\alpha^+, f(x))$ and hence $(\alpha^+, f(x)) \in u'$.

- u' satisfies (3′).

($\because$) Let α be a limit ordinal and assume that $(\gamma, x_\gamma) \in u'$ for all $\gamma < \alpha$. Let $x = \sup\{x_\gamma \mid \gamma < \alpha\}$.

$\Rightarrow (\gamma, x_\gamma) \in u$ for all $\gamma < \alpha$.

$\Rightarrow (\alpha, x) \in u$ by (3′).

$\Rightarrow$ Since β is a nonlimit ordinal, $(\alpha, x) \neq (\beta, y)$.

$\Rightarrow (\alpha, x) \in u'$.

$\Rightarrow u' = u \setminus \{(\beta, y)\}$ is perfect, which is impossible because of the way we defined u. Hence $(\beta, y) \notin u$.

$\Rightarrow f(x_\delta)$ is a unique element of A such that $(\beta, f(x_\delta)) \in u$ so that $\beta \in S$ for a nonlimit ordinal β.

(b) β is a limit ordinal.

$\Rightarrow$ Since $(\gamma, x_\gamma) \in u$ for all $\gamma < \beta$ by the hypothesis of induction, $(\beta, x) \in u$ by (3′), where $x = \sup\{x_\gamma \mid \gamma < \beta\}$.

Suppose now that $(\beta, y) \in u$ where $y \neq x$ and let $u' = u \setminus \{(\beta, y)\}$.

$\Rightarrow$ • u' satisfies (1′).

($\because$) Since $\beta \neq 0$, $(\beta, y) \neq (0, a)$.

$\Rightarrow (0, a) \in u'$.

- u' satisfies (2′).

($\because$) Let $(\alpha, x) \in u'$.

$\Rightarrow (\alpha, x) \in u$ and hence $(\alpha^+, f(x)) \in u$ by (2′).

$\Rightarrow$ Since β is a limit ordinal, $\alpha^+ \neq \beta$.

$\Rightarrow (\alpha^+, f(x)) \neq (\beta, y)$ and hence $(\alpha^+, f(x)) \in u'$.

- u' satisfies $(3')$

$(\because)$ Let α be a limit ordinal and assume that $(\gamma, x_\gamma) \in u'$ for all $\gamma < \alpha$.

$\Rightarrow (\gamma, x_\gamma) \in u$ for all $\gamma < \alpha$.

$\Rightarrow (\alpha, x) \in u$, where $x = \sup\{x_\gamma \mid \gamma < \alpha\}$.

$\Rightarrow$ If $\alpha \neq \beta$, then $(\alpha, x) \neq (\beta, y)$ and hence $(\alpha, x) \in u'$.

If $\alpha = \beta$, then $(\alpha, x) = (\beta, x)$ but since $y \neq x$ by assumption, $(\alpha, x) \neq (\beta, y)$ and hence $(\alpha, x) \in u'$.

$\Rightarrow (\alpha, x) = (\alpha, \sup\{x_\gamma \mid \gamma < \alpha\}) \in u'$ in any case.

$\Rightarrow u' = u \setminus \{(\beta, y)\}$ is perfect, which is impossible because of the way we defined u.

$\Rightarrow (\beta, y) \notin u$ for a limit ordinal β.

$\Rightarrow (\beta, x) \in u$ for exactly one $x = \sup\{x_\gamma \mid x_\gamma < \beta\} \in A$ so that $\beta \in S$ for a limit ordinal β.

We have proven that u is a function which satisfies the conditions (1), (2) and (3). The proof of the uniqueness of u is left as an exercise for the reader. □

Just as finite recursion was used in Chapter 6 to define the addition and multiplication of natural numbers, transfinite recursion may be used to give alternative definitions for the addition and multiplication of ordinal numbers.

Definition 10.2. For an arbitrary $a \in \mathrm{OR}$, we define a function $\sigma_\alpha : \mathrm{OR} \to \mathrm{OR}$ as follows:

(1) $\sigma_\alpha(0) = \alpha$,

(2) $\sigma_\alpha(\beta^+) = (\sigma_\alpha(\beta))^+$,

(3) $\sigma_\alpha(\beta) = \sup\{\sigma_\alpha(\gamma) \mid \gamma < \beta\}$ if β is a limit ordinal.

Theorem 10.1 guarantees the existence of a unique function $\sigma_\alpha : \mathrm{OR} \to \mathrm{OR}$ satisfying the conditions (1), (2), (3) of Definition 10.2.

Theorem 10.3. *For arbitrary ordinals α and β, $\sigma_\alpha(\beta) = \alpha + \beta$.*

Proof. If $\beta = 0$, then $\sigma_\alpha(0) = \alpha = \alpha + 0$.

By induction, let us suppose now that the theorem holds for all $\gamma < \beta$.

$\Rightarrow$ If β is a limit ordinal, then

$$\begin{aligned}\sigma_\alpha(\beta) &= \sup\{\sigma_\alpha(\gamma) \mid \gamma < \beta\} \text{ (by Definition 10.2(3))}\\ &= \sup\{\alpha + \gamma \mid \gamma < \beta\} \text{ (by the hypothesis of induction)}\\ &= \alpha + \beta \text{ (by Exercise 9.3.12 (2)).}\end{aligned}$$

If β is a nonlimit ordinal with $\beta = \delta^+$, then

$$\begin{aligned}\sigma_\alpha(\beta) &= \sigma_\alpha(\delta^+) = (\sigma_\alpha(\delta))^+ \text{ (by Definition 10.2(2))}\\ &= (\alpha + \delta)^+ \text{ (by the hypothesis of induction)}\\ &= \alpha + \delta^+ \text{ (by Theorem 9.14(1))}\\ &= \alpha + \beta.\end{aligned}$$

□

Theorem 10.3 tells us that if we define the addition of ordinal numbers by

$$\alpha + \beta = \sigma_\alpha(\beta) \tag{10.1}$$

where $\sigma_\alpha : \mathrm{OR} \to \mathrm{OR}$ is given by Definition 10.2, then Equation (10.1) is equivalent to Definition 9.13.

Equation (10.1) may also be written in the following form:

$$\begin{aligned}&(1)\ \alpha + 0 = \alpha,\\ &(2)\ \alpha + \beta^+ = (\alpha + \beta)^+,\\ &(3)\ \alpha + \beta = \sup\{\alpha + \gamma \mid \gamma < \beta\} \text{ if } \beta \text{ is a limit ordinal.}\end{aligned} \tag{10.2}$$

Since Equation (10.2) is equivalent to Definition 9.13, we will henceforth consider Equation (10.2) to define the addition of ordinal numbers.

Definition 10.4. For an arbitrary $\alpha \in \mathrm{OR}$, we define a function $\pi_\alpha : \mathrm{OR} \to \mathrm{OR}$ as follows:

(1) $\pi_\alpha(0) = 0$,

(2) $\pi_\alpha(\beta^+) = \pi_\alpha(\beta) + \alpha$,

(3) $\pi_\alpha(\beta) = \sup\{\pi_\alpha(\gamma) \mid \gamma < \beta\}$ if β is a limit ordinal.

Theorem 10.1 guarantees the existence of a function $\pi_\alpha : \text{OR} \to \text{OR}$ satisfying the conditions (1), (2) and (3) of Definition 10.4.

Theorem 10.5. *For arbitrary ordinals $\alpha, \beta \in \text{OR}$, $\pi_\alpha(\beta) = \alpha\beta$.*

Proof. If $\beta = 0$, then $\pi_\alpha(0) = \alpha 0$. By induction, let us suppose that the theorem holds for every $\gamma < \beta$; that is, $\pi_\alpha(\gamma) = \alpha\gamma$ for all $\gamma < \beta$.

$\Rightarrow$ If β is a limit ordinal, then

$$\begin{aligned}\pi_\alpha(\beta) &= \sup\{\pi_\alpha(\gamma) \mid \gamma < \beta\} \text{ (by Definition 10.4(3))}\\ &= \sup\{\alpha\gamma \mid \gamma < \beta\} \text{ (by the hypothesis of induction)}\\ &= \alpha\beta \text{ (by Exercise 9.3.13).}\end{aligned}$$

If β is a nonlimit ordinal with $\beta = \delta^+$, then

$$\begin{aligned}\pi_\alpha(\beta) &= \pi_\alpha(\delta^+) = \pi_\alpha(\delta) + \alpha \text{ (by Definition 10.4(2))}\\ &= \alpha\delta + \alpha \text{ (by the hypothesis of induction)}\\ &= \alpha(\delta + 1) \text{ (by Theorem 9.14(3))}\\ &= \alpha\beta.\end{aligned}$$

$\Rightarrow$ $\pi_\alpha(\beta) = \alpha\beta$ for all ordinals $\alpha, \beta \in \text{OR}$, by Theorem 4.57(transfinite mathematical induction). □

Theorem 10.5 tells us that if we define the multiplication of ordinal numbers by

$$\alpha\beta = \pi_\alpha(\beta) \tag{10.3}$$

where $\pi_\alpha(\beta)$ is given by Definition 10.4, then Equation (10.3) is equivalent to Definition 9.13.

Equation (10.3) may also be written in the following form:

$$\begin{aligned} &(1)\ \alpha 0 = 0, \\ &(2)\ \alpha\beta^{+} = \alpha\beta + \alpha, \\ &(3)\ \alpha\beta = \sup\{\alpha\gamma \mid \gamma < \beta\} \text{ if } \beta \text{ is a limit ordinal.} \end{aligned} \tag{10.4}$$

Since Equation (10.4) is equivalent to Definition 9.13, we will henceforth consider Equation (10.4) to be the definition of ordinal multiplication.

10.2 Properties of ordinal exponentiation

Equations (10.2) and (10.4) provide us with an alternative way of defining the addition and multiplication of ordinal numbers, using transfinite recursion. We will use this new (and in many ways, more convenient) approach to define ordinal exponentiation.

Definition 10.6. If α is any nonzero ordinal number, we define a function $\eta_\alpha : \mathrm{OR} \to \mathrm{OR}$ as follows:

(1) $\eta_\alpha(0) = 1$,

(2) $\eta_\alpha(\beta^{+}) = \eta_\alpha(\beta)\alpha$,

(3) $\eta_\alpha(\beta) = \sup\{\eta_\alpha(\gamma) \mid \gamma < \beta\}$ if β is a limit ordinal.

Theorem 10.1 guarantees the existence of a function $\eta_\alpha : \mathrm{OR} \to \mathrm{OR}$ which satisfies conditions (1), (2), (3) of Definition 10.6.

Definition 10.7. We define *ordinal exponentiations* as follow:
If α and β are arbitrary ordinals, we let

$$\alpha^\beta = \eta_\alpha(\beta) \text{ if } \alpha \neq 0, \quad \text{and } 0^\beta = 0.$$

In view of Definition 10.6, Definition 10.7 may therefore be written as follows:

$$\begin{aligned}&(1)\ \alpha^0 = 1,\\ &(2)\ \alpha^{\beta^+} = \left(\alpha^\beta\right)\alpha,\\ &(3)\ \alpha^\beta = \sup\{\alpha^\gamma \mid \gamma < \beta\}, \text{ if } \beta \text{ is a limit ordinal,}\\ &(4)\ 0^\beta = 0.\end{aligned} \tag{10.5}$$

We will now develop the fundamental properties of ordinal exponentiation.

Lemma 10.8. *If γ is a limit ordinal number and $\epsilon < \alpha^\gamma$, then there exists an ordinal number π such that $\pi < \gamma$ and $\epsilon < \alpha^\pi$.*

Proof. Suppose, on the contrary, that $\alpha^\pi \leqslant \epsilon$ for all ordinal numbers $\pi < \gamma$.

$\Rightarrow$ ϵ is an upper bound of the set $\{\alpha^\pi \mid \pi < \gamma\}$.

$\Rightarrow$ Since $\alpha^\gamma = \sup\{\alpha^\pi \mid \pi < \gamma\}$, $\alpha^\gamma \leqslant \epsilon$,

which contradicts our assumption that $\epsilon < \alpha^\gamma$.

$\Rightarrow$ $\epsilon < \alpha^\pi$ for some $\pi < \gamma$. □

Theorem 10.9. *For any ordinal numbers α, β and γ with $\alpha > 1$, we have the following.*

(1) *If $\beta < \gamma$, then $\alpha^\beta < \alpha^\gamma$.*

(2) *If $\alpha^\gamma < \alpha^\beta$, then $\gamma < \beta$.*

Proof. (1) Assume that $\beta < \alpha$. The proof is by induction on γ.

First, if $\gamma = 0$, the condition is satisfied vacuously.

Next, assume inductively that (1) holds for all ordinal number $\delta < \gamma$.

i.e., If $\beta < \delta$, then $\alpha^\beta < \alpha^\delta$ for every ordinal number $\delta < \gamma$.

$\Rightarrow$ $\alpha^\beta < \alpha^\delta$ for every ordinal number δ with $\beta < \delta < \gamma$.

($\because$) (i) Assume first that γ is a limit ordinal.

$\Rightarrow$ Since $\beta < \gamma$, $\beta+1 < \gamma$ and hence $\alpha^\beta < \alpha^{\beta+1}$ by the hypothesis of induction.

$\Rightarrow$ Since $\alpha^\gamma = \sup\{\alpha^\beta \mid \beta < \gamma\}$, $\alpha^\beta < \alpha^{\beta+1} \leqslant \alpha^\gamma$.

(ii) Now assume that γ is a nonlimit ordinal with $\gamma = \delta + 1$.

$\Rightarrow$ Since $\beta < \gamma$, $\beta = \delta$ or $\beta < \delta$.

$\Rightarrow$ If $\beta = \delta$, then we have $\alpha^\gamma = \alpha^{\delta+1} = \alpha^\delta\alpha > \alpha^\delta = \alpha^\beta$ by Equation (10.5) (2), and Theorem 9.25(5) (note that $\alpha > 1$).

If $\beta < \delta$, then $\alpha^\beta < \alpha^\delta$ by the hypothesis of induction, so we have $\alpha^\beta < \alpha^\delta < \alpha^\delta\alpha = \alpha^{\delta+1} = \alpha^\gamma$ by Theorem 9.25(5) and Equation (10.5) (2).

$\Rightarrow$ $\beta < \gamma$ implies $\alpha^\beta < \alpha^\gamma$ for all ordinal numbers α, β and γ with $\alpha > 1$, by Theorem 4.57(transfinite mathematical induction).

(2) Assume that $\alpha^\gamma < \alpha^\beta$.

$\Rightarrow$ $\gamma \leqslant \beta$ by the contrapositive of (1).

$\Rightarrow$ If $\gamma = \beta$, then $\alpha^\gamma = \alpha^\beta$ which contradicts the assumption, and hence $\gamma < \beta$. □

Theorem 10.10. *For any ordinal numbers α, β, and γ, we have the following.*

(1) *If $\alpha \leqslant \beta$, then $\alpha^\gamma \leqslant \beta^\gamma$.*

(2) *If $\beta^\gamma < \alpha^\gamma$, then $\beta < \alpha$.*

Proof. (1) The proof is by induction on γ.

If $\gamma = 0$, the condition is satisfied trivially.

Now assume inductively that (1) holds for all ordinal number $\delta < \gamma$.

$\Rightarrow$ If $\alpha \leqslant \beta$, then $\alpha^\delta \leqslant \beta^\delta$ for every ordinal number $\delta < \gamma$.

(i) Assume first that γ is a nonlimit ordinal.

$\Rightarrow$ $\gamma = \delta + 1$ for some ordinal number δ.

We assume that $\alpha \leqslant \beta$ and hence $\alpha^\delta \leqslant \beta^\delta$ by the induction hypothesis.

$\Rightarrow$ By Equation 10.5(2) and Theorem 9.25(5) and (7), we have $\alpha^\gamma = \alpha^{\delta+1} = \alpha^\delta\alpha \leqslant \alpha^\delta\beta \leqslant \beta^\delta\beta = \beta^{\delta+1} = \beta^\gamma$.

(ii) Next, assume that γ is a limit ordinal.

$\Rightarrow$ For every ordinal number $\delta < \gamma$, since $\beta^\delta < \beta^\gamma$ by Theorem 10.9(1) and since $\alpha^\delta \leqslant \beta^\delta$ by the induction hypothesis, $\alpha^\delta < \beta^\gamma$.

$\Rightarrow$ β^γ is an upper bound of $\{\alpha^\delta \mid \delta < \gamma\}$.

$\Rightarrow$ $\alpha^\gamma = \sup\{\alpha^\delta \mid \delta < \gamma\} \leqslant \beta^\gamma$.

$\Rightarrow$ $\alpha \leqslant \beta$ implies $\alpha^\gamma \leqslant \beta^\gamma$ for all ordinal numbers α, β, and γ by Theorem 4.57(transfinite mathematical induction).

(2) This is simply the contrapositive of (1). □

Theorem 10.11. *$\alpha^\beta \alpha^\gamma = \alpha^{\beta+\gamma}$ for any ordinal numbers α, β, and γ.*

Proof. Let α, β, and γ be ordinal numbers. The proof is by induction on γ. The theorem holds trivially if $\gamma = 0$. Assume that $\gamma > 0$ and that $\alpha^\beta \alpha^\delta = \alpha^{\beta+\delta}$ for every ordinal number $\delta < \gamma$.

$\Rightarrow$ If $\alpha = 0$, then $\alpha^\beta \alpha^\gamma = 0 = \alpha^{\beta+\gamma}$.

Assume that $\alpha \neq 0$ and hence $1 = \alpha^0 \leqslant \alpha^\gamma$ by Theorem 10.9.

First, suppose that γ is a nonlimit ordinal.

$\Rightarrow$ $\gamma = \delta + 1$ for some ordinal number δ.

$\Rightarrow$ $\alpha^{\beta+\gamma} = \alpha^{\beta+\delta+1} = \alpha^{\beta+\delta}\alpha = \alpha^\beta\alpha^\delta\alpha = \alpha^\beta\alpha^{\delta+1} = \alpha^\beta\alpha^\gamma$.

Next, suppose that γ is a limit ordinal. We shall prove the two inequalities (i) $\alpha^{\beta+\gamma} \leqslant \alpha^\beta\alpha^\gamma$ and (ii) $\alpha^\beta\alpha^\gamma \leqslant \alpha^{\beta+\gamma}$.

(i) $\alpha^{\beta+\gamma} \leqslant \alpha^\beta\alpha^\gamma$

($\because$) Since γ is a limit ordinal, clearly $\beta + \gamma$ is a limit ordinal.

$\Rightarrow$ $\alpha^{\beta+\gamma} = \sup\{\alpha^\delta \mid \delta < \beta + \gamma\}$ by Equation (10.5) (3).

$\Rightarrow$ $\alpha^\delta \leqslant \alpha^\beta\alpha^\gamma$ for every ordinal number $\delta < \beta + \gamma$.

Let $\delta < \beta + \gamma$ so that either $\delta \leqslant \beta$ or $\delta > \beta$.

$\Rightarrow$ If $\delta \leqslant \beta$, then $\alpha^\delta \leqslant \alpha^\beta \leqslant \alpha^\beta\alpha^\gamma (\because 1 \leqslant \alpha^\gamma)$ by Theorem 10.9(1) and Theorem 9.25(5).

If $\delta > \beta$, then, $\delta = \beta + \rho$ for some $\rho < \gamma$ by Theorem 9.24 and hence $\alpha^\rho < \alpha^\gamma$ by Theorem 10.9(1).

$\Rightarrow$ Since $\alpha^\delta = \alpha^{\beta+\rho} = \alpha^\beta\alpha^\rho < \alpha^\beta\alpha^\gamma$ by the hypothesis of induction and Theorem 9.25(5) and hence $\alpha^{\delta} \leqslant \alpha^{\beta}\alpha^{\gamma}$ in either of the two cases.

$\Rightarrow \alpha^{\beta}\alpha^{\gamma}$ is an upper bound of $\{\alpha^{\delta} \mid \delta < \beta + \gamma\}$.

$\Rightarrow \alpha^{\beta+\gamma} = \sup\{\alpha^{\delta} \mid \delta < \beta + \gamma\} \leqslant \alpha^{\beta}\alpha^{\gamma}$.

(ii) $\alpha^{\beta}\alpha^{\gamma} \leqslant \alpha^{\beta+\gamma}$.

($\because$) Since γ is a limit ordinal, α^{γ} is a limit ordinal by Exercise 10.2.1 (1).

$\Rightarrow \alpha^{\beta}\alpha^{\gamma} = \sup\{\alpha^{\beta}\epsilon \mid \epsilon < \alpha^{\gamma}\}$ by Equation (10.4) (3).

Let $\epsilon < \alpha^{\gamma}$.

$\Rightarrow$ There exists $\pi < \gamma$ such that $\epsilon < \alpha^{\pi}$ by Lemma 10.8.

$\Rightarrow \alpha^{\beta}\epsilon < \alpha^{\beta}\alpha^{\pi} = \alpha^{\beta+\pi} < \alpha^{\beta+\gamma}$ by Theorem 9.25(5), Theorem 10.9(1), and the hypothesis of induction.

$\Rightarrow \alpha^{\beta+\gamma}$ is an upper bound of $\{\alpha^{\beta}\epsilon \mid \epsilon < \alpha^{\gamma}\}$.

$\Rightarrow \alpha^{\beta}\alpha^{\gamma} = \sup\{\alpha^{\beta}\epsilon \mid \epsilon < \alpha^{\gamma}\} \leqslant \alpha^{\beta+\gamma}$.

$\Rightarrow \alpha^{\beta}\alpha^{\gamma} = \alpha^{\beta+\gamma}$ by (i) and (ii) above.

$\Rightarrow \alpha^{\beta}\alpha^{\gamma} = \alpha^{\beta+\gamma}$ for any ordinal numbers α, β, and γ by the transfinite mathematical induction(Theorem 4.57). □

Theorem 10.12. $(\alpha^{\beta})^{\gamma} = \alpha^{\beta\gamma}$ *for any ordinal numbers* α, β, *and* γ.

Proof. Let α, β, and γ be ordinal numbers. The proof is by induction on γ.

If $\gamma = 0$, then $(\alpha^{\beta})^{\gamma} = 1 = \alpha^{\beta\gamma}$ by Equation (10.5) (1).

Let us assume that $(\alpha^{\beta})^{\delta} = \alpha^{\beta\delta}$ for every ordinal number $\delta < \gamma$.

First assume that γ is a nonlimit ordinal.

$\Rightarrow \gamma = \delta + 1$ for some ordinal number δ.

$\Rightarrow (\alpha^{\beta})^{\gamma} = (\alpha^{\beta})^{\delta+1} = (\alpha^{\beta})^{\delta}\alpha^{\beta} = \alpha^{\beta\delta}\alpha^{\beta} = \alpha^{\beta\delta+\beta} = \alpha^{\beta(\delta+1)} = \alpha^{\beta\gamma}$ by Equation (10.5) (2), the hypothesis of induction, and Theorem 10.11.

Next, assume that γ is a limit ordinal and we will prove the two inequalities (i) $(\alpha^{\beta})^{\gamma} \leqslant \alpha^{\beta\gamma}$ and (ii) $\alpha^{\beta\gamma} \leqslant (\alpha^{\beta})^{\gamma}$.

$\Rightarrow$ (i) $(\alpha^{\beta})^{\gamma} \leqslant \alpha^{\beta\gamma}$.

($\because$) For any ordinal $\delta < \gamma$, since $\gamma > 0$, we have $\beta\delta < \beta\gamma$ by

Theorem 9.25(5) and hence $\alpha^{\beta\delta} < \alpha^{\beta\gamma}$ by Theorem 10.9(1).

$\Rightarrow \alpha^{\beta\gamma}$ is an upper bound of $\{\alpha^{\beta\delta} \mid \delta < \gamma\}$.

$\Rightarrow (\alpha^{\beta})^{\gamma} = \sup\left\{(\alpha^{\beta})^{\delta} \mid \delta < \gamma\right\} = \sup\{\alpha^{\beta\delta} \mid \delta < \gamma\} \leqslant \alpha^{\beta\gamma}$

by Equation (10.5)(3) and the hypothesis of induction.

(ii) $\alpha^{\beta\gamma} \leqslant (\alpha^{\beta})^{\gamma}$.

($\because$) Since γ is a limit ordinal, $\beta\gamma$ is a limit ordinal by Exercise 9.3.6.

$\Rightarrow \alpha^{\beta\gamma} = \sup\{\alpha^{\delta} \mid \delta < \beta\gamma\}$ by Equation (10.5) (3).

$\Rightarrow$ For $\delta < \beta\gamma$, $\delta = \beta\xi + \epsilon$ with $\xi < \gamma$ and $\epsilon < \beta$ by Lemma 9.27.

$\Rightarrow \delta = \beta\xi + \epsilon < \beta\xi + \beta = \beta(\xi + 1)$ by Theorems 9.25(1), 9.14(3).

$\Rightarrow$ Since $\xi < \gamma$ and γ is a limit ordinal, $\xi + 1 < \gamma$ and hence

$\alpha^{\delta} < \alpha^{\beta(\xi+1)} = (\alpha^{\beta})^{\xi+1} < (\alpha^{\beta})^{\gamma}$ by the hypothesis of induction

and Theorem 10.9(1).

$\Rightarrow (\alpha^{\beta})^{\gamma}$ is an upper bound of $\{\alpha^{\delta} \mid \delta < \beta\gamma\}$.

$\Rightarrow \alpha^{\beta\gamma} \leqslant \sup\{\alpha^{\delta} \mid \delta < \beta\gamma\} \leqslant (\alpha^{\beta})^{\gamma}$.

$\Rightarrow \alpha^{\beta\gamma} = (\alpha^{\beta})^{\gamma}$ by (i) and (ii), when γ is a limit ordinal.

$\Rightarrow \alpha^{\beta\gamma} = (\alpha^{\beta})^{\gamma}$ for any ordinals α, β, and γ by Theorem 4.57(transfinite mathematical induction). □

Note that by Equation (10.5) (3), $2^{\omega} = \sup\{2^{n} \mid n < \omega\}$. We noted in the preceding chapter that the "usual" arithmetic laws do not *all* apply to transfinite ordinal numbers; in particular, the commutative law for multiplication does not hold, nor does the right distributive law. As we shall see now, the law $(\alpha\beta)^{\gamma} = \alpha^{\gamma}\beta^{\gamma}$ does not apply generally to ordinal numbers.

Example 10.13. (1) $(2 \cdot 2)^{\omega} = 4^{\omega} = \omega$ because $4^{\omega} = \sup\{4^{n} \mid n < \omega\}$.

(2) $2^{\omega}2^{\omega} = \omega\omega = \omega^{2}$. Since $\omega > 1$, it follows by Theorem 9.25(5) that $\omega\omega > \omega 1 = \omega$; thus $(2 \cdot 2)^{\omega} \neq 2^{\omega}2^{\omega}$.

Exercise 10.2

1. Prove the following.

   (1) If γ is a limit ordinal and $\alpha > 1$, then α^γ is a limit ordinal.

   (2) If α is a limit ordinal and $\gamma \neq 0$, then α^γ is a limit ordinal.

2. Prove that for any ordinal numbers $\alpha > 1, \beta$ and γ, if $\alpha^\beta = \alpha^\gamma$, then $\beta = \gamma$.

3. Use Equations 10.2, 10.4 and 10.5 to prove the following for any finite ordinal number n.

   (1) $n + \omega = \omega$. (2) $n\omega = \omega$. (3) $n^\omega = \omega$.

4. Use induction to prove that for every ordinal number β, $2^\beta \geqslant \beta$. Conclude that for any ordinal numbers $\alpha > 0$ and β, $\alpha^\beta \geqslant \beta$.

5. Prove that for all ordinal numbers α and β with $\alpha > 1$, $\alpha\beta \leqslant \alpha^\beta$.

6. Prove that if α is a limit ordinal and p and q are finite ordinal numbers, then $(\alpha p)^q = \alpha^q p$. [Hint: Use Exercise 9.3.8.]

7. An ordinal number γ is said to be *simple* if it cannot be written $\gamma = \delta + \omega$ for any ordinal number δ. Prove that γ is simple if and only if for each ordinal number $\epsilon < \gamma$, there exists a limit ordinal λ such that $\epsilon < \lambda < \gamma$.

8. If α is a denumerable ordinal number (that is, if $\sharp\alpha = \aleph_0$), use Definition 9.13 to prove that $\alpha\omega$ is denumerable. Conclude that the ordinal numbers $\omega, \omega^2, \cdots, \omega^n (n \text{ finite})$ are all denumerable.

9. (1) Let $\{\gamma_i \mid i \in I\}$ be a set of ordinal numbers. Prove that

   $$\sup\{\gamma_i \mid i \in I\} = \bigcup_{i \in I} \gamma_i.$$

   [Hint: Use Definition 9.37(4) and Lemma 9.39(2).]

   (2) Prove that if $\{\gamma_n \mid n \in \alpha\}$ is a set of denumerable ordinal numbers, where α is denumerable, then $\sup\{\gamma_n \mid n \in \alpha\}$ is a denumerable ordinal number. [Hint: Note that it follows immediately from Exercises 8.5.2 and 8.5.9 that $\sharp(\bigcup_{i \in I} \gamma_i) \leqslant \sum_{i \in I} \sharp\gamma_i \leqslant \aleph_0$.]

   (3) Use the result of Exercise 10.2.8 above to prove that ω^ω is a denumerable ordinal number. [Hint: Note that $\omega^\omega = \sup\{\omega^n \mid n \in \omega\}$.]

   (4) Conclude that if α and β are denumerable ordinal numbers, then α^β is a denumerable ordinal number. [Hint: Use Exercise 10.2.9(2) above.]

10.3 Normal form

It is a well-known fact of elementary number theory that every natural number n has a uniquely determined decimal representation. That is, given n, there exist unique natural numbers $k, m_0, m_1, \cdots, m_k$ (each $m_i < 10$) such that $n = m_k \cdot 10^k + \cdots + m_1 \cdot 10 + m_0$.

The decimal representation of n is also called its *representation with base* 10; it can easily be shown that every natural number n also has a unique representation with base b, for any $b > 1$.

The idea of giving every number a representation with base b can easily be extended to ordinal numbers generally. In the sequel we will only consider base ω, but it should be clear to the reader that any other base will do.

Definition 10.14. Let γ be an ordinal number. Suppose that there are nonzero natural numbers $a_1, \cdots, a_n$ and ordinals $\alpha_1 > \alpha_2 > \cdots > \alpha_n$ such that

$$\gamma = \omega^{\alpha_1} a_1 + \omega^{\alpha_2} a_2 + \cdots + \omega^{\alpha_n} a_n. \tag{10.6}$$

Then Equation 10.6 is called a *normal form representation* of γ.

Theorem 10.15. *Every nonzero ordinal number has a normal form representation.*

Proof. Let γ be a nonzero ordinal number. The proof will be by induction on γ. If $\gamma = 1$, then $\gamma = \omega^0 1$ is a normal form representation of γ. Now assume the theorem is true for all ordinal number $\rho < \gamma$.

Let $A = \{\mu \mid \omega^\mu > \gamma\}$.

$\Rightarrow$ A is nonempty, as may easily be seen by using Exercise 10.2.4.

$\Rightarrow$ A has a least element υ such that $\omega^\upsilon > \gamma$.

$\Rightarrow$ υ is a nonlimit ordinal.

($\because$) Suppose that υ is a limit ordinal.

$\Rightarrow \omega^\upsilon = \sup\{\omega^\delta \mid \delta < \upsilon\}$.

$\Rightarrow$ Since υ is the least element of A, $\omega^\delta \leqslant \omega^\upsilon \leqslant \gamma$ for each ordinal number $\delta < \upsilon$.

$\Rightarrow \gamma$ is an upper bound of $\{\omega^\delta \mid \delta < \upsilon\}$.

$\Rightarrow \omega^\upsilon \leqslant \gamma$. This is contrary to our choice of υ.

$\Rightarrow \upsilon$ is a nonlimit ordinal.

$\Rightarrow \upsilon = \alpha_1 + 1$ for some ordinal number α_1.

$\Rightarrow$ There exist ordinal numbers ζ and ρ such that $\gamma = \omega^{\alpha_1}\zeta + \rho$, where $\rho < \omega^{\alpha_1}$, by Theorem 9.28.

$\Rightarrow \zeta < \omega$.

($\because$) Suppose that $\zeta \geqslant \omega$.

$\Rightarrow \omega^{\alpha_1}\zeta \geqslant \omega^{\alpha_1}\omega = \omega^\upsilon > \gamma = \omega^{\alpha_1}\zeta + \rho$ which is impossible by Theorem 9.23(3) and hence $\zeta < \omega$.

$\Rightarrow \zeta$ is a natural number a_1.

$\Rightarrow \gamma = \omega^{\alpha_1}\zeta + \rho = \omega^{\alpha_1}a_1 + \rho$, where $\rho < \omega^{\alpha_1} \leqslant \gamma$.

$\Rightarrow$ By the hypothesis of induction, there exist nonzero natural numbers $a_2, \cdots, a_n$ and ordinal numbers $\alpha_2 > \cdots > \alpha_n$ such that $\rho = \omega^{\alpha_2}a_2 + \cdots + \omega^{\alpha_n}a_n$.

$\Rightarrow \alpha_1 > \alpha_2$.

($\because$) Suppose that $\alpha_1 \leqslant \alpha_2$.

$\Rightarrow \omega^{\alpha_1} \leqslant \omega^{\alpha_2} \leqslant \omega^{\alpha_2}a_2 \leqslant \rho$ which is false because $\rho < \omega^{\alpha_1}$.

$\Rightarrow \gamma = \omega^{\alpha_1}\zeta + \rho = \omega^{\alpha_1}a_1 + \omega^{\alpha_2}a_2 + \cdots + \omega^{\alpha_n}a_n$, where $a_1, \cdots, a_n$ are nonzero natural numbers and $\alpha_1 > \alpha_2 > \cdots > \alpha_n$.

$\Rightarrow$ Every nonzero ordinal number has a normal form representation by Theorem 4.57(transfinite mathematical induction). $\square$

We will prove next that the normal form representation of γ is unique.

Lemma 10.16. *If $\gamma = \omega^{\alpha_1}a_1 + \cdots + \omega^{\alpha_n}a_n$ is a normal form representation of an ordinal number γ, then $\gamma < \omega^{\alpha_1+1}$.*

Proof. Let γ be an ordinal number. Assume that $\gamma = \omega^{\alpha_1}a_1 + \cdots + \omega^{\alpha_n}a_n$

is a normal form representation of γ.

$\Rightarrow \alpha_i < \alpha_1$ for all $i = 2, \cdots, n$ by Definition 10.14.

$\Rightarrow \omega^{\alpha_i} < \omega^{\alpha_1}$ for all $i = 2, \cdots, n$ Theorem 10.9(1).

$\Rightarrow \gamma = \omega^{\alpha_1} a_1 + \omega^{\alpha_2} a_2 + \cdots + \omega^{\alpha_n} a_n \leqslant \omega^{\alpha_1} a_1 + \omega^{\alpha_1} a_2 + \cdots + \omega^{\alpha_1} a_n$

$= \omega^{\alpha_1}(a_1 + \cdots + a_n) < \omega^{\alpha_1}\omega = \omega^{\alpha_1 + 1}$. □

Theorem 10.17. *The normal form representation of any ordinal number is unique.*

Proof. Suppose that an ordinal γ has two normal form representation;

$$\gamma = \omega^{\alpha_1} a_1 + \omega^{\alpha_2} a_2 + \cdots + \omega^{\alpha_n} a_n = \omega^{\beta_1} b_1 + \omega^{\beta_2} b_2 + \cdots + \omega^{\beta_m} b_m,$$

where $a_1, \cdots, a_n, b_1, \cdots, b_m$ are nonzero natural numbers, and $\alpha_1 > \alpha_2 > \cdots > \alpha_n$ and $\beta_1 > \beta_2 > \cdots > \beta_n$.

Let us write $\rho_a = \omega^{\alpha_2} a_2 + \cdots + \omega^{\alpha_n} a_n$ and $\rho_b = \omega^{\alpha_2} b_2 + \cdots + \omega^{\alpha_m} b_m$.

$\Rightarrow \gamma = \omega^{\alpha_1} a_1 + \rho_a = \omega^{\beta_1} b_1 + \rho_b$.

$\Rightarrow \alpha_1 = \beta_1$.

(∵) Suppose that $\alpha_1 < \beta_1$.

$\Rightarrow \alpha_1 + 1 \leqslant \beta_1$.

$\Rightarrow$ Since $1 \leqslant b_1$, $\omega^{\alpha_1 + 1} \leqslant \omega^{\beta_1} \leqslant \omega^{\beta_1} b_1 \leqslant \gamma$.

But $\gamma < \omega^{\alpha_1 + 1}$ by Lemma 10.16. We have a contradiction.

$\Rightarrow \alpha_1 \geqslant \beta_1$. By symmetry, we have $\beta_1 \geqslant \alpha_1$.

$\Rightarrow \alpha_1 = \beta_1$.

$\Rightarrow \gamma = \omega^{\alpha_1} a_1 + \rho_a = \omega^{\alpha_1} b_1 + \rho_b$.

$\Rightarrow a_1 = b_1$.

(∵) Suppose that $a_1 < b_1$.

$\Rightarrow a_1 + 1 \leqslant b_1$.

$\Rightarrow \omega^{\alpha_1}(a_1 + 1) \leqslant \omega^{\alpha_1} b_1$ by Theorem 9.25(7).

$\Rightarrow \omega^{\alpha_1} a_1 + \omega^{\alpha_1} \leqslant \omega^{\alpha_1} b_1$ by Theorem 9.14(3).

$\Rightarrow \omega^{\alpha_1} a_1 + \omega^{\alpha_1} + \rho_b \leqslant \omega^{\alpha_1} b_1 + \rho_b = \gamma$ by Theorem 9.25(3).

$\Rightarrow \omega^{\alpha_1} a_1 + \omega^{\alpha_1} + \rho_b \leqslant \omega^{\alpha_1} a_1 + \rho_a = \gamma$.

$\Rightarrow \omega^{\alpha_1} + \rho_b \leqslant \rho_a$ by Theorem 9.25(2).

$\Rightarrow \omega^{\alpha_1} \leqslant \rho_a$. But this is impossible.

($\because$) Since $\alpha_2 < \alpha_1$, $\omega^{\alpha_2+1} \leqslant \omega^{\alpha_1}$ and hence $\rho_a < \omega^{\alpha_2+1} \leqslant \omega^{\alpha_1}$ by Lemma 10.16, which is contrary to $\omega^{\alpha_1} \leqslant \rho_a$.

$\Rightarrow a_1 \geqslant b_1$. By symmetry, we have that $a_1 \leqslant b_1$.

$\Rightarrow a_1 = b_1$.

$\Rightarrow \gamma = \omega^{\alpha_1} a_1 + \rho_a = \omega^{\alpha_1} a_1 + \rho_b$.

$\Rightarrow \rho_a = \rho_b$ by Theorem 9.26(1).

$\Rightarrow$ By induction hypothesis, the normal form representation of $\rho_a = \rho_b$ is unique, and hence $m = n$, $\alpha_i = \beta_i$, $a_i = b_i$ for all $i = 2, 3, \cdots, n$.

$\Rightarrow m = n$, $\alpha_i = \beta_i$ and $a_i = b_i$ for all $i = 1, 2, \cdots, n$.

$\Rightarrow$ The normal form representation of any ordinal number is unique. □

The theorem which follows makes it easy to add and multiply ordinal numbers when they are written in normal form.

Theorem 10.18. *Let α and β be any ordinal numbers.*

(1) *If $\alpha < \beta$, then $\omega^\alpha a + \omega^\beta b = \omega^\beta b$ for any natural numbers a and b.*

(2) *If $\gamma = \omega^{\alpha_1} a_1 + \cdots + \omega^{\alpha_n} a_n$ is the normal form representation of an ordinal number γ and if $\beta \neq 0$, then $\gamma \omega^\beta = \omega^{\alpha_1 + \beta}$.*

(3) *If $\gamma = \omega^{\alpha_1} a_1 + \cdots + \omega^{\alpha_n} a_n$ is the normal form representation of γ and if b is a finite nonzero ordinal number, then*
$\gamma b = \omega^{\alpha_1} a_1 b + \omega^{\alpha_2} a_2 + \cdots + \omega^{\alpha_n} a_n$.

Proof. Let α and β be any ordinal numbers.

(1) Let a, b be natural numbers. Assume that $\alpha < \beta$.

$\Rightarrow \beta = \alpha + \delta$ for some ordinal number $\delta > 0$ by Theorem 9.24.

$$\Rightarrow \omega^\alpha a + \omega^\beta b = \omega^\alpha a + \omega^{\alpha+\delta} b = \omega^\alpha a + (\omega^\alpha \omega^\delta) b = \omega^\alpha a + \omega^\alpha (\omega^\delta b)$$
$$= \omega^\alpha (a + \omega^\delta b)$$

by applying Theorem 10.11 and Theorem 9.14(2) and (3).

$\Rightarrow$ Since a is finite, $a + \omega^\delta b = \omega^\delta b$. (It can be proved easily. See, for example, Exercise 10.2.3.)

$\Rightarrow \omega^\alpha a + \omega^\beta b = \omega^\alpha \left(a + \omega^\delta b\right) = \omega^\alpha \left(\omega^\delta b\right) = \omega^{\alpha+\delta} b = \omega^\beta b.$

(2) Let $\gamma = \omega^{\alpha_1} a_1 + \cdots + \omega^{\alpha_n} a_n$ be the normal form representation of an ordinal number γ and let β be a nonzero ordinal number.

$\Rightarrow n\omega = \omega$ for every ordinal number $n \in \omega$ by Exercise 10.2.3.

$\Rightarrow n\omega^\beta = \omega^\beta$ for every ordinal number $\beta > 0$ and $n \in \omega$.

($\because$) If $\beta = 1$, then $n\omega^\beta = n\omega^1 = n\omega = \omega$.

Assume that $\beta > 1$.

$\Rightarrow \beta = 1 + \delta > 0$ by Theorem 9.23(1).

$\Rightarrow n\omega^\beta = n\omega^{1+\delta} = n\left(\omega\omega^\delta\right) = (n\omega)\omega^\delta = \omega\omega^\delta = \omega^{1+\delta} = \omega^\beta$

by applying $n\omega = \omega$, Theorem 10.11 and Theorem 8.7(4).

$\Rightarrow \gamma\omega^\beta = \left(\omega^{\alpha_1} a_1 + \cdots + \omega^{\alpha_n} a_n\right)\omega^\beta \leqslant \left(\omega^{\alpha_1} a_1 + \cdots + \omega^{\alpha_1} a_n\right)\omega^\beta$

$= \left(\omega^{\alpha_1}(a_1 + \cdots + a_n)\right)\omega^\beta$ (by applying Theorem 9.14(3))

$= \omega^{\alpha_1}\left((a_1 + \cdots + a_n)\omega^\beta\right)$ (by applying Theorem 8.7(4))

$= \omega^{\alpha_1}\left(m\omega^\beta\right)$ (where $m = a_1 + \cdots + a_n$ is finite)

$= \omega^{\alpha_1}\omega^\beta$ (because $m\omega^\beta = \omega^\beta$, as above)

$= \omega^{\alpha_1+\beta}$ (by applying Theorem 10.11) and

$\omega^{\alpha_1+\beta} = \omega^{\alpha_1}\omega^\beta \leqslant \gamma\omega^\beta$ because $\omega^{\alpha_1} = \omega^{\alpha_1} 1 \leqslant \omega^{\alpha_1} a_1 \leqslant \gamma$

by Exercise 9.2.8 and Theorem 9.25(5) ($\because 1 \leqslant a_1$).

$\Rightarrow \gamma\omega^\beta = \omega^{\alpha_1+\beta}$.

(3) Let $\gamma = \omega^{\alpha_1} a_1 + \cdots + \omega^{\alpha_n} a_n$ be the normal form representation of γ and let b be a finite nonzero ordinal number.

The proof is by finite induction on $b \geqslant 1$.

If $b = 1$, then $\gamma b = \omega^{\alpha_1} a_1 b + \omega^{\alpha_2} a_2 + \cdots + \omega^{\alpha_n} a_n$

($\because$) $\alpha 1 = \alpha$ for any ordinal number α by Exercise 9.2.8.

Now suppose that the statement (3) holds for b.

$\Rightarrow$ We have, by applying the induction hypothesis,

$$\begin{aligned}
&(\omega^{\alpha_1} a_1 + \cdots + \omega^{\alpha_n} a_n)(b+1) \\
&= (\omega^{\alpha_1} a_1 + \cdots + \omega^{\alpha_n} a_n)\, b + (\omega^{\alpha_1} a_1 + \cdots + \omega^{\alpha_n} a_n) \\
&= (\omega^{\alpha_1} a_1 b + \omega^{\alpha_2} a_2 + \cdots + \omega^{\alpha_n} a_n) + (\omega^{\alpha_1} a_1 + \cdots + \omega^{\alpha_n} a_n).
\end{aligned}$$

The reader should note that in the above sum, the terms $\omega^{\alpha_2}a_2, \cdots,$ $\omega^{\alpha_n}a_n$ all precede the term $\omega^{\alpha_1}a_1$, and that $\alpha_1 > \alpha_2, \cdots, \alpha_1 > \alpha_n$ and hence by Theorem 10.18(1) above, they disappear from the sum.

$$\begin{aligned}\Rightarrow \gamma(b+1) &= (\omega^{\alpha_1}a_1 + \cdots + \omega^{\alpha_n}a_n)(b+1) \\ &= (\omega^{\alpha_1}a_1b + \omega^{\alpha_2}a_2 + \cdots + \omega^{\alpha_n}a_n) + (\omega^{\alpha_1}a_1 + \cdots + \omega^{\alpha_n}a_n) \\ &= \omega^{\alpha_1}a_1b + \omega^{\alpha_1}a_1 + \cdots + \omega^{\alpha_n}a_n \\ &= \omega^{\alpha_1}a_1(b+1) + \omega^{\alpha_2}a_2 + \cdots + \omega^{\alpha_n}a_n.\end{aligned}$$

$\Rightarrow$ By finite mathematical induction(Corollary 4.59),
$\gamma b = \omega^{\alpha_1}a_1b + \omega^{\alpha_2}a_2 + \cdots + \omega^{\alpha_n}a_n$ for all finite ordinal b. □

When adding or multiplying ordinal numbers in normal form, it is also useful to remember that if β is an infinite ordinal number and n is a finite ordinal number, then $n + \beta = \beta$ by Exercise 10.2.3.

Example 10.19. Let $\alpha = \omega^{\omega 2}8 + \omega^{\omega+1}2 + \omega^3 4$ and $\beta = \omega^{\omega+3}7 + \omega^{10}5 + 2$. We shall form the sums $\alpha + \beta$ and $\beta + \alpha$ and the products $\alpha\beta$ and $\beta\alpha$.

$$\begin{aligned}\alpha + \beta &= \omega^{\omega 2}8 + \omega^{\omega+1}2 + \omega^3 4 + \omega^{\omega+3}7 + \omega^{10}5 + 2 \\ &= \omega^{\omega 2}8 + \omega^{\omega+3}7 + \omega^{10}5 + 2.\end{aligned}$$

Note that the terms $\omega^{\omega+1}2$ and $\omega^3 4$ precede $\omega^{\omega+3}7$, and hence they disappear from the sum by Theorem 10.18(1). Thus

$$\begin{aligned}\beta + \alpha &= \omega^{\omega+3}7 + \omega^{10}5 + 2 + \omega^{\omega 2}8 + \omega^{\omega+1}2 + \omega^3 4 \\ &= \omega^{\omega 2}8 + \omega^{\omega+1}2 + \omega^3 4 = \alpha.\end{aligned}$$

$$\begin{aligned}\alpha\beta &= \alpha(\omega^{\omega+3}7 + \omega^{10}5 + 2) \\ &= \alpha\omega^{\omega+3}7 + \alpha\omega^{10}5 + \alpha 2 \text{ (by Theorem 9.14(3))} \\ &= \left(\omega^{\omega 2}8 + \omega^{\omega+1}2 + \omega^3 4\right)\omega^{\omega+3}7 + \left(\omega^{\omega 2}8 + \omega^{\omega+1}2 + \omega^3 4\right)\omega^{10}5 \\ &\quad + \left(\omega^{\omega 2}8 + \omega^{\omega+1}2 + \omega^3 4\right)2 \\ &= \omega^{\omega 2+\omega+3}7 + \omega^{\omega 2+10}5 + \omega^{\omega 2}16 + \omega^{\omega+1}2 + \omega^3 4 \\ &\quad \text{(by Theorem 10.18(2) and (3)).}\end{aligned}$$

$$\begin{aligned}\beta\alpha &= \beta\left(\omega^{\omega 2}8 + \omega^{\omega+1}2 + \omega^3 4\right) \\ &= \beta\omega^{\omega 2}8 + \beta\omega^{\omega+1}2 + \beta\omega^3 4 \text{ (by Theorem 9.14(3))} \\ &= \left(\omega^{\omega+3}7 + \omega^{10}5 + 2\right)\omega^{\omega 2}8 + \left(\omega^{\omega+3}7 + \omega^{10}5 + 2\right)\omega^{\omega+1}2\end{aligned}$$

$$+\left(\omega^{\omega+3}7+\omega^{10}5+2\right)\omega^{3}4$$
$$=\omega^{\omega+3+\omega2}8+\omega^{\omega+3+\omega+1}2+\omega^{\omega+3+3}4 \text{ (by Theorem 10.18(2))}$$
$$=\omega^{\omega3}8+\omega^{\omega2+1}2+\omega^{\omega+6}4.$$

Note that $3+\omega2=\omega2$, $3+\omega=\omega$ and $3+3=6$.

Exercise 10.3

1. In each of the following, compute $\alpha+\beta, \beta+\alpha, \alpha\beta$ and $\beta\alpha$.

   (1) $\alpha=\omega^{\omega3}2+\omega^{\omega}4+\omega^{10}5,\quad \beta=\omega^{\omega+1}7+\omega^{2}9+14.$

   (2) $\alpha=\omega^{\omega^{\omega}}9+\omega^{\omega}7+\omega2,\quad \beta=\omega^{\omega5}8+\omega^{7}2+1.$

   (3) $\alpha=\omega^{\omega^{\omega}}22+\omega^{\omega18}4+71,\quad \beta=\omega^{\omega2+3}12+100.$

2. Let γ be an ordinal number. Prove that γ is irreducible(See Exercise 9.3.2) if and only if $\gamma=\omega^{\beta}$ for some ordinal number β.

   In each of the following exercises, we will assume that $\gamma=\omega^{\alpha_1}a_1+\cdots+\omega^{\alpha_n}a_n$ is the normal form representation of γ.

3. Prove that γ is a limit ordinal if and only if $\alpha_n\neq 0$.

4. Let us define the *magnitude* of γ to be the ordinal number α_1. Prove that $\alpha+\beta=\beta$ if and only if magnitude of $\alpha <$ magnitude of β.

5. Prove that $\omega\gamma=\gamma$ if and only if $\alpha_1,\cdots,\alpha_n$ are all infinite ordinal numbers. Conclude that $\omega\gamma=\gamma$ if and only if $\gamma=\omega^{\omega}\beta$ for some ordinal number β.

6. Let γ be a limit ordinal number and let b be a finite ordinal number. Use finite induction on b to prove that $\gamma^{b}=\omega^{\alpha_1 b}a_1+\omega^{\alpha_1(b-1)}\left(\omega^{\alpha_1}a_2+\cdots+\omega^{\alpha_n}a_n\right)$.

7. Use the result of Exercise 10.3.6 above to compute α^{6} and β^{15}, where α and β are given in Exercise 10.3.1(1).

10.4 Epsilon numbers

Cantor investigated the properties of an interesting class of limit ordinals which he called *epsilon numbers*. These numbers shed some light on the structure of the well-ordered class OR and have useful applications in analysis and elsewhere. We shall give a brief review of their properties in this section.

Definition 10.20. An ordinal number α is called an *epsilon number* if $\alpha = \omega^\alpha$.

It is immediate that every epsilon number is necessarily a limit ordinal. Now, the first question we are led to ask is: Are there any epsilon numbers? To answer this question, we first need a lemma.

Lemma 10.21. *Let* $\{\beta_i \mid i \in I\}$ *be a set of ordinal numbers, and let* $\beta = \sup\{\beta_i \mid i \in I\}$. *Then* $\alpha^\beta = \sup\{\alpha^{\beta_i} \mid i \in I\}$.

Proof. Let $\beta = \sup\{\beta_i \mid i \in I\}$.

(1) Suppose that $\beta = \beta_i$ for some $i \in I$.

$\Rightarrow \beta_j \leqslant \beta_i$ for all $j \in I$.

$\Rightarrow \alpha^{\beta_j} \leqslant \alpha^{\beta_i}$ by Theorem 10.9(1).

$\Rightarrow \alpha^\beta = \alpha^{\beta_i} = \sup\left\{\alpha^{\beta_j} \mid j \in I\right\}$.

(2) Now suppose that $\beta \neq \beta_i$ for all $i \in I$.

$\Rightarrow \beta$ must be a limit ordinal.

($\because$) Suppose that $\beta = \delta + 1$.

$\Rightarrow$ Since $\beta = \sup\{\beta_i \mid i \in I\}$, there exists $i \in I$ such that $\beta_i > \delta$.

$\Rightarrow \beta_i = \delta + 1 = \beta$ which is contrary to our assumption.

$\Rightarrow$ (i) $\sup\{\alpha^{\beta_i} \mid i \in I\} \leqslant \alpha^\beta$.

($\because$) Since $\alpha^\beta = \sup\{\alpha^\gamma \mid \gamma < \beta\}$ and $\beta_i < \beta$ for each $i \in I$, $\alpha^{\beta_i} < \alpha^\beta$.

$\Rightarrow \sup\left\{\alpha^{\beta_i} \mid i \in I\right\} \leqslant \alpha^\beta$.

(ii) $\alpha^\beta \leqslant \sup\{\alpha^{\beta_i} \mid i \in I\}$.

($\because$) Note that $\alpha^\beta = \sup\{\alpha^\gamma \mid \gamma < \beta\}$ and let $\gamma < \beta$.

$\Rightarrow \beta_i > \gamma$ for some $i \in I$.

$\Rightarrow \alpha^\gamma < \alpha^{\beta_i} \leqslant \sup\left\{\alpha^{\beta_i} \mid i \in I\right\}$.

$\Rightarrow \alpha^{\beta} = \sup\{\alpha^{\gamma} \mid \gamma < \beta\} \leqslant \sup\left\{\alpha^{\beta_i} \mid i \in I\right\}.$

$\Rightarrow \alpha^{\beta} = \sup\left\{\alpha^{\beta_j} \mid j \in I\right\}$ by (i) and (ii). □

We now return to the question: Are there any epsilon numbers?

The answer is "yes", and this can easily be shown as follows:

We define a function $f_0 : \omega \to \mathrm{OR}$ by

$$f_0(0) = 1, f_0(n+1) = \omega^{f_0(n)}.$$

The existence of f_0 is guaranteed by the finite recursion(Theorem 6.14). Clearly, we have

$$f_0(0) = 1, f_0(1) = \omega, f_0(2) = \omega^{\omega}, f_0(3) = \omega^{\omega^{\omega}}, \text{ and so on.}$$

Now, let $\epsilon_0 = \sup\{f_0(n) \mid n \in \omega\}$. Then ϵ_0 is an epsilon number.

Indeed, by Lemma 10.21, we have

$$\omega^{\epsilon_0} = \sup\left\{\omega^{f_0(n)} \mid n \in \omega\right\} = \sup\{f_0(n+1) \mid n \in \omega\} = \epsilon_0. \qquad (10.7)$$

Thus there is at least one epsilon number, namely ϵ_0. We can easily show, in fact, that ϵ_0 is the least epsilon number.

Theorem 10.22. *ϵ_0 is the least epsilon number.*

Proof. Let α be an epsilon number.

$\Rightarrow f_0(0) = 1 \leqslant \alpha$($\because$ Clearly 0 is not an epsilon number).

Now suppose that $f_0(n) \leqslant \alpha$.

$\Rightarrow f_0(n+1) = \omega^{f_0(n)} \leqslant \omega^{\alpha} = \alpha.$

$\Rightarrow f_0(n) \leqslant \alpha$ for every $n \in \omega$ by finite induction(Corollary 4.59).

$\Rightarrow \epsilon_0 \leqslant \alpha$ and hence ϵ_0 is the least epsilon number. □

Is ϵ_0 the only epsilon number or are there others? It is easy to answer this question, for the method we used to construct ϵ_0 may be used to construct infinitely many other epsilon numbers. Indeed, we have the following:

Theorem 10.23. *If α is any ordinal number, let $f_\alpha : \omega \to \mathrm{OR}$ be the function defined inductively by the pair of conditions*

$$f_\alpha(0) = \alpha + 1, \quad \textit{and} \quad f_\alpha(n+1) = \omega^{f_\alpha(n)}$$

and let $S(\alpha) = \sup\{f_\alpha(n) \mid n \in \omega\}$. Then

(1) *$S(\alpha)$ is an epsilon number greater than α.*

(2) *$S(\alpha)$ is the least epsilon number greater than α.*

Proof. (1) $\omega^{S(\alpha)} = \sup\left\{\omega^{f_\alpha(n)} \mid n \in \omega\right\} = \sup\{f_\alpha(n+1) \mid n \in \omega\} = S(\alpha)$ by Lemma 10.21 and clearly $\alpha < S(\alpha)$.

(2) Now let γ be an epsilon number such that $\alpha < \gamma$.

$\Rightarrow f_\alpha(0) = \alpha + 1 \leqslant \gamma$.

Furthermore, if $f_\alpha(n) \leqslant \gamma$, then $f_\alpha(n+1) = \omega^{f_\alpha(n)} \leqslant \omega^\gamma = \gamma$.

$\Rightarrow f_\alpha(n) \leqslant \gamma$ for every $n \in \omega$, by finite induction(Corollary 4.59).

$\Rightarrow S(\alpha) \leqslant \gamma$.

$\Rightarrow S(\alpha)$ is the least epsilon number greater than α. □

It is easy to see that ϵ_0 is $S(0)$. Thus, the first few epsilon numbers are $S(0), S(1), S(2), S(3)$, and so on.

The next question we are led to ask is: Are there any epsilon numbers greater than all the $S(n), (n \in \omega)$? The question is easily answered in the following theorem.

Theorem 10.24. *Let $\{\alpha_i \mid i \in I\}$ be a set of epsilon numbers; if $\beta = \sup\{\alpha_i \mid i \in I\}$, then β is an epsilon number.*

Proof. $\omega^\beta = \sup\{\omega^{\alpha_i} \mid i \in I\} = \sup\{\alpha_i \mid i \in I\} = \beta$ by Lemma 10.21. □

Let $\epsilon : \mathrm{OR} \to \mathrm{OR}$ be the function defined recursively as follows [we agree to write ϵ_i for $\epsilon(i)$]:

$$\begin{aligned} \epsilon_0 &= S(0), \\ \epsilon_{\alpha+1} &= S(\epsilon_\alpha) \text{ for all } \alpha \in \mathrm{OR}, \\ \epsilon_\beta &= \sup\{\epsilon_\gamma \mid \gamma < \beta\} \text{ if } \beta \text{ is a limit ordinal.} \end{aligned} \tag{10.8}$$

Theorem 10.25. *ϵ is an isomorphism from* OR *to the class of all the epsilon numbers.*

Proof. Note that ϵ_β is an epsilon number for every $\beta \in$ OR.

($\because$) We argue by transfinite induction on $\beta \geqslant 0$:

First, we have already seen that ϵ_0 is an epsilon number.

Next, assume that ϵ_α is an epsilon number for every $\alpha < \beta$.

$\Rightarrow$ If β is a nonlimit ordinal, $\beta = \delta + 1$, and then $\epsilon_\beta = \epsilon_{\delta+1} = S(\epsilon_\delta)$ is an epsilon number by Theorem 10.23.

If β is a limit ordinal, then $\epsilon_\beta = \sup\{\epsilon_\pi \mid \pi < \gamma\}$ is an epsilon number by Theorem 10.24.

$\Rightarrow$ ϵ_β is an epsilon number for all $\beta \in$ OR by Theorem 4.57(transfinite mathematical induction).

$\Rightarrow$ Since $\epsilon :$ OR $\to$ OR is strictly increasing function by Theorem 10.23, $\epsilon :$ OR $\to$ OR is injective.

To show that the range of ϵ is the class E of the epsilon numbers, let $\delta \in E$ and let ϵ_γ be the least ϵ_ς such that $\delta < \epsilon_\varsigma$.

$\Rightarrow$ γ is a nonlimit ordinal.

($\because$) Suppose that γ is a limit ordinal.

$\Rightarrow$ $\epsilon_\gamma = \sup\{\epsilon_\pi \mid \pi < \gamma\}$ by Equation 10.8.

$\Rightarrow$ Since $\delta < \epsilon_\gamma$, there exists an ordinal number $\pi < \gamma$ such that $\delta < \epsilon_\pi$, which contradicts our choice of γ.

$\Rightarrow$ γ cannot be a limit ordinal.

$\Rightarrow$ $\gamma = \pi + 1$ for some ordinal number π.

$\Rightarrow$ Since $\epsilon :$ OR $\to$ OR is injective, $\epsilon_\pi < \epsilon_{\pi+1} = \epsilon_\gamma$.

$\Rightarrow$ $\epsilon_\pi \leqslant \delta < \epsilon_\gamma = \epsilon_{\pi+1}$ by the choice of ϵ_γ.

$\Rightarrow$ $\delta = \epsilon_\pi$ by Theorem 10.23.

$\Rightarrow$ $\epsilon :$ OR $\to$ OR is an isomorphism by Theorem 4.49. $\square$

It turns out, then, that there are "as many" epsilon numbers as there are ordinals. The class of the epsilon numbers is

$$\epsilon_0,\ \epsilon_1,\ \epsilon_2,\ \cdots,\ \epsilon_\omega,\ \epsilon_{\omega+1},\ \cdots,\ \epsilon_{\omega 2},\ \cdots,\ \epsilon_{\omega 3},\ \cdots,\ \epsilon_{\omega\omega},\ \cdots,\ \epsilon_\alpha,\ \cdots,$$

as α ranges over all the ordinals.

A few easy arithmetic rules simplify computations which involve epsilon numbers. They are given in the next theorem.

Theorem 10.26. *Let ϵ designate an arbitrary epsilon number. Then*

(1) *If $\alpha < \epsilon$, then $\alpha + \epsilon = \epsilon$.*

(2) *If $\alpha < \epsilon$, then $\alpha\epsilon = \epsilon$.*

(3) *If $\alpha < \epsilon$, then $\alpha^\epsilon = \epsilon$.*

Proof. Let $\alpha < \epsilon$, and write α in normal form: $\alpha = \omega^{\alpha_1}a_1 + \cdots + \omega^{\alpha_n}a_n$.

(1) Note that $\alpha + \epsilon = \alpha + \omega^\epsilon = \omega^{\alpha_1}a_1 + \cdots + \omega^{\alpha_n}a_n + \omega^\epsilon$.

$\Rightarrow$ Since $\alpha < \epsilon$, $\omega^{\alpha_i} < \omega^\epsilon = \epsilon$ for each $i = 1, \cdots, n$.

$\Rightarrow$ $\alpha_i < \epsilon$ for each $i = 1, \cdots, n$ by Theorem 10.10(2).

$\Rightarrow$ $\alpha + \epsilon = \omega^{\alpha_1}a_1 + \cdots + \omega^{\alpha_n}a_n + \omega^\epsilon = \omega^\epsilon = \epsilon$ by Theorem 10.18(1).

(2) Since $\alpha_1 + \epsilon = \epsilon$ by Theorem 10.26(1) above,

$$\alpha\epsilon = \alpha\omega^\epsilon = (\omega^{\alpha_1}a_1 + \cdots + \omega^{\alpha_n}a_n)\,\omega^\epsilon = \omega^{\alpha_1+\epsilon} = \omega^\epsilon = \epsilon.$$

(3) Since
$$\alpha = \omega^{\alpha_1}a_1 + \cdots + \omega^{\alpha_n}a_n \leqslant \omega^{\alpha_1}a_1 + \cdots + \omega^{\alpha_1}a_n = \omega^{\alpha_1}(a_1 + \cdots + a_n) < \omega^{\alpha_1}\omega = \omega^{\alpha_1+1},$$

$$\alpha^\epsilon \leqslant \left(\omega^{\alpha_1+1}\right)^\epsilon = \omega^{(\alpha_1+1)\epsilon}.$$

$\Rightarrow$ Since every epsilon number is a limit ordinal, $\alpha_1 < \epsilon$ implies $\alpha_1 + 1 < \epsilon$ and hence $(\alpha_1 + 1)\epsilon = \epsilon$ by Theorem 10.26(2).

$\Rightarrow$ $\alpha^\epsilon \leqslant \omega^{(\alpha_1+1)\epsilon} = \omega^\epsilon = \epsilon$ and hence $\alpha^\epsilon \leqslant \epsilon$.

$\Rightarrow$ Since $\alpha^\epsilon \geqslant \epsilon$ by Exercise 10.2.4, $\alpha^\epsilon = \epsilon$. $\square$

It is worth nothing the following property.

Proposition 10.27. *An ordinal number α is an epsilon number if and only if α satisfies the following condition:*

$$\text{If } \beta < \alpha \text{ and } \gamma < \alpha, \text{ then } \beta^\gamma < \alpha. \tag{10.9}$$

Proof. The proof is left as an exercise for the reader (Exercise 10.4.4). □

Exercise 10.4

1. Prove that if α is an epsilon number, then α is a limit ordinal.
2. If α is an epsilon number, prove that $\beta^{\omega^\alpha} = \beta^{\omega\alpha}$ for every ordinal number β.
3. Prove that if α is an infinite ordinal and $\alpha^\beta = \beta$, then β is an epsilon number.
4. (1) Prove that if $\alpha(\alpha > \omega)$ satisfies Condition 10.9 of Proposition 10.27, then α is a limit ordinal.

   (2) Prove that $\alpha(\alpha > \omega)$ is an epsilon number if and only if α satisfies the condition 10.9 of Proposition 10.27.
5. Prove that $\alpha(\alpha > \omega)$ is an epsilon number if and only if $2^\alpha = \alpha$.

10.5 Inaccessible ordinals and cardinals

Inaccessible ordinal numbers and cardinal numbers play an important role in current investigations on the axiomatic foundations of set theory. They also have applications in functional analysis, topology, algebra, mathematical logic and other areas of advanced mathematics. In this section we will introduce them and give a few of their basic properties.

Definition 10.28. Let α be an ordinal number, and let $Z(\alpha)$ designate the class of all the ordinal numbers which are equipotent with α. The least element of $Z(\alpha)$ is called the *initial ordinal belonging to* α, and is denoted by $\mathrm{Io}(\alpha)$.

Proposition 10.29. *Let* α *be an ordinal number.*

(1) $\mathrm{Io}(\alpha) \leqslant \alpha$.

(2) $\mathrm{Io}(\alpha)$ *is the largest initial ordinal less than* α.

Proof. (1) follows immediately from Definition 10.28.

(2) Assume that $\mathrm{Io}(\alpha) < \gamma \leqslant \alpha$.

$\Rightarrow \mathrm{Io}(\alpha) \subseteq \gamma \subseteq \alpha$.

$\Rightarrow \mathrm{Io}(\alpha) \approx \alpha \approx \gamma$ by Definition 10.28. □

Remark 10.30. We have seen that the class of all the initial ordinals satisfies the conditions K1 and K2 of Axiom 10 of Cardinality, hence we are justified in making the following definition (see Definition 9.47): By a *cardinal number* we mean an initial ordinal. Thus, the class CD of all the cardinal numbers is the class of all the initial ordinals.

Notation 10.31. We have noted that if α is any ordinal number, it is common practice to write $\aleph_\alpha = \omega_\alpha$, treating $\aleph_\alpha$ as a cardinal number and ω_α as an ordinal number. (See Remark 9.48(3).)

Proposition 10.32. *Let α be an ordinal number and let A be any well-ordered set.*

(1) $\mathrm{Io}(\alpha) = \#\alpha$.

(2) $\oslash A \geqslant \omega_\alpha$ *if and only if* $\#A \geqslant \aleph_\alpha$.

Proof. Let α be an ordinal number and let A be any well-ordered set.

$\Rightarrow$ (1) is clear.

(2) follows easily from Proposition 10.29 and (1) above.

In fact, if we write $\gamma = \oslash A$, then $\mathrm{Io}(\gamma) = \#\gamma = \#A$.

The simple details are left as an exercise for the reader. □

We will now begin the process of defining inaccessible ordinals and cardinals.

Definition 10.33. Let A be a well-ordered class and let $B \subseteq A$. We say that B is a *cofinal subclass* of A if for arbitrary $x \in A$, there exists $y \in B$ such that $y > x$.

Lemma 10.34. *If C is a cofinal subclass of B and B is a cofinal subclass of A then C is a cofinal subclass of A.*

Proof. Let C be a cofinal subclass of B and B a cofinal subclass of A.

To show that C is a cofinal subclass of A, let $x \in A$.

$\Rightarrow$ Since B is a cofinal subclass of A, there exists $y \in B$ such that $y > x$.

$\Rightarrow$ Since C is a cofinal subclass of B, there exists $z \in C$ such that $z > y$.

$\Rightarrow$ $z \in C$ and $z > x$.

$\Rightarrow$ C is a cofinal subclass of A. □

Lemma 10.35. *Let α be an ordinal number and let $B \subseteq \alpha$. Then B is a cofinal subset of α if and only if* $\sup B = \alpha$.

Proof. Let α be an ordinal number and let $B \subseteq \alpha$.

$\Rightarrow$ Every element of an ordinal number is an ordinal number, and $\gamma < \alpha$ if and only if $\gamma \in \alpha$ by Lemma 9.39(2) and Definition 9.37.

(only if) Let B be a cofinal subset of α and assume that $\gamma < \alpha$.

$\Rightarrow$ $\gamma \in \alpha$, and $x \in \alpha$ for all $x \in B$ and hence $x < \alpha$ for all $x \in B$.

$\Rightarrow$ Since B is a cofinal subset of α, there exists $\beta \in B$ such that $\beta > \gamma$.

$\Rightarrow$ γ is not an upper bound of B for all $\gamma < \alpha$.

$\Rightarrow$ Since α is an upper bound of B, α is the least upper bound of B.

$\Rightarrow$ $\alpha = \sup B$.

(if) Assume that $\sup B = \alpha$ and let $\gamma \in \alpha$.

$\Rightarrow$ $\gamma < \alpha$ and hence γ is not an upper bound of B.

$\Rightarrow$ There exists $\beta \in B$ such that $\beta > \gamma$.

$\Rightarrow$ B is a cofinal subset of α. □

Definition 10.36. Let α be a limit ordinal. By the *cofinality* of α we mean the least ordinal number β such that α has a cofinal subset similar to β. If β is the cofinality of α, we write $\beta = \mathrm{cf}(\alpha)$.

Lemma 10.37. *If $\beta = \mathrm{cf}(\alpha)$ for some $\alpha \in \mathrm{OR}$, then $\beta = \mathrm{cf}(\beta)$.*

Proof. Assume that $\beta = \mathrm{cf}(\alpha)$ for some $\alpha \in \mathrm{OR}$ and let $\gamma = \mathrm{cf}(\beta)$.

$\Rightarrow$ Since α has a cofinal subset similar to β and β has a cofinal subset similar to γ, by Lemma 10.34, α has a cofinal subset similar to γ.

$\Rightarrow$ Since β is the least ordinal number η such that α has a cofinal subset similar to η, $\beta \leqslant \gamma$.

On the other hand, since γ is similar to a subset of β, $\gamma \leqslant \beta$.

$\Rightarrow$ $\beta = \gamma = \mathrm{cf}(\beta)$. □

Theorem 10.38. *Let β be a limit ordinal. If $\beta = \mathrm{cf}(\beta)$, then β is an initial ordinal.*

Proof. Let β be a limit ordinal such that $\beta = \mathrm{cf}(\beta)$.

Let $\omega_\alpha = \mathrm{Io}(\beta)$, and let $f : \omega_\alpha \to \beta$ be a bijective function.

Let $A = \{\gamma \in \omega_\alpha \mid f(\delta) < f(\gamma) \text{ for all } \delta < \gamma\}$.

$\Rightarrow$ $f(A)$ is a cofinal subset of β.

($\because$) Assume that $\upsilon \in \beta$.

$\Rightarrow$ Since β is a limit ordinal and a limit ordinal has no greatest element, there are many elements $\zeta \in \omega_\alpha$ such that $f(\zeta) > \upsilon$. (Note that f is bijective!)

Let π be the least such element.

$\Rightarrow$ $f(\zeta) \leqslant \upsilon < f(\pi)$ for all $\zeta < \pi$.

$\Rightarrow$ $\pi \in A$. This proves that for all $\upsilon \in \beta$, there exists $\pi \in A$ such that $f(\pi) > \upsilon$.

$\Rightarrow$ $f(A)$ is a cofinal subset of β.

$\Rightarrow$ Because of the way we have defined A, it is easy to see that $f|_A : A \to f(A)$ is an isomorphism. Let $\delta = \oslash A$.

$\Rightarrow$ δ is similar to a cofinal subset of β.

$\Rightarrow$ $\mathrm{cf}(\beta) \leqslant \delta$.

$\Rightarrow$ Since A is a subset of ω_α, $\delta \leqslant \omega_\alpha$ and hence $\beta = \mathrm{cf}(\beta) \leqslant \delta \leqslant \omega_\alpha$.

On the other hand,

since ω_α is the least ordinal number equipotent with β, $\omega_\alpha \leqslant \beta$.

$\Rightarrow \omega_\alpha = \beta$ and hence β is an initial ordinal number. □

Definition 10.39. A limit ordinal β is called *regular* if $\mathrm{cf}(\beta) = \beta$.

It follows, by Lemma 10.37, that the class of the regular ordinals is the range of the function $\mathrm{cf} : \mathrm{OR} \to \mathrm{OR}$. By Theorem 10.38, *every regular ordinal is an initial ordinal* ω_α. Thus, in particular, every regular ordinal is a cardinal number.

Definition 10.40. If ω_α is a regular ordinal number, then $\aleph_\alpha$ is called a *regular cardinal.*

The significance of regular cardinals in set theory will become apparent once we have given an alternative definition for them. We begin with the following two lemmas:

Lemma 10.41. *Let* $\{\gamma_i \mid i \in I\}$ *be a set of ordinal numbers and let* $\#\gamma_i = \aleph_{\delta_i}$ *for each* $i \in I$. *If* $\omega_\alpha = \sup\{\gamma_i \mid i \in I\}$, *then* $\alpha = \sup\{\delta_i \mid i \in I\}$.

Proof. Let $\{\gamma_i \mid i \in I\}$ be a set of ordinal numbers and let $\#\gamma_i = \aleph_{\delta_i}$ for each $i \in I$. Assume that $\omega_\alpha = \sup\{\gamma_i \mid i \in I\}$.

$\Rightarrow \gamma_i \leqslant \omega_\alpha$ for all $i \in I$.

$\Rightarrow \aleph_{\delta_i} = \#\gamma_i \leqslant \aleph_\alpha$ for all $i \in I$, by Proposition 10.32(2).

$\Rightarrow \delta_i \leqslant \alpha$ for all $i \in I$, by Theorem 9.36.

Now let π be an upper bound of $\{\delta_i \mid i \in I\}$.

$\Rightarrow \delta_i < \pi$ for all $i \in I$.

$\Rightarrow \#\gamma_i = \aleph_{\delta_i} < \aleph_\pi$ for all $i \in I$, by Theorem 9.36.

$\Rightarrow \gamma_i < \omega_\pi$ for all $i \in I$, by Proposition 10.32(2).

$\Rightarrow$ Since $\omega_\alpha = \sup\{\gamma_i \mid i \in I\}$, $\omega_\alpha \leqslant \omega_\pi$.

$\Rightarrow \alpha \leqslant \pi$ by Theorem 9.36.

$\Rightarrow \alpha$ is the least upper bound of $\{\delta_i \mid i \in I\}$.

$\Rightarrow \alpha = \sup\{\delta_i \mid i \in I\}$. □

Lemma 10.42. *Let $\{\delta_i \mid i \in I\}$ be a set of ordinal numbers and let α be any ordinal number such that $\#I < \aleph_\alpha$. Then $\sum_{i\in I} \aleph_{\delta_i} = \aleph_\alpha$ if and only if $\alpha = \sup\{\delta_i \mid i \in I\}$.*

Proof. Let $\{\delta_i \mid i \in I\}$ be a set of ordinal numbers and let α be any ordinal number such that $\#I < \aleph_\alpha$.

If I is finite, the result follows trivially. Assume that I is infinite.

(only if) Assume that $\sum_{i\in I} \aleph_{\delta_i} = \aleph_\alpha$.

$\Rightarrow$ For each $j \in I$, $\aleph_{\delta_j} \leqslant \sum_{i\in I} \aleph_{\delta_i} = \aleph_\alpha$.

$\Rightarrow$ $\delta_j \leqslant \alpha$ for all $j \in I$ by Theorem 9.36.

$\Rightarrow$ α is an upper bound of $\{\delta_i \mid i \in I\}$.

Before going on, note that there exists an $i \in I$ such that $\#I < \aleph_{\delta_i}$.

($\because$) Suppose that $\aleph_{\delta_i} \leqslant \#I$ for every $i \in I$.

$\Rightarrow$ Since I is infinite $(\sharp I)(\sharp I) = \sharp I$ by Theorem 8.23 and hence $\sum_{i\in I} \aleph_{\delta_i} \leqslant (\#I)(\#I) = \#I < \aleph_\alpha$ by Corollary 8.33 which is contrary to our assumption.

Now let π be an upper bound of $\{\delta_i \mid i \in I\}$.

$\Rightarrow$ $\delta_i < \pi$ for all $i \in I$.

$\Rightarrow$ $\sum_{i\in I} \aleph_{\delta_i} \leqslant (\#I)\aleph_\pi$ by Corollary 8.33.

$\Rightarrow$ Since $\#I < \aleph_{\delta_i}$ for some $i \in I$, $\#I < \aleph_\pi$.

$\Rightarrow$ $\aleph_\alpha = \sum_{i\in I} \aleph_{\delta_i} \leqslant (\#I)\aleph_\pi = \aleph_\pi$.

$\Rightarrow$ $\alpha \leqslant \pi$ by Theorem 9.36.

$\Rightarrow$ α is the least upper bound of $\{\delta_i \mid i \in I\}$.

$\Rightarrow$ $\alpha = \sup\{\delta_i \mid i \in I\}$.

(if) Assume that $\alpha = \sup\{\delta_i \mid i \in I\}$.

$\Rightarrow$ $\delta_i \leqslant \alpha$ for all $i \in I$.

$\Rightarrow$ $\aleph_{\delta_i} \leqslant \aleph_\alpha$ for all $i \in I$, by Theorem 9.36.

$\Rightarrow$ $\sum_{i\in I} \aleph_{\delta_i} \leqslant (\#I)\aleph_\alpha = \aleph_\alpha$ by Corollary 8.33.

Now to show that $\sum_{i\in I} \aleph_{\delta_i} = \aleph_\alpha$, let $\sum_{i\in I} \aleph_{\delta_i} = \aleph_\delta$.

$\Rightarrow \aleph_{\delta_j} \leqslant \sum_{i\in I} \aleph_{\delta_i} = \aleph_\delta$ for all $j \in I$

$\Rightarrow \delta_j \leqslant \delta$ for all $j \in I$, by Theorem 9.36.

$\Rightarrow$ Since $\alpha = \sup \delta_i$, $\alpha \leqslant \delta$ and hence $\aleph_\alpha \leqslant \aleph_\delta = \sum_{i\in I} \aleph_{\delta_i}$.

$\Rightarrow$ Since $\sum_{i\in I} \aleph_{\delta_i} \leqslant \aleph_\alpha$, $\sum_{i\in I} \aleph_{\delta_i} = \aleph_\alpha$. □

Theorem 10.43. *Let α be a limit ordinal. Then $\aleph_\alpha$ is a regular cardinal number if and only if it satisfies the following condition:*

> *If $\{a_i \mid i \in I\}$ is any set of cardinal numbers such that $a_i < \aleph_\alpha$ for each $i \in I$ and $\#I < \aleph_\alpha$, then $\sum_{i\in I} a_i < \aleph_\alpha$.*

Proof. Let α be a limit ordinal.

(only if) Assume that $\aleph_\alpha$ is a regular cardinal number and let $\{a_i \mid i \in I\}$ be any set of cardinal numbers such that $a_i < \aleph_\alpha$ for all $i \in I$ and $\#I < \aleph_\alpha$.

$\Rightarrow$ Since any finite $a_i \in \{a_i \mid i \in I\}$ would not affect our result, we may assume the a_i's are all infinite cardinal numbers and hence can write $a_i = \aleph_{\delta_i}$ for each $i \in I$.

$\Rightarrow \aleph_{\delta_i} = a_i < \aleph_\alpha$ for all $i \in I$.

$\Rightarrow \sum_{i\in I} a_i = \sum_{i\in I} \aleph_{\delta_i} \leqslant \aleph_\alpha(\sharp I) \leqslant \aleph_\alpha\aleph_\alpha = \aleph_\alpha$ by applying Corollary 8.33, Theorem 9.25(5) and Theorem 8.23.

$\Rightarrow \sum_{i\in I} a_i < \aleph_\alpha$.

($\because$) Suppose that $\sum_{i\in I} a_i = \sum_{i\in I} \aleph_{\delta_i} = \aleph_\alpha$.

$\Rightarrow \alpha = \sup\{\delta_i \mid i \in I\}$ by Lemma 10.42.

$\Rightarrow \omega_\alpha = \sup\{\omega_{\delta_i} \mid i \in I\}$ by Theorem 9.36.

$\Rightarrow \{\omega_{\delta_i} \mid i \in I\}$ is a cofinal subset of ω_α by Lemma 10.35.

This is impossible.

($\because$) Since $\#I < \aleph_\alpha$, $\#\{\omega_{\delta_i} \mid i \in I\} < \aleph_\alpha$.

$\Rightarrow$ By Proposition 10.32(2), $\oslash\{\omega_{\delta_i} \mid i \in I\} < \omega_\alpha$ which contradicts to the fact that ω_α is a regular ordinal.

$\Rightarrow \sum_{i \in I} a_i \neq \aleph_\alpha$.

$\Rightarrow$ Since $\sum_{i \in I} a_i \leqslant \aleph_\alpha$, $\sum_{i \in I} a_i < \aleph_\alpha$ and the condition holds.

(if) Suppose that $\aleph_\alpha$ satisfies the condition in Theorem 10.43, and let $\{\gamma_i \mid i \in I\}$ be a cofinal subset of ω_α. We assume that the γ_i's are all distinct and hence $\#\{\gamma_i \mid i \in I\} = \#I$.

$\Rightarrow \omega_\alpha = \sup\{\gamma_i \mid i \in I\}$ by Lemma 10.35. Set $\aleph_{\delta_i} = \#\gamma_i$ for each $i \in I$.

$\Rightarrow \alpha = \sup\{\delta_i \mid i \in I\}$ by Lemma 10.41.

$\Rightarrow \sum_{i \in I} \aleph_{\delta_i} = \sum_{i \in I} (\#\gamma_i) = \aleph_\alpha$ by Lemma 10.42 and $\aleph_{\delta_i} < \aleph_\alpha$ for all $i \in I$.

$\Rightarrow \#I \geqslant \aleph_\alpha$ by the condition of Theorem 10.43.

$\Rightarrow \oslash\{\gamma_i \mid i \in I\} \geqslant \omega_\alpha$ by Proposition 10.32(2).

$\Rightarrow$ Any cofinal subset of ω_α has ordinality $\geqslant \omega_\alpha$ and hence $\omega_\alpha = \mathrm{cf}(\omega_\alpha)$.

$\Rightarrow \omega_\alpha$ is a regular ordinal number. □

Corollary 10.44. *If α is a nonlimit ordinal, then $\aleph_\alpha$ is a regular cardinal.*

Proof. Assume that α is a nonlimit ordinal.

$\Rightarrow \alpha = \delta + 1$ for some ordinal δ.

Set $a_i = \aleph_\delta$ for each $i \in I$, where $\#I = \aleph_\delta$.

$\Rightarrow \sum_{i \in I} a_i = (\aleph_\delta)^2 = \aleph_\delta < \aleph_{\delta+1} = \aleph_\alpha$ by Theorem 8.31 and Theorem 8.23.

$\Rightarrow \aleph_\alpha$ is a regular cardinal number by Theorem 10.43. □

Definition 10.45. A cardinal number a is called an *inaccessible cardinal*(more precisely, a *strongly inaccessible* cardinal) if

(1) a is a regular cardinal, and

(2) $b < a$ and $c < a$ imply $b^c < a$.

The ordinal number ω_α is called an *inaccessible ordinal* if $\aleph_\alpha$ is an inaccessible cardinal.

In view of Theorem 10.43, we have the following proposition.

Proposition 10.46. *An infinite cardinal number b is inaccessible if and only if it satisfies the following pair of conditions:*

(1) *If $\{a_i \mid i \in I\}$ is a set of cardinal numbers such that $a_i < a$ for each $i \in I$ and $\#I < a$, then $\sum_{i \in I} a_i < a$.*

(2) *If $b < a$ and $c < a$, then $b^c < a$.*

In other words, an infinite cardinal number b is inaccessible if and only if it cannot be obtained *either as a sum of fewer than b cardinal numbers smaller than b, or by raising a cardinal smaller than b to a power smaller than b.* This explains the use of the word "inaccessible". It can be shown, furthermore, that if b is an inaccessible cardinal, then b cannot be obtained as a product of fewer than b cardinal numbers smaller than b (see Exercise 10.5.4 below).

Definition 10.47. We may call a nonempty set A *inaccessible* if it cannot be constructed from smaller sets by using the set-theoretical operations of union, intersection, product, or power set. To be technical, this means that

(1) A is not equal to $\bigcap_{i \in I} B_i$, $\bigcup_{i \in I} B_i$, or $\prod_{i \in I} B_i$ for any set of sets $\{B_i \mid i \in I\}$, where $\{B_i \mid i \in I\} \prec A$ and $B_i \prec A$ for each $i \in I$;

(2) A is not the power set of any set B such that $B \prec A$.

For example, if we begin with the empty set $\emptyset$ and start to construct sets such as

$$\{\emptyset\}, \{\emptyset, \{\{\emptyset\}\}, \{\emptyset\}\}, \{\{\emptyset\}, \{\emptyset, \{\emptyset\}\}\}, \cdots,$$

and if we proceed to construct larger and larger sets from these by using the operations of union (including infinite union), product (including infinite product) and power set, we will never end up with an inaccessible set. Now, in view of what we have said in the preceding paragraph, it is clear that inaccessible cardinals are the cardinals of inaccessible sets.

A rather obvious question which arises in axiomatic set theory is, "Are there any inaccessible sets?" In other words, do inaccessible cardinals exist? The existence of inaccessible cardinals cannot be proved by means of the axioms we have already introduced. However, a new axiom may be added to set theory, called the *axiom for inaccessible cardinals*, asserting their existence. It has been proven in recent years that this axiom is not a consequence of the other axioms of set theory. Whether it is consistent with the other axioms of set theory is still an open question.

Axiom 12 (Axiom of Inaccessible Cardinal). There exist inaccessible cardinals.

The following definition is useful:

Definition 10.48. $\aleph_\alpha$ is called a *weakly inaccessible cardinal* if

(1) $\aleph_\alpha$ is a regular cardinal,

(2) α is a limit ordinal.

By Corollary 10.44, if α is a nonlimit ordinal, then $\aleph_\alpha$ is necessarily regular. Thus, it is natural to ask whether there are any regular cardinals $\aleph_\alpha$ where α is a limit ordinal. Again, we cannot prove the existence of such cardinals from the usual axioms of set theory, but their existence follows immediately from Axiom 12 of Inaccessible Cardinal. Indeed, we have the following

Theorem 10.49. *If α is strongly inaccessible, then α is weakly inaccessible.*

Proof. Assume that α is a strongly inaccessible ordinal number.

$\Rightarrow \aleph_\alpha$ is a strongly inaccessible cardinal number.

Now suppose that α is a nonlimit ordinal.

$\Rightarrow \alpha = \delta + 1$ for some ordinal δ.

$\Rightarrow \aleph_\delta^{\aleph_\delta} = 2^{\aleph_\delta} > \aleph_\delta$ by Theorem 8.28.

$\Rightarrow \aleph_\delta^{\aleph_\delta} \geqslant \aleph_\alpha$,

which is contrary to our hypothesis that $\aleph_\alpha$ is strongly inaccessible.

$\Rightarrow$ α is a limit ordinal.

$\Rightarrow$ Since $\aleph_\alpha$ is a regular cardinal number, α is weakly inaccessible. □

There is another interesting connection between strong and weak inaccessibility:

Theorem 10.50. *Assuming the generalized continuum hypothesis to be true, if α is weakly inaccessible then α is strongly inaccessible.*

Proof. Assume the generalized continuum hypothesis, let $\aleph_\alpha$ be weakly inaccessible, and suppose that $\aleph_\beta < \aleph_\alpha$.

$\Rightarrow$ $\beta < \alpha$ by Theorem 9.36.

$\Rightarrow$ Since α is a limit ordinal, $\beta + 1 < \alpha$, so $\aleph_{\beta+1} < \aleph_\alpha$.

$\Rightarrow$ Since $2^{\aleph_\beta} = \aleph_{\beta+1}$ by the generalized continuum hypothesis, $2^{\aleph_\beta} < \aleph_\alpha$.

$\Rightarrow$ Since $\aleph_\alpha$ is regular, $\aleph_\alpha$ is strongly inaccessible by Exercise 10.5.3. □

Thus, if we assume the generalized continuum hypothesis, then the notions of strongly inaccessible and weakly inaccessible are equivalent.

Exercise 10.5

1. Prove Proposition 10.29(1) and (2).

2. Prove Proposition 10.32(1) and (2).

3. Prove that a is strongly inaccessible if and only if

   (1) a is regular, and (2) $2^b < a$ for all $b < a$. [See Exercise 8.4.3.]

4. Prove that if b is an inaccessible cardinal, then b satisfies the following condition:

If $\{a_i \mid i \in I\}$ is a set of cardinals such that $a_i < b$ for each $i \in I$ and $\sharp I < b$, then $\prod_{i \in I} a_i < b$.

5. Prove that if an infinite cardinal b satisfies the condition in Exercise 10.5.4 above, then b is an inaccessible cardinal.

Chapter 0

역사적 소개

0.1 집합론의 배경

집합론(集合論)이 비록 "**새로운**" 수학의 초석으로 인식되지만, 집합에 대한 직관적인 개념에 있어서는 본질적으로 새로운 것이 없다. 초창기부터 수학자들은 이런 저런 종류의 대상에 대한 집합을 생각해 왔으며, 현대 집합론의 기본적인 개념들은 고전적인 수학논리에 아주 많이 함축되어 나타나고 있다. 그러나 집합론이 주요한 수학이론으로서 독자적인 영역을 구축하게 된 것은 19세기 말에 칸토르(G. Cantor, 1845 ~ 1918)가 일련의 논문들을 발표하면서부터 시작되었다.

칸토르가 처음으로 집합의 성질을 연구하게 된 동기는 이상하게도 매우 전문적인 연구분야인 삼각급수에서의 그의 논문이었다. 처음 그는 급수의 수렴과 관련해서 나타나는 어떤 특정한 실수들의 집합에 국한하여 연구하였다. 그러나 칸토르는 재빨리 그의 발견이 아주 일반적인 집합에 적용될 수 있음을 이해하고, 1872년과 1897년 사이에 연속적으로 발표한 주목할 만한 연구 논문에서, 집합에 관한 그의 초기생각을 이끌어 낸 구체적인 문제들에서부터 오늘날 집합론의 기초가 되는 중요한 일반 개념들로 발전시켜 나아갔다.

칸토르와 동시대 사람들의 눈에 비친 칸토르가 취한 가장 대담한 개혁은 무한 집합을 유한 집합처럼 자연스럽게 다루는 것이었다. "**무한**"(無限, infinity)이라는 문제는 오랫동안 수학에서 아주 민감한 문제였다. 독자는 아마 제논(Zeno, B.C 490 ~ B.C. 425)의 유명한 역설(逆說, paradox)을 알고 있을 것이다. 단위길이의 선분이 $\frac{1}{2}, \frac{1}{4}, \frac{1}{8}, \frac{1}{16}, \cdots$ 등의 점들에 의해 부분구간으로 나누어 진다. 분할된 각 부분구간은, 아무리 짧을지라도, 확실히 0이 아닌 길이를 가지고 있다. 즉, 0이 아닌 무한히 많은 길이를 합해서 유한의 길이가 되는 외관상 역설로 보이는 결론에 이른다. 이런 덫을 피하기 위해서, 고대 수학자들은 무한을 "**현실적 무한**"(現實的 無限, actual infinity)과 "**가상적 무한**"(假想的 無限, virtual infinity)의 두 가지로 구별하여 현실적 무한은 무한히 많은 대상이 한꺼번에 존재하는 것으로, 가상적 무한은 임의로 주어진 유한한 크기를 단순히 초월할 수 있는 잠재력을 가진 것으로 생각하였다. **가상적 무한**은 안전한 개념으로 허용되었고, **현실적 무한**은 금기의 대상이었다.

그러나 칸토르의 이론에서는 무한집합을 무제한으로 다루었으며 무한의 개념은 분명히 **현실적 무한**의 의미로 이해되었다. 현실적 무한을 금기의 대상으로 간주하였던 그 시대 사람들이 이러한 칸토르의 이론을 즉시 인정하지 않았던 것은 당연한 일이었다. 그들은 칸토르의 이론을 처음에는 매우 회의적으로 여겼으며, 때때로 공공연하게 비판도 하였다. 그러나, 1890년대에 이르러 이 이론들이 다양한 수학이론에 훌륭한 기초를 마련해 주었기 때문에, 칸토르의 이론 중에서 비교적 덜 비판받는 부분이 널리 이용되었다. 그리하여 20 세기로 넘어 올 무렵에는 집합론의 가장 혁신적인 관점까지도 아주 많은 수학자들에 의해서 인정되었다. 이렇게 된 주된 이유는 수학의 여러 분야에서, 특히 해석학에서, 칸토르의 이론이 매우 가치있는 도구로 밝혀졌기 때문이었다.

한편, 데데킨트(J. W. R. Dedekind, 1831 ~ 1916)를 비롯한 몇몇 뛰어난 수학자들의 업적은 집합론이 수학의 기본으로서 여러 분야의 수학을 **통일하는**(unifying) 가장 바람직한 역할을 하게 되는 전기가 되었다. 초창기부터 수학자들은 수학의 모든 분야를 몇 개의 기본적 원리로 통일할 가능성에 대하여 관심을 갖고 있었다. 유크리트(Euclid, BC325? ~ BC265?)시대부터 중세에 이르기까지 많은 고전학파들은 수학의 여러 분야가 기하학에 포함된다고 생각하였다. (숫자도 기하적 비율로 생각할 수 있다.)

바이어슈트라스(K. T. W. Weierstrass, 1815 ~ 1897)와 데데킨트, 등 그 밖의 다른 수학자들의 연구에서, 모든 고전수학은 자연수(양의 정수)의 산

술로부터 유도될 수 있다는 사실이 밝혀져서 수학통일에 있어서 아주 많은 성공적인 진전이 19세기에 나타났다. 모든 실수는 "코시 수열(Cauchy sequence)"이라고 부르는 유리수의 수열로 볼 수 있으며, 따라서 실수의 연구는 유리수의 연구로 귀착되게 된다. 한편 유리수는 정수의 순서쌍으로서 생각할 수 있으므로 결국 미적분학과 해석기하를 비롯한 모든 기하학을 포함하는 실수에 대한 수학이 자연수에 그 뿌리를 두고 있다는 것이다.

수학기초론에 대한 사상의 전환기에 있는 이 결정적인 시점에 데데킨트는 그의 작은 책 "**수란 무엇이며 무엇이어야만 하는가?**"(Was sind und was sollen die Zahlen, 1888)에서 자연수의 개념은 집합론의 기본원리로부터 유도될 수 있다고 발표하였다. 이것을 현대적으로 설명하면 다음과 같다:

> "0"을 공집합 $\emptyset$로 두고 "1"은 하나의 원소 $\emptyset$만을 포함하는 집합 $\{\emptyset\}$로 정의한다. 그리고 나서, "2"는 집합 $\{0, 1\}$로 정의하고, "3"은 $\{0, 1, 2\}$로 정의한다. 이와 같은 방법으로 계속해서 정의해 나간다. 따라서 자연수의 모든 성질들은 이 정의와 기초적인 집합론을 사용하여 증명될 수 있다.

20세기로 바뀔 무렵에는, 집합론은 수학계의 모든 분야에서 없어서는 안 될 도구로 인정 받았을 뿐만 아니라, 수리과학 분야에서 첫째 자리를 차지할 중요한 후보가 되었다.

그러나 칸토르의 생각이 이제 인정 받았다고 여겨지는 바로 이 시점에, 이러한 분위기와는 정반대로, 최초의 역설이 발표되었는데, 이는 마침내 칸토르가 생각한 집합론의 기본적인 건전성에 대하여 심각한 의문을 던지게 되었다. 이 역설은 아주 광범위한 반향을 불러 일으켰기 때문에 그것들을 자세히 살펴보기로 하겠다.

0.2 역설

1895년에서 1910년 사이에 집합론의 여러 부분에서 몇가지 모순점들이 발견되었다. 처음에 수학자들은 이 모순들을 "**역설**"(逆說, paradox)이라고 부르며 단지 수학적 호기심 정도로 치부하여 별로 관심을 두지 않았다. 최초의 역설은 1897년에 부랄리-포르티(C. Burali-Forti, 1861 ~ 1931)에 의해서 발표되었으나, 사실 이 역설은 이미 그보다 2년 전에 칸토르 자신이 발견한 것이었다.

부랄리-포르티의 역설은 집합론의 다소 기술적인 영역에서 나타났기 때문에 처음에는 기초적인 정의만 약간 수정하면 충분히 바로 잡을 수 있으리라고 생각되었다. 그러나, 1902년에 럿셀(B. Russell, 1872 ~ 1970)이 발표한 역설은 집합론의 가장 기본적인 개념과 관련되어 있어서 무시할 수 없는 것이었다. 그 후 계속해서 몇 년 동안 또 다른 모순점들이 발견되자 완전무결하다고 생각되었던 수학적인 개념들이 의심을 받게 되었다.

집합론의 역설은 두가지 종류로 나누는데 그 하나는 "**논리적**"(論理的, logical) 역설이라 하고, 다른 하나는 "**어의적**"(語義的, semantic) 역설이라고 한다. "논리적"과 "어의적"으로 부르는 이유는 이들 역설들의 예를 보면 쉽게 이해된다. 본질적으로, 논리적 역설은 잘못된 논리로부터 나오고, 어의적 역설은 잘못된 언어 사용에서 나온다.

이제 가장 유명한 역설 두 가지를 소개하고자 한다. 이 역설들은 집합론의 단지 기본적인 개념만을 포함하고 있는 것들이다. 첫 번째는 논리적 역설이고, 두번째는 어의적 역설로서 모두 다 전형적인 예이다.

논리적 역설 중에서 가장 단순한 것은 럿셀의 역설(Russell's paradox)인데, 이것은 다음과 같다.

「A를 집합이라고 하면, 그 원소들도 집합이 될 수 있다. 이와 같은 상황은 수학에서 자주 일어난다. 예를 들면, 각 직선을 점들의 집합이라고 생각하고 A를 직선들의 집합이라고 하면, A 자신이 A의 원소가 될 가능성이 있다. 예를 들면, 모든 집합들의 집합은 이와 같은 성질을 갖는다.
S를 「자기 자신이 자기 자신의 원소가 되지 않는 모든 집합들의 집합」이라고 하자. 그러면, S는 그 자신의 원소가 될 수 있을까?
만약 S가 S의 원소라고 한다면 S는 S의 정의에 따라서 S의 원소가 아니다.
만약 S가 S의 원소가 아니라면, S는 S의 정의에 따라서 S의 원소이다. 결과적으로 S가 S의 원소일 필요충분조건은 S가 S의 원소가 아닐 때이다.」

이것이 가장 기본적인 종류의 모순이다. 보통, 수학에서 이런 종류의 모순에 도달할 경우 우리는 가정 중의 어느 하나에 잘못이 있다고 인정해야만 한다. 이 경우에 있어서는「집합이 그 자신의 원소가 된다는 것은 무의미하다」라든가 또는「자기 자신의 원소가 되지 않는 모든 집합들의 집합이라는 것이 존재하지 않는다」라는 결론을 내리게 된다. 잠시 후 이 문제로 다시 돌아오기로 하고 어의적 역설에 관하여 조금 더 설명하기로 하자. 어의적 역설의 전형적인 예는 베리(G. G. Berry, ? ∼ ?)의 역설(Berry's paradox)이다.

「논리전개를 위하여, 모든 영어단어가 어떤 표준사전에 수록되어 있다고 가정하고 T를 영어 단어 20개 미만을 사용하여 설명될 수 있는 자연수들 전체의 집합이라고 하자. 영어단어들의 수는 유한이기 때문에 20개 미만의 단어들로 결합될 수 있는 문장 전체의 개수는 유한이다. 따라서 T는 유한 집합이다. 따라서 T의 원소들 전체의 개수보다 더 큰 자연수들이 분명히 존재한다. 즉, 영어단어 20개 미만으로는 설명될 수 없는 최소의 자연수가 존재한다. T의 정의에 의해서, 이 자연수는 T에 속하지 않는다. 그러나 우리는 이 자연수를 16개의 영어단어로 나타내었다. (이 자연수 = a least number which can be described in fewer than twenty words of the English language) 따라서 이 자연수는 T 안에 있어야 한다.」

다시 우리는 명백한 모순에 직면하였다. 위의 논리는 결점이 없기 때문에, 만약 집합 T의 존재를 인정한다면 이와 같은 집합 T는 존재할 수 없게 된다.

역설이 발표되기 전에는, 집합의 **존재성**에 대한 의문은 전혀 제기되지 않았다. 칸토르는 집합을「우리의 지각에 의해서 총괄적으로 파악할 수 있는 분명하고, 구별이 가능한 대상들의 모임」으로 "**정의**"(定義)하였다. 좀더 상세하게 설명하면, 칸토르와 그의 추종자들은「대상들의 성질을 설명할 수 있다면, 그 성질을 소유하는 모든 대상들 전체의 집합을 또한 말할 수 있다」는 "상식수준"의 개념을 믿었다. 그러나 바로 이러한 성질로 부터 역설적인 집합을 만들어 내기 때문에 역설들은 칸토르가 정의한 이 소박한 집합의 정의를 받아 들일 수 없도록 반박하는 특별한 장점을 갖고 있다.

1900년대 초에 집합론의 기초를 수정할 목적으로 나온 여러가지 동향에서 가장 중심적인 주제는 집합의 존재성이었다. 즉, 어떤 성질들이 집합을 모순없이 정의하는가? 어떤 조건하에서 성질들이 집합을 정의하는가? 어떤 방법으로 주어진 집합들로부터 새로운 집합을 만들 수 있는가?

0.3 공리적 방법

수학기초론에 대한 위기는 역설의 출현으로 시작되었으며 그 위기는 오늘날까지도 아직 완전히 해소되지 않고 있다. 칸토르의 집합에 대한 "정의"에서 나타났던 것처럼 집합에 관한 직관적인 개념으로는 만족스러운 집합론의 기초를 확립할 수 없으며 따라서 수학 전체에 대한 기초도 확립할 수 없다는 것이 아주 명백해졌다. 특별한 형태의 개념과 정의를 제외함으로서 역설을 극복하려는 사소한 시도는 결국 실패로 돌아 갔고 이제 완전히 새롭게 접근하는 수 밖에 없게 되었다. 1905년을 기점으로 수학자들은 이 문제를 다룰 여러 가지 방법을 제시했고, 또 발전시켰다. 그들은 주로 세 개의 큰 부류로 나눌 수 있는데 공리주의, 논리주의, 직관주의로 불리는 학파들이다. 이 절의 나머지 부분을 통하여 이 세 방향의 사상을 소개하고자 한다. 먼저 공리적 방법의 발전과정을 간단히 살펴보기로 하자.

수학에서의 공리적 방법은 기원전 300년 경 유크리트의 "**원론**"(原論, Elements)과 함께 고도로 발전된 형식으로 나타났다. 유크리트에 의해서 널리 알려진 이 공리적 방법은 오늘날 수학의 모든 분야에서의 특징적인 모습이 되었지만 기하학 이외의 수학분야에 적용된 것은 비교적 최근의 일이다. 즉, 공리 체계에 대한 우리의 현대식 이해와 일반적인 연역적 사고는 대부분 기하학 분야의 연구에서 비롯되었다. 따라서 공리적 방법의 개발에 영향을 미친 기하학에 있어서의 주요한 발전자취는 살펴 볼 가치가 있다.

유크리트와 그가 살던 시대에는 공리(公理, axioms)와 공준(公準, postulates)은 의심의 여지가 없는 진리를 뜻하였다. 예를 들면, 유크리트의 평행선의 공리에 관하여 천년에 걸친 논쟁이 계속 된 것은「기하학적 명제는 절대적인 진리」라는 신념 때문이었다. 이 "**평행선(平行線)의 공준(公準)**"은「두 직선 A와 B가 제 3의 직선 C와 만날때 A와 B가 C와 이루는 내각의 합이 두 직각보다 작으면 두 직선 A와 B는 반드시 만난다」라는 것이다. 이 명제는 분명히 진리인 것 같이 보였지만 다른 공리와 공준들처럼 간결하지 않았기 때문에, 유크리트시대부터 1700년대에 이르기까지 많은 기하학자들이 대대로 이 평행선의 공리를 나머지 가설들로부터 증명해 보려고 노력했었지만 모두 허사였다. 결국 19세기 중엽에 이르러서야 이 문제는 볼리야이(J. Bolyai, 1802

$\sim$ 1860)와 로바체프스키(N. I. Lobachevski, 1792 $\sim$ 1856)에 의해 해결되었다. 즉, 두사람 모두 각자 평행선의 공리 대신에 이와 모순되는 가정으로 바꾸어 비유크리트 기하학(non-Euclidean geometry)을 창안하였다. 그 후 비유크리트 기하학도 유크리트적인 해석이 가능하였기 때문에 유크리트 기하학만큼 합리적이란 것이 증명되었다. 즉, 점, 선, 각 등을 적당히 다시 해석함으로써 볼리야이 또는 로바체프스키의 공준이 유크리트 기하학에서도 성립하게 되었다. 따라서 평행선의 공리는 유크리트 체계 안에서 다른 공리와 공준에 대하여 독립적일 뿐만 아니라, 또한 유크리트기하 만큼 모순이 없으면서 우리의 일상 생활에서 경험하는 공간을 설명하는 것과는 다른 독립된 기하학이 성립한다는 것이 밝혀지게 된 것이다. 이 비유크리트 기하학의 발견으로 공리는 더 이상 절대적인 진리가 아니라 어떤 명제이든 논리전개를 위한 수학적 이론의 **전제**(前提, premise)로서 사용하고자 하는 명제라는 인식에 이르게 되었다.

아마도 유크리트 원론에 있어서의 최대의 결함은 유크리트가 공리에 의하여 제시하지 못하고 암암리에 가정한 몇 가지 가설들일 것이다. 예를 들면 어떤 증명에서는 두 개의 원은 각각 다른 쪽의 중심을 통과할 경우 두 개의 공통점을 가진다고 했지만, 공리에는 이러한 점의 존재를 가정하지 않았다. 또 다른 한 가지는 유크리트가 두 점 사이에 있는 한 점에 대해서 "**사이**"라는 개념을 정의하지 않았으며 또한 그 성질을 공리로 세우지도 않았다. 또 다른 주장에서 원론은 강체운동을 말하고 있지만, 그 개념을 정의하지 않았고 공리에서 그런 개념에 대해서 언급하지도 않았다. 따라서 유크리트 원론에는 논리적 추론의 순차적인 연결이 종종 단절되는데, 그 까닭은 시각적인 경험에 암암리에 호소하기 때문이다. 19세기에 이르러 이 결함이 발견되었을 때, 수학적 이론은 공간적 또는 기타의 직관적 매개를 거치지 않고 전개될 수 있어야 한다는 데 수학자들의 공감대가 형성되었다. 즉, 점, 선, 사이, 같은 특정한 대상들과 그들 사이의 특정한 관계들은 "**무정의 개념**"(無定義 槪念, undefined notions)으로 인식되어야 하며 또한 이 특정한 것들에 대한 성질들이 완전히 명시되어 있어야 하고, 연역적인 방법은 개념의 뜻과는 본질적으로 독립이어야 한다. 1882년 파쉬(M. Pasch, 1843 $\sim$ 1930)는 기하학의 구성을 처음으로 출판하면서 직관에 의한 호소를 제거하는 것이 그 목적이라고 명백히 언급하고 이를 체계적으로 실천하였다.

19세기 말이 되어서야 공리론적 방법의 현대적 개념이 나타나기 시작했다. 그 대체적인 윤곽에서는 유크리트의 생각과 다를 바가 없었다.

즉, 어떤 수학적 이론에서 특정 명제들을 "**공리**"(公理, axiom)로 채택하고 그 이론이 그 밖의 다른 모든 명제들은 논리적 추론(論理的 推論)을 통해 이 공리들로 부터 유도(誘導)될 수 있을 때 이 이론을 "**공리적**"(公理的, axiomatic)이라고 한다. 그러나 유크리트의 생각과 달리 수학적 "**증명**"(證明, proof)의 **형식적**(形式的, formal) 성격에 대해서는 새로운 개념이 추가되었다. 「공리는 가능한 한 충분히 자세해야 하며, 논리적 연역의 법칙은 충분히 명시적이어야 하며, 증명 과정에서 직관이나 지식이 잠입할 여지가 전혀 없어야 한다.」 이상적(理想的)으로 말한다면 증명이 맞았느냐 틀렸느냐 하는 것을 컴퓨터가 검증해 줄 수 있어야 한다.

수학이 영어와 같은 일상 용어로 구성되는 한, 명제를 설명하고 복잡한 문장의 구조를 파악하기 위해 사용하는 언어에 대한 사람들의 이해가 불가피하다. 따라서, 수학적 증명에서 직관을 완전히 제거하기 위해서는 "**형식적인**" 수학언어(數學言語, mathematical language)의 개발이 필수적이다. 이 언어의 **규칙**은 엄격하게 조문화되어 모든 명제는 그 뜻이 애매하지 않고 또한 그 구조가 명백해야 한다. 형식적이고 상징적인 언어의 창조는 현대수학에 있어서 가장 중요한 발전 중의 하나였다. 그러한 언어가 어떤 형태인지 여기서 살펴 보자.

가장 기본적인 수학적 명제는 다음과 같은 형태이다.

X는 Y에 평행이다.

y는 x와 z 사이에 있다.

X는 열린 집합이다. 등등.

이 명제들은 하나의 대상, 또는 두 대상, 혹은 더 일반적으로 n 순서쌍의 대상들에 관한 명제들이다. 이와 같은 명제들을 **기본적 술어**(基本的 述語, elementary predicates)라고 하고, X, Y, x, y, z 같은 문자는 **변항**(變項, variables)이라고 한다.

술어(述語, predicates)를 나타내는 방법으로 술어를 하나의 문자로 나타내고 그 문자 다음에 변항을 열거하면 편리하다. 예를 들어 「X는 Y에 평행이다」를 $A(X, Y)$로 「y는 x와 z사이에 있다」를 $B(x, y, z)$로 나타낸다. 오늘날 수학을 잘 이해하고 있는 독자들이라면 「기하논리의 추론과정에서는 술어의 뜻이 중요하지 않다」는 것을 알고 있을 것이다. 예를 들면, **평행**(平行, parallel)이란 말의 뜻은 기하논리의 추론과정에서는 아무런 영향을 주지 않는다. 영향을 미치는 것은 「X는 Y에 평행이다」라는 명제와 다른 명제 사이의 관계이다. 여기서 다른 명제란 예를 들어 「X는 V와 만난다」라든가 「Y는 Z에 수직이다」라든가 하는 명제이다. 이와 같은 이유로 기본적 술어를 **원자적 공식**(原子的 公式, atomic formula)이라고도 한다. 즉, 술어는 분리할 수 없는 완전한 것으로서 더 이상 분석할 수도 없으며, 다만 다른 것들과 구별만 가능할 뿐이다.

수학의 모든 알려져 있는 분야에서는 단지 유한개(보통은 매우 적은 개수이다)의 서로 다른 기본적 술어만을 필요로 한다는 것은 놀랄만한 일이다. 예를 들어서 평면 유크리트 기하학은 다음 7개의 기본적 술어만으로 표현된다.

기본적 술어(基本的 述語, elementary predicates) (1)

$P(x)$ (x는 점이다.)

$L(x)$ (x는 직선이다.)

$B(x, y, z)$ (y는 x와 z 사이에 있다.)

$E(x, y)$ (x는 y와 같다.)

$I(x, y)$ (x는 y에 속한다.)

$C(u, v, x, y)$ (선분 uv는 선분 xy와 일치한다.)

$D(u, v, w, x, y, z)$ (각 uvw는 각 xyz와 일치한다.)

집합론은 앞으로 소개 하겠지만, 하나의 술어 $x \in A$(x는 A의 원소이다)에 의해서 집합론 전체가 완전히 구성된다.

명사만으로는 영어문장을 쓴다는 것이 불충분한 것처럼 술어만으로는 수학의 모든 명제를 나타내기에 충분하지 않다. 예를 들어 「**만약** x가 y에 평행**이고**, y가 z에 수직**이면** x는 z에 수직이다」라는 명제를 보면 이 명제는 술어가 **논리적 연결사**(論理的 連結詞, logical connectives)와 결합되어 있다. 일반적으로 만약 P와 Q가 수학이론에서의 명제라면 논리적 연결사로 결합된

아래 문장 또한 명제이다.

논리적 연결사(論理的 連結詞, logical connectives)

$\neg P$ (P의 부정)

$P \wedge Q$ (P 그리고 Q)

$P \vee Q$ (P 또는 Q)

$P \rightarrow Q$ (P이면 Q)

$P \leftrightarrow Q$ ([P이면 Q] 그리고 [Q이면 P])

마지막으로 예를 들어 「"x가 점"이고 "y가 점"이면 "z는 x와 y 사이"를 만족하는 "점 z가 **있다**"」라는 명제를 보면 여기서는 **한정기호**(限定記號, quantifier)를 필요로 한다. 즉, 만약 $P(x)$가 변수 x를 가진 명제라면 다음도 명제이다.

한정기호(限定記號, quantifier)

$\forall x P(x)$ (모든 x에 대하여, $P(x)$)

$\exists\, x \cdot\!\ni\!\cdot$[1] $P(x)$ (어떤 x에 대하여, $P(x)$)

혹은 ($P(x)$를 만족하는 x가 있다.)

혹은 ($P(x)$를 만족하는 x가 존재한다.)

이상이 형식적 수학언어에 대한 설명의 전부이다. 모든 알려진 수학이론은 기본적 술어와 논리적 연결사 그리고 한정기호를 사용하여 나타낼 수 있다. 이 **형식적 수학언어**(形式的 數學言語, formal mathematical language)가 어떻게 사용되느냐 하는 것을 설명하기 위해 간단한 예를 들어 보자. 명제 「만약 x와 y가 서로 다른 점이라면 x와 y 사이에 점 z가 있다」를 기본적 술어(1)에서의 술어를 사용하여 다음과 같이 기호로 나타낼 수 있다.

형식적 언어(形式的 言語, formal languages) (2)

$$[P(x) \wedge P(y) \wedge \neg E(x,y)] \rightarrow [\exists\, z \cdot\!\ni\!\cdot (P(z) \wedge B(x,z,y))].$$

형식적 언어를 사용함으로서 얻을 수 있는 많은 이점 중의 하나는 이 언어에서는 연역과정을 엄밀하고 명시적으로 나타낼 수 있다는 점이다. 명백하고

---

[1] 기호 $\cdot\!\ni\!\cdot$은 영어의 such that의 줄여서 기호로 나타낸 것이다.

분명한 다음과 같은 몇 가지 규칙에 의해 명제 T가 명제 S로 부터 추론될 수 있는 지, 아닌 지를 결정한다.

추론 규칙(推論規則, rules of inference) (3)

규칙 A. P와 $(P \to Q)$로 부터 Q를 추론한다.

규칙 B. P와 Q로 부터 $P \wedge Q$를 추론한다.

규칙 C. $\neg(\neg P)$로 부터 P를 추론한다.

규칙 D. $P(c)$로 부터 $[\exists\, x \cdot\!\ni\!\cdot\, P(x)]$를 추론한다.

이 규칙들과 몇 가지 다른 규칙들[2]을 **추론 규칙**(推論規則, rules of inference)이라고 한다. 주어진 전제에서 출발하는 형식적인 논증은 형식적 언어에 의한 표현들의 나열이며, 여기에서 표현은 전제이거나 혹은 추론규칙을 적용하여 그 앞에서의 표현(또는 표현들)으로 부터 유도된 표현이다.

보 기 0.1. 술어가 $L(x, y)$인 형식적 언어를 생각해 보자. $L(x, y)$는 $x < y$를 나타낸다고 하자. 다음은 이 형식적 언어의 아주 간단한 형식적인 논증이다.

전제(前提, premises)

전제 1. $a < b$.

전제 2. $b < c$.

전제 3. $[(a < b) \wedge (b < c)] \to (a < c)$.

정리(定理, Theorem) $\exists\, x \cdot\!\ni\!\cdot\, [(a < x) \wedge (b < x)]$.

증명(證明, proof)

단계 1. $a < b$ (전제 1)

단계 2. $b < c$ (전제 2)

단계 3. $[(a < b) \wedge (b < c)]$ (단계 1, 2, 추론 규칙 B)

단계 4. $[(a < b) \wedge (b < c)] \to (a < c)$ (전제 3)

단계 5. $a < c$ (단계 3, 4, 추론 규칙 A)

---

[2] Slupecki와 Borkowski는 책[10]에서 위 (3)에서의 네 개의 규칙에 일곱 개의 규칙을 추가하여 자연스러운 연역체계(system of natural deduction)를 소개하였는데 이 열 한 개의 규칙이면 모든 타당한 논증을 추론할 수 있다. Quine[8]과 Suppes[11]는 다른 연역체계를 서술하였다.

단계 6. $[(a < c) \wedge (b < c)]$ (단계 5, 2, 추론 규칙 B)
단계 7. $\exists\, x \cdot\!\ni\!\cdot [(a < x) \wedge (b < x)]$ (단계 6, 추론 규칙 D)

여기서 주의깊게 살펴 볼 점은 각 표현에서 추론규칙이 완전히 기계적인 방식으로 적용되었다는 점이며 표현은 무의미한 기호의 나열에 불과하다. 우리에게 어떤 것을 의미한다는 것은 증명하는 것과는 아무 관련이 없다. 따라서 형식적 수학증명에서는 직관이 개입할 여지가 전혀 없다.

공리적 이론이 **형식화**(形式化) 되었다(formalized)라는 말은 모든 공리들이 형식적 언어로 표현되어 있고 모든 증명 또한 형식적 증명으로 되어 있는 것을 말한다. (예를 들면, 형식적 언어 (2)는 힐베르트(David Hilbert, 1862 ~ 1943)의 평면 기하학의 공리이다.) 모든 공리적 이론은 형식적으로 전개되는 것이 이상적이지만 실제 이렇게 하는 것은 너무나 번거로운 일이다. 기호만을 사용하는 기호명제는 해독하기 어려우며 또한 형식적 증명은 지나치게 길어지는 경향이 있다. 따라서 수학자들은 공리적 이론이 형식화될 수 있다는 사실에 만족하고 공리적 논리를 형식화 하지 않고 논리를 전개하는 것이 보통이다. 이 책도 이와 같은 통상적인 방법을 따르기로 한다.

0.4 공리적 집합론

1900년대 초에 이르러 많은 수학자들은 역설을 극복하는 열쇠가 공리적 기초 위에 집합론을 세우는 일이라고 생각하였다. 마치 **점**(點)과 **선**(線)이 기하에서 무정의 개념(無定義 概念)이 되는 것과 같이, "**집합**"(集合)이란 용어(用語)와 "**의 원소이다**(속한다)"라는 용어(用語)가 여기에서는 무정의 개념(無定義 概念)이 되었다. 무정의 개념들의 **뜻**은 아무 관계가 없으며, 그 성질은 공리에 의해서 형식적으로 주어져야 했다. 특히,「칸토르의 집합론에서 모든 의미 있는 결과는 증명이 가능하도록 되어야 했으며 역설들은 만들어 질 수 없도록 공리가 선택되어야 했다.」

집합론의 첫 공리화는 1908년에 체르멜로 (Ernst F. F. Zermelo, 1871 ~ 1953)에 의하여 이루어졌다. 체르멜로의 체계는 스코렘(T. Skolem, 1887 ~ 1963)과 프랜켈(A. A. H. Fraenkel, 1891 ~ 1961)에 의해 약간의 수정을 거

쳐 오늘날까지 널리 쓰이고 있다. 형식적 방법이 널리 이해되고 인정되기 이전에 체르멜로는 그의 논문을 썼다. 따라서 그의 집합론은 형식적 언어로 쓰여지지 않았다. 그러나 그 체계는 기하학의 낡은 공리적 방법에 가까운 것이었다. 체르멜로의 체계에 있어서는 한 개의 원시적 관계가 있는데 기호로는 $\in$로 적는다. $x \in Y$란 표현을 「x는 Y의 원소이다」라고 읽는다. 변항 $x, y, z, X, Y, \cdots$은 기호 $\in$의 오른쪽 또는 왼쪽에 자리하며 이 변항이 우리가 집합이라고 부르는 대상이다.

독자들은 아마도 두 종류의 대상 즉, 집합과 원소가 있어야만 한다고 생각할 것이다. 사실 이 구별은 필요 없다. 왜냐하면 한편으로는 원소와 집합의 관계는 절대적인 것이라기 보다 상대적인 것이며 사실, 원소와 집합 사이의 관계가 정확히 $\in$이다. 다른 한편, 수학에서 거의 모든 집합은 **집합들**의 집합이기 때문이다. 예를 들면, 평면 해석기하에서 선은 점들의 집합이며, 점은 두 실수의 짝이며, 실수는 유리수의 열이다. 따라서 공리적 집합론에서 이룩한 유용한 단순화가 모든 집합의 원소를 다시 집합으로 간주하게 하는 것이다. 다시 말하여 모든 집합은 실제로 집합들의 집합인 것이다. 이렇게 단순화하여도 잃을 것은 아무 것도 없고, 오히려 원시적 개념과 공리의 수를 줄이는 장점이 있다.

기호에 대해서도 이와 같은 관점에서 볼 수 있다. 소문자와 대문자로 $x \in Y$와 같이 쓰는 습관이 있지만, 사실 그럴 필요는 조금도 없다. 우리는 때때로 $x \in y$ 또는 $X \in Y$라고도 쓴다. 이 표현에서는 모든 변항은 집합을 나타낸다.

우리가 생각하는 거의 모든 집합은 특정한 종류의 대상 전체로 구성되어 있다. 즉, 주어진 조건을 만족하는 대상 전체로 구성되어 있다. 이것이 집합을 만드는 가장 자연스러운 방법이다. x에 관한 조건을 서술할 수 있어서 이것을 상징하여 $S(x)$와 같이 적고, $S(x)$를 만족하는 모든 대상 x의 집합을 생각하게 된다.

보 기 0.2. (i) "x는 무리수이고 $0 \leq x \leq 1$이다"라는 조건을 만족하는 대상 x 전체의 집합. (즉, 0와 1 사이의 모든 무리수들의 집합.)

(ii) "x는 사람이다"라는 문장으로 나타낼 수 있는 대상 x 전체의 집합.

(즉, 모든 사람들의 집합.)

이것이 집합이 형성되는 가장 자연스러운 방법이기 때문에, 이 방법이 가능하도록 집합론에 어떤 원칙을 가지는 것이 분명 바람직하다. 즉, 어떤 조건 $S(x)$가 주어졌을 때, $S(x)$를 만족하는 대상 x 전체의 집합을 만드는 원칙이 있어야 한다. 그러나 우리가 2절에서 지적한 바와 같이, 만약 이러한 원칙을 **아무런 제한 없이** 체택한다면, 예를 들어 자기 자신을 원소로 하지 않는 모든 집합들의 집합을 만드는 것과 같이, 역설에 빠지게 된다. 따라서 이 원칙에 역설들을 배제할 수 있는 제한장치를 고안해야만 한다. 체르멜로는 다음과 같은 제한을 생각하였다.

「$S(x)$를 x에 관한 조건이라고 하면, $S(x)$를 만족하는 모든 x들의 집합을 만들 수 없다. 그러나 만약 A라는 집합이 주어졌다면, **$S(x)$를 만족하는 A에 속하는 모든 x들의 집합**은 만들 수 있다. 따라서 간단히 말하면, 대상의 성질이 어떤 "새로운" 집합을 만드는데 쓰여질 수는 없다. 그러나 이미 그 존재가 보장된 집합 A로부터 주어진 성질을 만족하는 모든 원소들을 "선택"하는 데에는 대상의 성질이 쓰여질 수 있다.」

체르멜로는 이 원칙을 그의 집합론 체계에 있어서 하나의 공리로 삼았다. 이 공리는 어떤 집합 안에 있는 원소를 선택하는 역할을 하기 때문에 이를 **선택의 공리**(選擇의 公理, axiom of selection)라고 불렀다. 그 내용은 다음과 같다.

「A가 어떤 집합이라 하고 A에 속하는 모든 대상 x에 대해 뜻을 가지는 명제를 $S(x)$라 하자. 그러면 A의 원소 중에서 $S(x)$를 만족하는 모든 원소 x로 구성된 집합이 존재한다.」

이 선택의 공리에 의해 존재성이 보장된 집합을 보통 $\{x \in A \mid S(x)\}$와 같이 나타내고 「$S(x)$를 만족하는 A의 모든 원소 x의 집합」이라고 읽는다. 여기서 주목할 점은 체르멜로의 체계에서는 「$S(x)$를 만족하는 모든 x의 집합」 $\{x \mid S(x)\}$와 같은 집합은 허용하지 않고 다만 임의의 집합 A에 대하여 집합 $\{x \in A \mid S(x)\}$를 만들 수 있도록 허용하는 것이다.

이 선택의 공리가 어떻게 역설을 피할 수 있는가? 먼저 럿셀의 역설에서 어떤 문제가 생겼는지 살펴 보자. 럿셀의 역설에서 문제의 집합은 「자기 자신을 원소로 가지지 않는 모든 집합들의 집합」이었다. 이것을 기호로 $\{x \mid x \notin$

$x\}$와 같이 적을 수 있다. 이미 지적한 바와 같이 체르멜로의 체계에서는 이와 같은 집합은 만들 수 없고 다만 우리가 할 수 있는 최선은 존재가 보장된 집합 A에 대하여, $\{x \in A \mid x \notin x\}$와 같은 것은 만들 수 있다. 만약 럿셀의 역설에서 나타나는 $\{x \mid x \notin x\}$을 $\{x \in A \mid x \notin x\}$로 대신한다면 그 결과는 완전히 달라진다. 즉, S를 $\{x \in A \mid x \notin x\}$라 하고 논리적 단계를 따라 검토해 보자.

「$S \in S$는 불가능하다. 왜냐하면 $S \in S$이면 $S \notin S$이니까 모순이다. 따라서 $S \notin S$이다. 따라서 $S \notin A$이다. 왜냐하면 $S \in A$이라면 $S \in S$이고 이는 모순이다.」

즉, 럿셀의 논법은, 만약 A가 임의의 집합일 경우 A가 집합 $\{x \notin A \mid x \in x\}$의 원소가 될 수 없다는 것을 말하고 있을 뿐이다.

다른 논리적 역설도 마찬가지로 극복할 수 있다. 모든 논리적 역설에서 문제되는 집합은 모두가 공통된 특성을 가지고 있다. 문제되는 집합들은 과도하게 포괄적이다. 즉, 문제되는 집합은 **너무 크며** 너무 많이 포함하고 있다. 럿셀의 역설에서는「자기 자신을 원소로 가지지 않는 집합들 전체의 집합」이었으며, 럿셀의 역설과 밀접하게 연관되어 있지만, 칸토르의 역설에서는「모든 집합들의 집합」이었다. 선택의 공리가 존재가 보장된 집합의 부분집합만을 만들 수 있도록 허용하기 때문에 이렇게 터무니없이 큰 집합을 만들 수 없다.

어의적 역설을 피하는 문제는 좀 더 어렵다. 베리의 역설과 같은 어의적 역설에서는 문제의 집합이 너무 큰 것은 아니었다. 문제는 오히려, 집합을 결정하는 조건 $S(x)$에 있는 듯하다. 이 경우에는 선택의 공리에 의한 제한도 효과적인 방파제 역할을 하지 못한다. 따라서 만약 $S(x)$가「x는 영어로 20개 미만인 단어로 설명할 수 있다」라는 문장이라고 하면 베리의 역설에서 문제의 집합은 $\{x \in \mathbb{N} \mid S(x)\}$이다. 여기서 $\mathbb{N}$은 자연수의 집합이다. 만약 $S(x)$를 x에 관하여 받아들일 수 있는 조건이라고 인정한다면 이 집합은 체르멜로의 체계에서도 만들 수 있다.

따라서 어의적 역설을 극복하려면 집합을 결정하는 **조건**(條件) $S(x)$의 형태에 제한을 두지 않으면 안된다. 체르멜로는 선택의 공리에서 $S(x)$가「A의 모든 원소 x에 대하여 의미가 있을 경우에 한하여 $\{x \in A \mid S(x)\}$를 만들 수 있다」라고 규정하여 이 문제를 해결하고자 시도하였다. 그러나 이렇게 하여 돌아 온 것은 그가 단지 다음과 같은 새로운 문제를 제기하였을 뿐이다.
즉「의미가 있을 경우」를 어떻게 이해해야 하는가?
$S(x)$의 의미가 있는지 어떤지를 어떻게 결정할 것인가의 새로운 문제이다.

이제 마지막으로 주의 깊은 독자라면 스스로에게 자문해 보았을 문제가 남았다. 즉,「대상 x에 관한 명제로서의 조건」 $S(x)$란 대체 무슨 뜻인가? "x에 대한 명제"란 개념을 그저 직관적으로 알 수 있다고 생각해서는 안된다. 왜냐하면 지금 우리의 목적은 집합론을 공리화하려는 것이며, 따라서 모든 직관으로부터 해방되려고 노력하고 있기 때문이다. 체르멜로의 체계가 형식적 언어(形式的 言語, formal languages)로 쓰여지지 않았기 때문에 그는 이 질문에 대하여 만족할 만한 답변을 하지 못하였다. 그러나 1922년에 스코렘과 프랜켈은 집합론의 형식적 공리화를 위한 연구를 하면서 이 딜레마로부터 벗어나는 자연스러운 방법을 발견하였다. 즉, "x에 관한 명제"를 단순히 **형식적 언어**에 의한 하나의 "**자유 변항**"(自由 變項, free variables) x를 가진 명제로 보는 방법이다. 여기서 x가 $S(x)$에서 자유 변항이라는 말은 x는 $\exists x$ 혹은 $\forall x$와 같은 한정기호에 의해서 영향을 받지 않는다는 것을 의미한다. 예를 들면, $\exists y \cdot\ni\cdot (x < y)$에서는 x는 자유 변항이지만 y는 자유 변항이 아니다.

체르멜로의 체계에서 기본적 술어는 하나뿐이며, 그것을 기호 $\in$로 적는다. 따라서 형식적 언어에 의한 명제는 $x \in Y, u \in V$와 같은 술어와 논리적 연결사 및 한정기호만을 써서 나타낼 수 있다.

만약 선택의 공리에서 쓸 수 있는 "명제" $S(x)$를 형식적 언어로 표현할 수 있는 것만으로 제한한다면 모든 어의적 역설들을 제거할 수 있게 된다. 예를 들어서「x는 영어로 20 단어 미만으로 설명할 수 있다」라는 문장을 형식적 언어로는 도저히 쓸 수 없다. 어의적 역설을 해결하는 이 풀이가 철학적으로는 이상적인 해결책이라고 하기는 어렵지만, 수학적 입장에서는 받아들일 수 있다. 수학적으로는 수학에서 필요한 모든 집합을 충분히 만들 수 있기 때문이다. 물론 넓은 뜻에서「모든 사람들의 집합」이나「모든 라틴어 동사들의 집합」과 같은 것들은 만들 수 없다. 지금까지도 이 보다 더 좋은 해결 방법은 마련되지 않았다.

지금까지 우리는 체르멜로의 체계가 스코렘과 프랜켈의 수정을 거쳐 어떻게 역설을 피하게 되었느냐 하는 것을 살펴 보았다. 체르멜로의 체계에서 그 밖의 다른 공리들은 이 책의 다음 장에서 전개하려는 것과 비슷하다.

본질적으로 그 공리들은 모든 자연수들의 집합의 존재성을 보장하며 이로부터 우리는 수학에 있어서의 다른 모든 수의 체계를 구성할 수 있게 되며, 또 합집합과 교집합, 곱집합의 존재성을 보장한다.

더 앞으로 나아가기에 앞서 공리화의 다른 방법을 간단히 살펴보기로 하자. 이 방법은 우리시대에 점점 더 많은 관심을 받고 있는 것으로서 그 본질적인 생각은 노이만에 의한 것이다.

노이만(John von Neumann, 1903 ~ 1957)은 다음 두 가지 사실이 결합하여 논리적 역설을 낳는다는 점에 주목하였다.

첫째 사실은 우리가 이미 보아 온 문제의 집합(예를 들어 럿셀의 역설에서의 집합 S)이 **너무 크다**는 것이다.

둘째 사실은 이 "큰" 집합이 「어떤 집합의 원소가 될 수 있다」고 하는 것이다. (예를 들어 럿셀의 집합 S가 그 자신의 원소가 될 수 있다고 하였다.)

이 두 가지 중에서 체르멜로는 첫째 사실만을 이용하였다. 즉, 그는 너무 큰 집합을 만들 수 없도록 제한하여 역설을 극복하였다.

노이만은 이 둘째 사실도 이용할 것을 제안하였다. 그는 터무니없이 큰 집합이 존재하는 것을 허용하되 이것이 집합의 원소가 되는 것은 허용하지 않았다. 간단히 말하여, 노이만의 체계는 다음과 같이 표현할 수 있다.

「체르멜로의 경우와 같이 기본 술어는 하나뿐이다. 즉, 술어 $x \in Y$가 그것이다. 변항 x, y, X, Y 등은 **클래스**(class)라고 부르는 것을 나타낸다. 그러나 클래스에는 두 가지가 있는데 어떤 클래스의 원소가 되는 클래스라고 정의하는 **원소**와 또 어떠한 클래스의 원소도 될 수 없는 클래스가 그것이다.」

따라서 체르멜로의 선택의 공리는 다음과 같은 "**클래스의 공리**"(類의 公理, axiom of class)로 대치할 수 있다.

「$S(x)$가 대상 x에 관한 임의의 명제라고 하면, $S(x)$를 만족하는 모든 원소 x들의 클래스가 존재한다.」

다시 말해서, 만약 $S(x)$가 x에 관한 임의의 명제라고 하면, $\{x \mid x$는 원소이며 $S(x)\}$와 같은 클래스를 만들 수 있다.

이 체계에서 럿셀의 역설이 성립하지 않는다는 것을 증명하기 위해 럿셀의 역설에서와 같이 여러 단계를 따라 분석해 보기로 하자. 여기서 S는 $\{x \mid x$는 원소이며 $x \notin x\}$이다.

「$S \in S$는 불가능하다. 왜냐하면 $S \in S$라면 $S \notin S$이므로 모순이기 때문이다. 따라서 $S \notin S$이다. 그래서 S는 원소가 아니다. 왜냐하면 S가 원소라면 $S \in S$일 것이기 때문에 그것은 모순이다.」

따라서 럿셀의 논법은 단지 위에서 정의한 클래스 S는 원소가 아니다는 것을 증명하고 있을 뿐이다. 개선된 체르멜로의 체계에서와 같이 노이만의 체계에서도 클래스의 공리에 형식적 언어로 표현되는 명제만을 허용함으로서, 어의적 역설도 제거할 수 있게 되었다.

노이만의 체계에 대한 변형들이 괴델 (K. Gödel, 1906 ~ 1978)과 베르네이즈 (P. I. Bernays, 1888 ~ 1977)에 의해 개발되었다. 이 노이만식의 체계가 체르멜로의 체계보다 좋은 점은 클래스의 공리가 선택의 공리보다 직관적인 집합론의 관점에 좀 더 가깝다는 점이다.

사실 $S(x)$가 x에 관한 임의의 명제라면, 클래스의 공리는 $S(x)$를 만족하는 모든 원소 x를 포함하는 클래스가 존재함을 보장한다.

수학을 다루는 데 있어서, 노이만 체계가 주는 편리한 점은 "모든 원소들의 클래스"에 관하여 말할 수 있게 해 주며 또한 원소가 아닌 클래스들 사이의 연산도 가능하게 해 준다. 이와 같은 원소가 아닌 클래스들이 고등 수학의 여러 분야에서 나타나는 경향이 있다. 이러한 것들은 피할 수는 있지만 그만한 대가를 지불하고서야 가능하다.

이 체계의 가장 큰 약점은 원소가 될 수 있는 클래스와 원소가 될 수 없는 클래스와의 구별(고도로 인공적인 것이지만)에 대해 항상 조심해야 한다는 점이다. 그러나 노이만식 체계에서의 풍부한 융통성과 그 자연스러움은 이와 같은 약점을 덮고도 확실히 남음이 있다. 이 교재에서는 약간 변형된 형태의 노이만의 공리체계를 채택하고자 한다.

0.5 공리적 집합론에 대한 반론과 다른 대안들

공리적 집합론의 주된 목적은 무엇이며, 이 목적이 어느 정도로 성공적으로 달성되었는가? 이 질문에 답하기 위해 우리는 20세기 초에 수학자들이 집합론에 대한 공리적 기초를 연구하던 환경을 살펴 봐야 한다. 칸토르의 사상은 이미 그 당시 현대수학의 구조 구석구석에 완전히 스며들어 있었으며 그 당시 열심히 공부하는 수학자들에게는 없어서는 안 될 도구가 되어 있었다. 대수학과 해석학은 집합론의 틀 안에서 체계화되었으며, 이 분야에서 가장 화려하고 강력한 새로운 연구결과들은 칸토르와 그의 후계자들이 도입한 방법에 의해 이루어졌다. 따라서 역설이 발견되고 칸토르 체계의 기초에 관한 타당성에 대하여 회의가 생겼을 때, 대부분의 수학자들은 그 체계를 버리기를 꺼려 하였다. 즉, 그들은 역설을 극복할 수 있는 어떤 방법이 발견되어 칸토르의 이론이 전부는 아니더라도 적어도 대부분은 그대로 보존될 것으로 믿고 있었다. 힐베르트는 이 점에 관해 다음과 같이 쓴 일이 있다. "칸토르가 우리에게 소개한 낙원에서 우리가 쫓겨나는 일은 없을 것이다."

또다시 새로운 역설들이 발견되고, 그것을 극복하려는 초기의 모든 노력이 실패로 돌아가자 직관적인 집합론이 그대로 보존될 수가 없게 될 것이 더욱 분명하게 되었다. 즉, 아주 많을지도 모르지만 어느 정도 버려져야만 할 것으로 보였다. 따라서 사람들이 기대 할 수 있었던 최선은 현대수학의 새로운 연구결과들을 유지하고 고전적인 수학을 적절히 체계화하는 데 필요한 만큼 직관적인 집합론을 유지하는 것이었다.

간단히 말해서 공리적 집합론은 이와 같은 제한된 목적을 달성하기 위하여 만들어 졌다. 즉, 직관적인 집합론에서와 같이 폭넓을 필요는 없지만 적어도 고전수학에서 필요로 했던 수체계나 함수 그리고 관계와 같은 구조를 포함하고 칸토르의 기본적인 내용들을 모두 포함하는 집합론 체계의 확고한 기초를 마련해야만 했었다.

체르멜로의 체계와 노이만의 체계는 둘 다 이 제한된 목적을 달성하는 데에 성공하였다. 그러나 직관적인 집합론도 상당히 많이 희생되었다. 예를 든다면 체르멜로의 체계에서는, 이미 살펴 본 바와 같이, 집합을 만드는 직관적인 방법인 대상의 성질을 서술하여 그 성질을 가진 모든 대상들의 집합을 만드는 방법은 더 이상 찾아 볼 수 없게 되었다. 이와 같은 집합을 만드는 직관적인 방법은 선택의 공리로 대치되었다. 선택의 공리에서는 성질이라는 것이 「주어진 집합의 부분집합」을 결정하는 데에만 허용되었으며 또한 이러한 허용될 수 있는 성질도 일곱 개의 기호 $\in$, $\wedge$, $\vee$, $\neg$, $\rightarrow$, $\forall$, $\exists$와 변수 $x, y, z, ...$(형식적 언어) 만으로 나타낼 수 있는 것이었다.

따라서 우리가 보통 집합이라고 생각하는 것의 대부분, 예를 들어 "모든 사과들의 집합"이나 "우주에 있는 모든 원자들의 집합"과 같은 것은 공리적 집합론에서는 집합으로 인정하지 않는다. 사실 공리적 집합론이 제공하는 "집합"은 우선 공집합 $\emptyset$, 그 다음에는 공집합으로부터 만들 수 있는 $\{\emptyset\}$, $\{\emptyset, \{\emptyset\}\}, \cdots$ 와 같은 것들 뿐이다. 수학 전체가 이와 같은 빈약한 집합론의 기초 위에 있다는 사실은 놀라운 일이다.

집합론에 관한 여러 가지 공리적 체계들이 수학을 급한 위기에서 구해 내었지만, 이 공리적 체계가 아주 많은 사람들을 만족시키지는 못했다. 특히 수학의 세련미와 보편성에 민감한 많은 사람들은 즉각 체르멜로와 노이만의 체계는 문제를 해결하기 위한 미봉책이며 조만간 완전히 직관적인 보편성에 입각하여 "집합"의 개념을 다룰 수 있는 더욱 포괄적인 수학이론으로 대치될 것이라고 지적하였다.

집합에 대한 직관적인 개념을 잘라 버린 반쪽을 취급하고 있다는 공리적 집합론에 대한 이와 같은 반론이 중요한 철학적인 파급효과를 가져왔다. 그것은 수학적 "진리"에 관한 보다 넓은 논의에 속하는 것으로서 다음과 같은 문제들을 중심으로 하고 있다.

「수학적 개념이 인간 마음의 산물(즉, **발명품**)인가? 그렇지 않으면 반대로 그것은 인간과는 관계없이 플라톤적인 개념의 영역에 존재하고 있는 것이어서 수학자들에 의하여 단순히 **발견되는** 것인가?」

이 후자의 견해를 "**플라톤적 실재론**"(platonic 實在論, platonic realism)이라고 하는데 이것이 고전적인 수학에 대한 지배적인 견해였다. 이 대립되는 두 견해가 하나의 구체적인 개념(자연수의 개념)에 대하여 어떻게 적용되는가를 살펴 봄으로서 이 두 대립되는 견해를 설명해 보자.

플라톤적 실재론의 입장에서 볼 때 "하나", "둘", "셋" 등은 자연에 존재하고 그것은 최초의 인간이 헤아리기 시작하기 전부터 존재하였다. 만약 우주의 어딘 가에 인간 이외의 지적인 존재가 있다면 그들이 우리와 어떻게 다를지라도 그들은 틀림없이 자연수를 발견했을 것이고, 그 성질은 우리가 발견한 것과 똑 같을 것이다.

한편, 반대편의 견해에 따르면 세 마리의 소, 세 개의 돌, 또는 세그루의 나무는 자연에 존재하지만, 자연수 3은 우리 마음의 창조물이며 우리가 자연수를 구성하는 과정을 발명한 것이다. 즉, 0에서 출발하여, 매 번 1을 더하여 감으로서 $1, 2, 3, \cdots$ 등을 만든다. 이렇게 하여 우리가 스스로 만든 개념적 도구를 유행시킨 것이다.

플라톤적 실재론은 공리적 집합론의 입장에 어떻게 영향을 주었는가? 플라톤적 실재론의 견지에서 볼 때 수학적 대상은 우리에게 그 모든 특징 및 그 모든 성질과 더불어 만들어 놓은 채로 미리 주어진 것이다. 따라서 수학적 정리가 "**참**"이다라는 말은 그 정리가 「관련된 수학적 대상에 대하여 바르게 표현했다」는 뜻이 된다.

예를 들어, $2 + 2 = 4$라는 정리는 산술에서 증명 가능한 형식적인 명제일 뿐만 아니라, 그것은 수에 관한 "실제적 사실"을 말하고 있는 것으로 본다.

그러면 가령 우리가 수학적 대상이 그 모든 성질과 더불어 우리에게 주어졌다는 것을 인정한다면, "집합"의 개념도 일정하게 잘 정의된 개념이며 우리의 편리에 따라 변경될 수 없는 것이 된다. 따라서 체르멜로와 노이만이 만든 집합은 존재하지 않아야 하며, 또한 이 존재하지 않는 대상을 표현하는 것을 의미하는 정리들은 거짓이 된다.

결국 우리가 플라톤적 실재론을 엄격히 받아들인다면 체르멜로와 노이만의 체계는 수학적으로 무효라고 배척하지 않을 수 없게 된다.

다행히도 얼마 전부터의 새로운 경향은, 플라톤 주의로 부터 벗어나 수학적 **진리**에 대해 보다 유연하고 보다 **불가지론**(不可知論)적인 쪽으로 흐르고 있다.

하나는 수학의 발달이 플라톤 철학이 지배했던 양상으로부터 점진적으로 벗어나고 있다는 점과 또 다른 하나는 「모든 수학적 대상은 확정적이며 서로 구별되는 직관의 대상에 대응한다」는 플라톤 주의의 근본적인 요청이 19세기

무렵의 창조적인 수학자들에게는 거의 견딜 수 없는 무거운 책임감이었다는 점을 들 수 있다. 그들은 구체적인 술어로 간단히 설명할 수 없는 무수히 많은 새로운 개념들, 예를 들어 복소수나 구성에서의 추상적인 규칙이나 함수의 일반적인 개념들 처럼 구체적인 용어로 쉽게 해석할 수 없는 개념들을 다루고 있었던 것이었다.

가령 복소수의 경우가 바로 그와 같은 사정을 잘 설명해 주고 있다. 고전적인 수학에서 복소수는 까다로운 것이었다. 왜냐하면 복소수는 구체적으로 해석할 수 없었으며, 따라서 이런 것이 과연 실제로 존재하느냐하는 데 대한 불만스러운 의심이 일어 나게 되었다. 실수는 길이나 양으로 설명될 수 있다. 그러나 음수의 제곱근은 현실 세계 또는 수에 대한 어떤 직관적인 대상에도 대응되지 않는 것같이 보였다. 그러나 복소수 체계는 가장 자연스러운 방법으로 창안되었다. 즉, 복소수는 실수를 포함하고 있으며, 실수를 계수로 하는 대수방정식의 모든 근을 포함하고 있는 최소의 수체계가 복소수이다. 따라서 복소수에 대응되는 실체가 물리적으로나 또는 심리적으로 존재하느냐 존재하지 않느냐 하는 물음은 촛점이 맞지 않다.

복소수의 경우는 공리적 집합론에서의 문제점들과 잘 대비되고 있다. 체르멜로와 노이만이 창안한 집합은 수학적인 상황에서 아주 자연스럽게 나타난 것이다. 체르멜로와 노이만은 수학에 적합하고 아주 간명하게 집합의 개념을 설정하여 모순이 없는 공리적 집합론을 확립하였다. 우리가 그것을 직관적으로 해석할 수 있느냐 없느냐는 상대적으로 중요하지 않다.

그러나 1900년대 초에 많은 수학자들은 공리적 집합론이 요구하는 정도만큼 전통적인 면을 포기할 것을 주저한 것이다. 더욱이 그들은 심미(審美)적 입장에서, 집합의 수학적 이론에서는 우리의 직관이 집합이라고 인정하는 것 모두와, 그리고 그것만을 포함해야 한다고 느끼고 있었다. 그 중 한 사람이 럿셀이다. 직관적인 집합론을 살리기 위한 노력을 하는 가운데 럿셀은 다음과 같은 착상이 떠올랐다.

즉, 집합이 **수준**(水準, level)이라는 계급에 의해 서열을 갖도록 하는 것으로, 「A와 B가 집합일 경우 A가 B의 원소라면 B는 A보다 한 수준 높은 것」으로 하였다.

예를 들어 평면기하에서 점의 집합으로서의 원은 원들의 족보다는 한 수준이 낮은 것이라 하였다. 이와 같은 기본적 착상은 럿셀에 의해서 하나의 이론으로 만들어 졌는데 이를 "**유형이론**(類型理論, Theory of types)"이라고 불렀다. 그 줄거리는 다음과 같다.

「모든 집합은 하나의 지정된 **수준**(水準, level)이라는 자연수를 가진다. 가장 단순한 집합, 즉, 수준이 0인 집합을 **개체**(個體, individuals)라고 하며, 그것은 원소를 가지지 않는다. 개체들의 모임이 수준 1인 집합이다. 수준이 1인 집합들의 모임이 수준이 2인 집합이 되는 등등이다.」

이 유형이론에서 $a \in B$라는 표현은 어떤 수 n에 대하여 a가 수준 n의 집합이며, B가 수준 $(n+1)$의 집합일 경우에만 의미를 가진다. 따라서 럿셀의 역설은 그것이 구성될 수 조차도 없다는 간단한 이유 때문에 사라져 버린다.

럿셀의 유형이론은 아주 간단한 생각에 의해서 만들어졌다. 그러나 불행하게도 이 이론이 "작용하기" 위해서 럿셀은 많은 새로운 가정들을 추가하지 않으면 안 되었다. 결국 모든 결점들을 보완한 이론은 사용하기에 너무도 거치장스럽고 만족하기에도 너무 복잡한 것이 되었다.

즉, 첫째로 집합의 "**수준**"(水準, level)이라는 계급에 대응하는 논리적 술어에 수준이라는 계급이 있어야 했다. 그 다음에는 어의적 역설을 피하기 위해 같은 수준에 있는 집합들은 "순서"에 의해 다시 나눠져야 했다. 끝으로 럿셀은 소위 "**약분 가능성의 공리**"(Axiom of Reduciblity)라는 공리를 인정하지 않을 수 없게 되었는데 이 공리는 지금까지 체르멜로가 세운 가정과 마찬가지로 매우 임의적인 것이고 직관에 근거를 두지 않은 것이었다. 유형이론은 현재까지도 여전히 흥미있고 아마도 유망한 연구 분야이기는 하지만 이러한 단점 때문에 수학자들 사이에서 널리 인정을 받지 못했다.

훨씬 더 급진적인 접근이 자칭 직관주의자라는 수학자들의 집단에 의해 취해졌다. 직관주의자들이 볼 때 거의 모든 칸토르의 집합론을 포함하여 대부분의 현대 수학은, 그들이 생각하기에 타당성이 없는, 논리의 규칙들을 무비판적으로 사용하는데 기초하고 있다는 것이다. 따라서 집합론에 대한 직관주의자들의 태도는 거의 총체적 거부라는 것으로 매우 쉽게 요약된다.

직관주의 철학을 완전히 이해하려면 먼저 논리에 대한 그들의 태도를 이해해야 한다. 직관주의자들은 수학자들이 쓰는 논리의 법칙은 경험적인 성격을 띤다고 본다.

어떤 증명방법들은 수학자들에 의하여 공통적으로 사용되게 되고, 여러 해를 거치는 동안 규칙이라는 틀 속에 규정 된다. 이런 규칙들은 그 원래의 상황에서는 뚜렷하게 바른 것이었다. 그러나 그것들이 규칙으로 규정되고 적용될 수 없는 전혀 다른 상황에서도 무비판적으로 사용되게 되었다.

더욱 구체적으로 말해 보자. 15세기 이전까지는 모든 수학의 근본이었던 유크리트의 기하에서의 모든 정리는 유한개의 대상만을 포함한다. 그리고 이 대상(기하적 도형)은 명백한 작도에 의하여 주어진다. 유크리트가 사용한 논리의 법칙은 이런 상황에서는 완전히 옳지만 무한을 포함하는 대상의 영역에 그것들을 옮겨 놓을 때 거기에서는 대상이 구체적인 작도로 주어질 수 없는 것이 되고 따라서 이들 규칙은 옳지 않다고 직관주의자들은 말한다.

예를 들어서 배중률을 생각해 보자. 이 규칙은「만약 S가 명제라면 S 혹은 S의 부정 중에서 어느 한 쪽만이 참이다」라고 하는 것이다.

특히 A가 어떤 집합이고 $P(x)$가 A의 모든 원소 x에 대하여 의미가 있는 어떤 명제라고 하자. **배중률**(排中律, law of the excluded middle)이란 $P(x)$가 참인 x가 A 안에 적어도 하나 있든지, 아니면 A에 속하는 모든 x에 대하여 $P(x)$가 거짓이든지 그 어느 한 쪽이어야만 한다. 만약 A가 유한집합이고 $P(x)$가 참인지 아닌지를 A의 각 원소에 대하여 조사할 수 있다면 직관주의자들은 이 규칙이 옳다고 받아들일 것이다. 그러나 직관주의자들이 배중율의 모태라고 주장하는 것으로서, $P(x)$가 참인지 아닌지를 A의 모든 원소 x에 대해서 조사할 수 있다면(그러기 위해서 A는 유한집합이어야만 한다) 우리들은 경험에 의해서 단지 아래 두 가지 경우가 결과로 나타날 수 있다는 것을 알 수 있다. 즉, 어떤 하나의 x에 대하여 $P(x)$가 참임을 우리가 발견하던지, 아니면 A의 모든 x에 대하여 $P(x)$가 거짓이든지의 두 가지이다.

따라서 우리의 경험은 유한집합인 경우에는 배중률을 확인하게 된다. 그러나 무한집합의 경우에는 우리가 경험하지 못한 영역이고 또 경험이 불가능한 영역이기 때문에 배중율이 옳다고 가정한다는 것은 전적으로 그 근거가 없는 것이라고 직관주의자들은 말한다. 이와 같은 입장에서 직관주의자들은 배중율을 거부한다.

고전 논리학에서의 다른 규칙들도 경험의 영역을 넘어 선다는 이유로 마찬가지로 거부한다.

직관주의자들은 수학의 근원은 주로 기하적 도형과 자연수의 단순한 구성에서와 같은 어떤 지적(知的, mental)인 구성(작도)의 연구라고 지적한다. 초기 수학의 정리들은 본질적으로「어떤 작도가 수행되면 어떤 결과가 얻어 진다」는 취지에서의 명제들이었다.

유크리트 기하학 및 일반적인 초기수학에서의 증명은 작도적 성격을 가졌다. 예를 들어 기하학에서의 다음 정리를 생각해 보자.

「주어진 두 삼각형에 대하여 만약 한 삼각형의 두 변과 그 사잇각이 다른 삼각형의 대응되는 두 변과 그 사잇각이 각각 서로 같으면 두 삼각형은 합동이다.」

여기에서 표현된 것을 살펴 보면 정리에서 말한 대로 두 변과 그 사잇각이 각각 같도록 두 삼각형을 작도 했다면 컴파스를 이용한다던지 하여 한 삼각형의 나머지 변이 다른 삼각형의 나머지 변과 같다는 것을 확인할 수 있게 된다는 사실을 이 정리로서 나타낸 것이다. 유크리트 기하학을 포함하여 일반적으로 초기 수학에서의 증명은 구성적(작도적) 성격을 지녔다. 예를 들면, 피타고라스의 정리는 대응하는 부분이 합동이기 때문에 면적이 같게 되는 도형의 작도를 통하여 증명된다. 작도가 완성되고 나면 남은 문제는 합동인 부분을 지적하고 원하는 결론에 도달하는 것 뿐이다. 따라서, 논리의 규칙이란 원래 이같은 작도와 결론에 이르는 내용에서 나타나는 상황을 서술하기 위한 것이라고 직관주의자들은 말한다.

예를 들면, 배중률이라는 것은, 만약 대상이 유한개의 집합으로 주어지고 자나 컴퍼스를 쓴다든지 해서 각 대상이 어떤 성질 P를 가지고 있느냐 않느냐를 검사하는 방법이 주어진다면, 우리는 각 대상에 대하여 그 검사방법을 수행해서, 대상 중의 하나가 요구조건(P)을 만족한다든지 아니면 모든 대상이 요구조건(P)을 만족하지 못한다는 것을 단순히 알 수 있다는 것을 의미할 뿐이다.

정리하면 직관 주의자들은 수학에서의 정리는 주어진 지적인 구성이 주어진 결과를 도출하게 된다는 취지에서의 사실적인 명제에 지나지 않는다고 주장한다. 모든 증명은 구성적(작도적)이어야만 한다. 우리가 어떤 수학적 대상이 존재한다고 말하려면, 실제로 그 대상을 구성(작도)하는 방법을 제시함으로서 증명해야 한다. 주어진 대상의 짝들 사이에서 어떤 관계가 성립한다고 말하려면 그 증명은 문제가 되고 있는 대상들의 모든 짝들에 대하여 검사하는 방법을 포함해야 한다.

"**논리의 규칙**"이란 수학적 구성을 실행하는 과정의 단순한 관찰에 불과하다. 구성적인 수학의 관련 범위를 넘어서까지 이 법칙이 적용된다고 믿을 근거는 없다. 사실 이와 같은 범위 밖에까지 이 규칙을 적용하는 것은 무의미한 일이다.「수학에 있어서 논리는 부수적인 것이지 본질적인 것이 아니다.」

이 직관주의자들의 집합 개념이 우리의 것과 다르다는 것은 분명하다. 예를 들어 우리가 대상의 성질을 서술하면 그 성질을 가진 집합이 존재한다는 칸토르의 원리를 생각해 보자. 지금 이 원리는, 체르멜로와 노이만의 제한된 변형과 함께, 직관주의자에게는 저주(詛呪)이다. 어떤 대상이 존재한다는 것은 그것을 구성(작도)할 수 있을 때에만 존재한다는 것이다. 따라서 집합은 우리가 그 집합을 구성할 방법이 설명될 때만 존재하게 되는 것이다.

직관주의자들의 집합론을 전부 소개하는 것은 이 책의 범위를 벗어난다. 그러나 직관주의적 수학에서 중요한 **확장**(擴張, spread)이라고 부르는 특수한 집합은 소개할 가치가 있다. 확장이란 것은 집합의 모든 원소를 만들어 내는 규칙이라고 할 수 있다. 따라서 확장이란 "전체적으로 이미 형성된" 것을 말하는 것이 아니고 "형성되어 가는 과정"을 말한다. 이 집합의 모든 원소는 충분히 지속적으로 이 규칙을 적용하기만 하면 형성할 수가 있다.

직관주의적 수학에서는 역설에서의 곤란한 집합이 만들어지지 않고, 또 본질적인 논의는 직관주의적 논리에 의해서 표현되어질 수 없기 때문에 직관주의적 집합론에서는 역설이 생길 수 없다.

0.6 맺음말

20세기 초에는, 우리가 앞에서 보았던 것 처럼, 여러 가지 무모순(無矛盾)적인 집합론이 수학의 다양한 "학파들(schools)"에 의해 제안되고 발전되었다. 우리는 공리적 집합론의 기본 원리와 럿셀의 유형이론 그리고 집합에 대한 직관주의자들(혹은 구성주의자들)의 접근을 요약해 보았다. 그러나 그 밖에도 여러가지 사상이 나타났지만 너무나 다양해 이 짧은 서론에서는 언급할 수가 없다.

집합을 다루는 모든 방법 중에서 공리론적인 방법이 현대수학의 욕구에 가장 잘 들어 맞는다고 생각된다. 체르멜로와 노이만의 체계에서 나타난 "**집합**"(集合, sets)의 개념은 수학의 목적을 위해서는 충분히 광범위한 것이고, 따라서, 수학적 영역에서는 실제로 그것은 칸토르의 집합개념과 별로 다르지 않다. 증명의 방법과 기호의 사용과 그 엄밀성, 이 모두가 현대수학의 사용법과 호응한다. 무엇보다도 가장 중요한 것은 공리적 집합론이 대부분의 열심히 공부하는 수학자들이 볼 때 자연스럽다는 점이다.

공리적 집합론을 배척하는 사람들은 어떤 철학적인 편견에 근거를 두고 있다. 그러나 공리적 집합론을 인정하기를 거부하는 철학적 입장은 그와 동시에 현대수학의 많은 부분에서의 타당성을 거부한다. 예를 들면 직관주의 학파는 현대해석학의 아주 많은 부분이 비구성적인 원리에 기초를 두고 있기 때문에 이를 거부한다. 이와 같은 비판에 대한 논란들은 새로운 요청이고 분명히 우리들에게 생각할 문제들을 제공하지만 세 세대(三世代)에 걸친 수학자들의 이 화려한 업적을 무너뜨리기에는 충분하지 않다.

과거 약 70년 동안 집합론은 기초적이고 모든 수학을 **단일화**(單一化, unify)하는 역할을 하는 수학 분야로서 널리 인정받게 되었다. 우리는 자연수가 어떻게 구성되느냐 하는 것과 집합론의 범위 안에서 유도된 자연수의 성질도 이미 살펴 보았다. 자연수로 부터 유리수, 실수, 복소수로 쉽게 발전할 뿐만 아니라 칸토르의 유명한 "**초한기수**"(超限基數, transfinite cardinals)로까지 발전할 수 있다. 함수, 관계, 연산 등의 개념도 집합으로 쉽게 정의된다. 따라서 모든 알려진 수학 분야가 집합론 안에서 만들어 질 수 있다. 따라서 다음과 같은 질문을 하는 것은 당연할 뿐 아니라 사실 극히 중요한 것이다.
"집합론이 수학 전체의 체제에 대하여 얼마나 안전한 토대를 제공하느냐?"

특히 공리적 집합론이 **일관성**(一貫性, consistency)이 있다고 즉, 모순(矛盾)이 없다고 절대적으로 확신할 수 있는가?

만약 일관성이 있다면 이는 그 안에서 발전시킨 모든 수학이 모순이 없게 되고 만약 일관성이 없다면 공리적 집합론에 바탕을 두고 건설한 것은 무엇이든 그것은 무가치한 것이 될 것이다.

사실 공리적 집합론의 일관성에 관해서는 아무 것도 증명되지 않았다. 현대 논리학의 몇 가지 결과에서 보면 이는 크게 놀랄 일은 아니다. 예를 들면, 1931년에 괴델은 「자연수의 정상적인 산술의 일관성을 유한적으로는 증명할 수 없음」을 증명하였다. 거의 모든 수학 분야에서의 상황이 마치 한 가지이다.

그래서 현 시점에서 공리적 집합론의 일관성에 대하여 현재 우리가 가지고 있는 최선의 보장은 잘 알고 있는 그러한 모순들은 정상적인 방법으로는 얻어질 수 없다는 사실이다. 현 시점에서는 이보다 더 나은 결론은 없다.

흥미를 끄는 **상대적 일관성**(相對的 一貫性, relative consistency)에 관한 다음과 같은 결과가 최근에 증명되었다.

「만약 체르멜로의 공리적 집합론이 일관성이 있다면 노이만의 공리도 또한 일관성이 있다.」

결국 수학에 대한 일관성의 보장은 기본적인 직관과 경험적 증거의 적당한 결합에 의지하게 될 것으로 추정된다.

Bibliography

[1] Bourbaki N., *Théorie des Ensembles*, Paris, Herman, 1963.

[2] Fraenkel, *A. A., Set Theory and Logic*, Reading, Mass., Addison-Wesley, 1966.

[3] Halmos, P., *Naive Set Theory* Princeton, Van Nostrand, 1960.

[4] Kamke, E., *Theory of Sets*, New York, Dover, 1950.

[5] Monk, J. D., *Introduction to Set Theory*, New York, McGraw-Hill, 1969.

[6] Pinter, Charles C., *Set Theory*, Addison-Wesley Publishing Company Inc., 1971.

[7] Pinter, Charles C., *Set Theory*, Kyungmoon Publishers, 2002.

[8] Quine, W. V., *Mathematical Logic*, Cambridge, Mass., Harvard University Press, 1951.

[9] Rubin, J. E., *Set Theory for the Mathematician*, San Francisco, Holden-Day, 1966.

[10] Slupecki, J. and Borkowski, L., *Elements of Mathematical Logic and Set Theory*, Oxford, Pergamon Press, 1967.

[11] Suppes, P., *Introduction to Logic*, Princeton, Van Nostrand, 1957.

[12] Suppes, P., *Axiomatic Set Theory*, Princeton, Van Nostrand, 1960.

[13] 임정대, *집합론*, 경문사, 1992.

Index

개정판

Lecture Notes in
Set Theory
집합론 해설

지은이 Charles C. Pinter
엮은이 박찬녕
펴낸이 조경희
펴낸곳 경문사
펴낸날 2006년 1월 20일 1판 1쇄
2022년 9월 1일 개정판 5쇄
등 록 1979년 11월 9일 제313-1979-23호
주 소 121-818, 서울특별시 마포구 와우산로 174
전 화 (02)332-2004 팩스 (02)336-5193
이메일 kyungmoon@kyungmoon.com

값 25,000원

ISBN 978-89-6105-685-4